GLENCOE MATH

PERFORMANCE TASKS

AUTHORS
Carter • Cuevas • Day • Malloy
Kersaint • Reynosa • Silbey • Vielhaber

McGraw Hill Education

Bothell, WA • Chicago, IL • Columbus, OH • New York, NY

Contents

Performance Tasks

NAME_____________________________ DATE _______________ PERIOD _________

Performance Task

Field Trip Snacks

Julio's class and Lupe's class are preparing to go on a field trip. In preparation for the trip, their teachers asked the students to select a snack item. Julio's class chose between apples and granola bars, and Lupe's class chose between oranges and bananas.

Based on the classes' choices, Julio's teacher purchased apples and granola bars in a ratio of 3:4, and Lupe's teacher purchased oranges and bananas in a ratio of 3:5.

Write your answers on another piece of paper. Show all your work to receive full credit.

Part A
Classes at Julio and Lupe's school have between 20 and 25 students. Based on the snacks the teachers purchased, how many students are in each of the classes?

Part B
How much will the snacks for Julio's class cost if apples and granola bars are \$0.35 and \$0.42 each, respectively?

Part C
The snacks for Lupe's class will cost \$6.75. How much is each banana if each orange costs \$0.40?

Performance Task *(continued)*

POINTS _______________

Suppose half the students in Julio's class who chose a granola bar now want to have an apple instead for their snack, and twice as many students as chose an orange in Lupe's class now want to have an orange instead of a banana.

Part D

What would be the new ratios of snacks in Julio's class, apples to granola bars, and in Lupe's class, oranges to bananas?

Part E

Without calculating, would the cost of snacks for Lupe's class be greater or less considering the change? Explain your reasoning.

Part F

If Lupe's teacher decides to buy the same number of oranges and bananas so that each student in the class receives one snack, what would be the new cost of snacks for her class? How does this compare to the original cost of snacks for her class?

Part G

Julio's teacher and Lupe's teacher also decide to make smoothies to bring on the field trip. The amounts of the ingredients for different-size servings are shown in the table. Complete the table to show all the ingredient amounts needed for 3, 6, and 12 servings. If each teacher makes enough smoothies for all students in his or her class, how much of each ingredient will the teachers need to purchase?

	3 Servings	6 Servings	12 Servings
Yogurt	3 cups		
Strawberries		18	
Apple Juice	6 ounces		
Honey			6 Tablespoons

Performance Task

Class Elections

The sixth-grade students at Audubon Middle School recently held an election to vote for their class president. All 160 students voted in the election. The student with the most votes will be the president, and the student with the second most votes will be the vice president. The table shows the results.

Candidate	Percent of Vote	Fraction of Vote	Number of Votes
Tia Williams	45		
Francisco Moreno		$\frac{2}{5}$	
Morgan Huffman			24

Write your answers on another piece of paper. Show all your work to receive full credit.

Part A
How many more students voted for Tia than Francisco?

Part B
The school is planning to have an assembly, and the principal would like the students to sit in sections according to their vote. Use a 10 × 10 grid to model the percent of votes for all three candidates.

Part C
Which candidate would be vice president if half of the students who voted for Francisco changed their votes and voted for Morgan? Explain your answer.

Performance Task *(continued)*

POINTS ________

The principal decided to have the students vote again because Morgan was not able to give her campaign speech the day before the election.

Part D

Francisco told one of his friends, "If I can get $\frac{1}{3}$ of Morgan's votes, I will win the election and become president." Is Francisco's statement correct? Explain.

Part E

Describe a possible scenario in which Morgan can become president by getting part of both Tia's and Francisco's votes.

Part F

During the revote, Morgan receives twice as many votes as she did in the original election. What percent of the vote does Morgan have now? Explain how to write Morgan's revote percent as a decimal.

Performance Task

Stuffed Animals

Happy Stuffs produces and distributes stuffed animals to stores across the United States. Their two most popular animals are Lance Lion and Olivia Owl. Each case of Lance Lion contains 75 stuffed animals. Each case of Olivia Owl contains 95 stuffed animals. It takes 0.95 kilogram of stuffing to make each Lance Lion and 1.15 kilograms of stuffing to make each Olivia Owl.

Write your answers on another piece of paper. Show all your work to receive full credit.

Part A

Happy Stuffs ships one case of each animal to 125 stores on a monthly basis. For January, 40 of the stores have ordered an additional case of each animal. For February, 24 stores have ordered an additional case of each animal. For March, 16 stores have ordered an additional case of each animal. Complete the table for the number of each stuffed animal ordered during the first quarter of the year.

Animal	January	February	March
Lance Lion			
Olivia Owl			

What is the monthly amount of stuffing needed for each of the three months?

Part B

Happy Stuffs receives stuffing in cases of 50 kilograms. Their production manager estimates that they will need 125 fewer cases of stuffing for this three-month period if they use only 0.9 kilogram and 1.05 kilograms of stuffing in Lance Lion and Olivia Owl, respectively. Is the production manager's estimate reasonable? Explain.

Performance Task *(continued)*

POINTS ________

Company A currently supplies stuffing for Happy Stuffs for $39 per case. Company Z has approached Happy Stuffs with an offer to provide stuffing for $56.25 per 75-kilogram case.

Part C
Happy Stuffs calculates the price per kilogram to decide which supplier to order from. What is the unit price for each supplier? Which supplier has the better price?

Part D
Using the unit prices from Part C, find the amount of money Happy Stuffs could save for each of the first three months of the year if they use the original amount of stuffing for each animal.

Part E
Company A does not want to lose the Happy Stuffs account. What is one counteroffer they could make to remain the supplier?

Performance Task

Banana Nut Muffins

James is using the recipe below for banana nut muffins. Each batch makes 18 muffins.

- 2 cups of all-purpose flour
- $1\frac{1}{2}$ teaspoons baking soda
- $\frac{1}{2}$ teaspoon salt
- $2\frac{1}{4}$ cups mashed bananas
- 1 cup brown sugar
- $\frac{3}{4}$ cup butter
- 2 eggs
- 1 teaspoon vanilla extract
- $\frac{2}{3}$ cup chopped pecans

Write your answers on another piece of paper. Show all your work to receive full credit.

Part A

James would like to make 27 muffins. How many batches of muffins will he make? Write equations to find the amount of baking soda, bananas, and butter James will need to make the muffins. Solve the equations. Explain.

Part B

James has 6 cups of chopped pecans. How many muffins can he can make with this amount of chopped pecans? Explain.

Performance Task *(continued)*

POINTS ______________

James does not have time to go to the grocery store before he starts to bake, so he needs to use the ingredients he has in his kitchen. James discovers he has a lot of all of the ingredients except for the butter and salt. He only has 9 cups of butter and 2 teaspoons of salt.

Part C
Use a drawing to find the number of batches of muffins he can make using his amount of salt.

Part D
Suppose James has enough of the other ingredients to make as many batches as he can with 9 cups of butter. How many batches of muffins can James make? Justify your answer with an equation.

Part E
James says to make $\frac{1}{2}$ dozen muffins, he needs is $\frac{1}{2}$ the amount of each ingredient. Is he correct? Explain.

Part F
James wants to serve ice tea with the muffins. The recipe calls for $\frac{1}{2}$-cup of powdered tea mix with 2 quarts of water. James only has a $\frac{1}{2}$-cup measuring cup. How many $\frac{1}{2}$-cups of water does he need to make a double batch of tea? What size container does he need for the tea? Explain.

Performance Task

Tracking Temperatures

Monique has been studying weather patterns for some cities during the month of January. Here are some things she noticed:

- The average temperatures for Mayes and Norwick are opposite integers.
- Toniville's average temperature is the same as the absolute value of Boone's average temperature.
- Willport's average temperature is the opposite of the absolute value of Hughes's average temperature.
- The absolute value of Mayes's average temperature is greater than that of Boone's average temperature.
- No city has an average temperature of zero.

Write your answers on another piece of paper. Show all your work to receive full credit.

Part A
Which city's average temperature is definitely a negative number? Explain why the other cities' average temperatures may or may not be negative.

Part B
Monique said it's possible that the average temperatures of Toniville and Boone are the same. Is she correct? Explain your answer.

Part C
Give a scenario in which the average temperature of Norwick is greater than that of Toniville.

NAME_______________ DATE _______________ PERIOD _______

Performance Task *(continued)*

POINTS _______

The average temperatures of Mayes, Boone, and Hughes are all positive integers. None of the average temperatures are greater than 15°C.

Part D
Give values to each city's average temperature so that all of the statements are true. Write your answer so that the average temperatures are in order from greatest to least.

Part E
Graph each city's average temperature on the thermometer. Use the first letter of each city's name to label the graph.

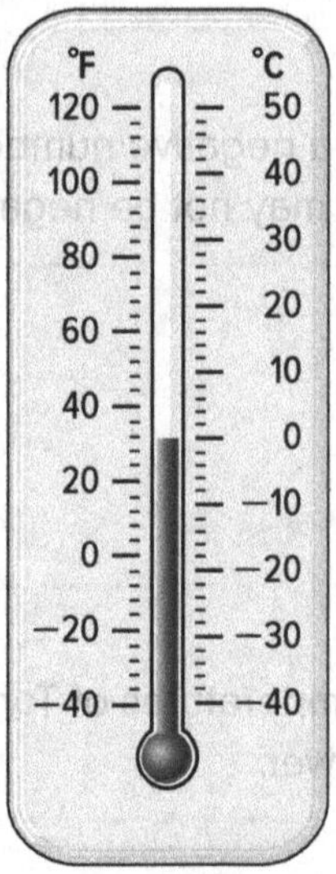

Part F
If each city's average temperature in February increased by 7°C, how would this affect the graph from Part E? How would that affect the relationship between the average temperature for Mayes and Norwick? Explain.

NAME____ DATE ____________________ PERIOD __________

Performance Task

Matt's Family

Matt is the youngest cousin in his family. His cousin, Benito, has written expressions for some of the cousins' ages.

- Matt's age is m.
- Benito's age is two more than three times Matt's age.
- Yolanda's age is one less than the square of Matt's age.
- Xavier's age can be written as $8(m - 1)$.
- Nicki's age is the difference of Xavier's and Benito's ages.
- Pamela is twice as old as Nicki.
- Julio is half the age of Yolanda.
- Luke's age is the sum of Xavier's and Benito's ages.

Write your answers on another piece of paper. Show all your work to receive full credit.

Part A
Write an expression for both Yolanda's and Benito's ages.

Part B
Write a simplified expression for Luke's age.

Part C
What is the sum of Julio's age and Nicki's age if Matt is five years old? Explain your answer.

Performance Task *(continued)*

POINTS __________

In four years, Matt will be nine years old. Benito is trying to modify the expressions he wrote for the cousins' ages.

Part D
How old will Yolanda be when Matt is nine years old?

Part E
Matt's grandmother Kathy is 43 years less than Matt's age cubed. Write an expression for Kathy's age in terms of Matt's age, *m*? Explain how you would calculate Kathy's age in four years.

Part F
Matt's uncle Anthony is currently 17 years old. Write five possible expressions for Anthony's age in four years in terms of Matt's current age, *m*. Write at least one expression using an exponent, and one expression using each operation.

Performance Task

Party Favors

Kyra and Austin are planning separate party events and would like to buy favors for all of the guests that will attend. Kyra has invited 42 people. Austin has invited 34 people. Party Central's offer to Austin is $204 for all of his favors. Great Party Suppliers will match the price per favor, but they will discount every favor purchased after 20 by $1.

Write your answers on another piece of paper. Show all your work to receive full credit.

Part A

Write and solve an equation to show how much Kyra can expect to pay for party favors if she buys them from Party Central. Explain your answer.

Part B

How much will Austin pay for his party favors if he chooses to purchase them from Great Party Suppliers? Write and solve an equation using *t* for the total cost.

Part C

Write and solve an equation using ratios that could be used to find the cost of purchasing 28 party favors from Party Central?

Performance Task *(continued)*

POINTS _______________

Favors for You sells the same party favors as Party Central and Great Party Suppliers. Each favor at Favors for You costs $7. However, they offer a $\frac{1}{4}$ discount for orders of 75 favors or more.

Part D

Austin asks Kyra if she wants to purchase the party favors together in order to save money. He says, "We can save $100 if we buy them together at Favors for You instead of separately from Party Central or Great Party Suppliers." Is Austin's statement correct? Explain.

Part E

Suppose Austin and Kyra decide they each want to have extra favors, so together they purchase 100 favors to share. Make a table showing an equation for the total cost, the actual total cost, and how much they will pay per favor at each of the three stores.

Performance Task

Pizza Party

Mrs. Alabi, the principal at Meadow Middle School, is planning a pizza party for all of the students who met their reading goals for the past semester. Pizza Palace sells large pizzas for $8.00, and each topping costs an additional $1.25. Mrs. Alabi needs to purchase one pizza for every four students who met their goal. The school has an enrollment of 350 students. Mrs. Alabi expects that 56% of the students reached their reading goal.

Mrs. Alabi has a budget of $475 to purchase the pizzas for the reading pizza party.

Write your answers on another piece of paper. Show all your work to receive full credit.

Part A

Create a function table and a graph to represent the price of a large cheese pizza with 0–5 extra toppings.

Part B

Mrs. Alabi asks her student assistant to write a function that can be used to determine the costs of a number of pizzas with 2 toppings. The student assistant wrote this function with *p* being the number of pizzas: $p(8.00 \times 2.25)$. Explain how to make the student assistant's function correct.

Part C

If Mrs. Alabi's expectation that 56% of the student population will reach the reading goal is correct, how many pizzas will Mrs. Alabi need to order to make sure everyone invited gets $\frac{1}{4}$ of a pizza?

Performance Task *(continued)*

POINTS ________

Part D

Keeping in mind Mrs. Alabi's budget, use inequalities to determine whether she can order all 1–topping pizzas or all 2–topping pizzas for the party. Write and solve the inequalities. Justify your answer.

Part E

Write a scenario in which Mrs. Alabi orders some 1–topping pizzas and some 2–topping pizzas but does not spend more than her budget of $475.

NAME_______________________________________ DATE _______________ PERIOD __________

Performance Task

Chapter 9

Home Addition

The Franco family built on an addition to their home so their daughters, Carmen and Sonia, will have their own rooms. On a grid of the additon, Carmen's bedroom is located at (2, 1), (5, 1), (10, 6), (2, 6), and Sonia's bedroom is located at (2, 7), (7, 12), (10, 12), (10, 7). The sisters share a rectangular closet in between their rooms.

Write your answers on another piece of paper. Show all your work to receive full credit.

Part A

Make a diagram of the girls' bedrooms on a coordinate grid.

Carmen's closet door is located at (2, 6), (3, 6). If Carmen's and Sonia's bedrooms are the congruent, where is Sonia's closet door located? Label the rooms and the closet.

Part B

Each section of the grid represents a length of 2 feet. If the girls are going to tile their closet with tiles that are each 0.5 square foot, how many tiles will the girls need to tile their entire closet?

Performance Task *(continued)*

POINTS ________

The girls are shopping around to find the best deal on flooring. Here are some of the deals they have found.

Mama's Flooring

Tiles: $3.50 per square foot

Carpet: $2.95 per square foot

Installation: $400

Happy Flooring

Tiles: $4.25 per square foot

Carpet: $3.10 per square foot

Installation for Carpet:
Free for first 200 square feet and then $0.50 per square foot

Installation for Tile:
$4.00 per square foot

Part C

Suppose each girl chooses carpet for their bedrooms and their closet will be tiled, which flooring store is the better deal?

Part D

The girls' parents said they can also add on a shared bathroom. The bathroom area must be 56 square feet. List possible coordinates for the bathroom.

How much will it cost to tile the bathroom if the girls use the same flooring store as in Part C?

Performance Task

Trophies

Unique Trophy Shop makes specialized trophies. A local university orders an exclusive crystal trophy each year for the student who has volunteered the most hours to a charitable organization.

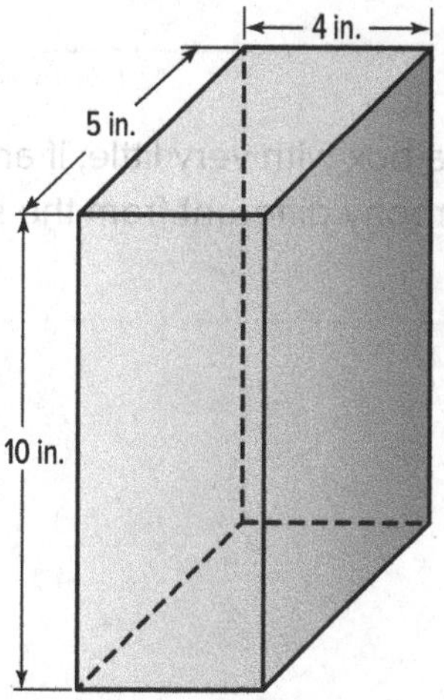

Write your answers on another piece of paper. Show all your work to receive full credit.

Part A

Create an orthogonal drawing of the trophy by drawing the front and back, sides, and top and bottom views. Label the drawing.

What is the surface area of the university's trophy?

Part B

Explain a way that the dimensions of the trophy could be changed without changing the volume.

Performance Task *(continued)*

POINTS ________

Unique Trophy Shop always wraps their trophies with 2 inches of bubble wrap all around the trophy. Each trophy is then packed in a box designed specifically for it.

Part C

Unique Trophy Shop sets the wrapped trophy in a box with very little, if any, extra space. How is the surface area of the wrapped trophy different from the surface area of the trophy itself?

Part D

Create a net of a box designed so that the wrapped trophy would fit snuggly inside. What is the volume of the box you designed for the wrapped trophy?

Performance Task

For this Performance Task, your teacher will ask you to do some research to provide the data you will use for your calculations.

How Much Sugar?

Do you stop to read nutrition facts on containers before eating or drinking what's in them? There are so many items to choose from it's difficult to make a decision about what to have for a meal or snack. The information on the nutrition fact labels is there to inform you and possibly help you make a decision about what to eat or drink.

For this task, research and record the number of grams of sugar in 12 ounces of ten different types of drinks. Make sure to include a variety of drinks, such as water, juices, soft drinks, sport drinks, and so on. Since drinks come in many different sizes, make sure you are recording the amount for 12 ounces.

Write your answers on another piece of paper. Show all your work to receive full credit.

Part A
Create a line plot for the number of grams of sugar in 12 ounces of the ten drinks you chose. What are the mean, median, and mode of the amount of sugar in the drinks?

Part B
What is the interquartile range of the amount of sugar in the drinks? Explain your answer.

Part C
Examine your data for any outliers. State the value of any outliers in your data. If your data set does not have any outliers, give an example of a value that would be an outlier. Explain how the outlier(s) affect your answers to Part A and Part B.

Performance Task *(continued)*

POINTS ________

Part D
What is the mean absolute deviation for your data? Explain how to find the mean absolute deviation and what it represents.

Part E
Compare your results with another classmate. Explain similarities and differences. Then combine both results and find the mean, median, and mode of the amount of sugar in all 20 drinks.

Suppose your classmate selected ten drinks with 0 grams of sugar. What would that do to your combined data?

Part F
Suppose you have an intern position with an advertising agency in charge of advertising for a new drink that has 10 grams of sugar in a 12-ounce serving. What advertising points would you make to boost sales when comparing this sugar total to the other drinks you researched? Give at least two points to sell the new product.

Performance Task

Track Results

Mr. Noyes and Mrs. Ramsey recorded the 100-yard dash times of their sixth-grade students. The table shows the time in seconds for each student.

100-Yard Dash																
Teacher	Student Times (s)															
Mr. Noyes	21	23	18	38	22	22	21	18	15	21	22	24	23	20	21	23
Mrs. Ramsey	15	27	23	14	28	24	23	22	24	18	22	26	23	24	24	31

Write your answers on another piece of paper. Show all your work to receive full credit.

Part A
Create a double box plot for the 100-yard dash times for each class. Use the interquartile range to determine which class had a greater variability in times.

Part B
Three students were absent the day Mr. Noyes's class recorded the results. When they returned to class, they were able to run the 100-yard dash. When their times were included, the median of the class data increased. What could their times have been? Explain your answer.

Part C
Mr. Noyes and Mrs. Ramsey are going to randomly select a student from one of their classes to represent their school in a 100-yard race against other schools. Mrs. Ramsey states that since the fastest runner is in her class, they should randomly select a student from her class. Use the data to make an argument to support randomly selecting a student from Mr. Noyes's class instead of Mrs. Ramsey's class. Give at least two reasons to support your answer.

Performance Task *(continued)*

POINTS _______________

Part D

Here are the times of the top ten runners in the 100-yard dash for a city-wide competition last year.

Time in seconds: 15, 13, 14.5, 14, 13.5, 14, 13.5, 14, 14.5, 14.5

Make a line plot of the times for the top ten city-wide competition. What is the mean time for the top ten race times last year?

Part E

How do the top ten sixth-graders in Mr. Noyes's and Mrs. Ramsey's classes combined compare to the top ten runners from last year's city competition? How would you expect the top ten sixth-graders from the combined classes to fare at this year's city-wide competition? Use a double box plot to explain your answer.

Page PT1 *Field Trip Snacks*

CCSS Content Standard(s)	6.RP.1, 6.RP.2, 6.RP.3, 6.RP.3a, 6.RP.3b
Mathematical Practices	MP1, MP2, MP3, MP7
Depth of Knowledge	DOK2, DOK3

Part	Max Points	Scoring Rubric
A	2	**Full Credit:** Julio: 3:4, 6:8, 9:12 = 21 Lupe: 3:5, 6:10, 9:15 = 24 Julio's class has 21 students, and Lupe's class has 24 students. Partial Credit will be given for correctly calculating the number of students in one class. No credit will be given for an incorrect answer.
B	1	**Full Credit:** Apples: $9 \cdot 0.35 = 3.15$; \$3.15 Granola bars: $12 \cdot 0.42 = 5.04$; \$5.04 Total cost: $3.15 + 5.04 = 8.19$; \$8.19 The snacks for Julio's class will cost \$8.19. No credit will be given for an incorrect answer.
C	1	**Full Credit:** Total cost for oranges: $9 \cdot 0.40 = 3.60$; \$3.60 Total cost for bananas: $6.75 - 3.60 = 3.15$; \$3.15 Each banana costs: $3.15 \div 15 = 0.21$; \$0.21 each No credit will be given for an incorrect answer.
D	2	**Full Credit:** Julio's class now needs 6 granola bars, so now they need 15 apples. 15:6 simplifies to 5:2. The new ratio of snacks, apples to granola bars, for Julio's class is 5:2. Lupe's class needs 18 oranges, meaning they need only 6 bananas. 18:6 simplifies to 3:1. The new ratio of snacks, oranges to bananas, for Lupe's class is 3:1. Partial Credit will be given for the correct ratio for one class. No credit will be given for an incorrect answer.
E	1	**Full Credit:** The cost would be greater because Lupe's teacher would have to buy more oranges, which are the more expensive snacks. No credit will be given for an incorrect answer.
F	1	**Full Credit:** Oranges: $12 \cdot 0.40 = 4.80$; \$4.80 Bananas: $12 \cdot 0.21 = 2.52$; \$2.52 Total Cost: $4.80 + 2.52 = 7.32$; \$7.32 Difference: $7.32 - 6.75 = 0.57$; \$0.57 The snacks for Lupe's class would now cost \$7.32. This is \$0.57 more than before. No credit will be given for an incorrect answer.
G	3	**Full Credit:** (see table below) For 21 students, Julio's teacher will need to purchase 21 cups of yogurt, 63 strawberries, 42 ounces of apple juice, and 10.5 tablespoons of honey. Or 24 students, Lupe's teacher will need to purchase 24 cups of yogurt, 72 strawberries, 48 ounces of apple juice, and 12 tablespoons of honey. Partial Credit will be given for each correct answer. One point will be given for completing the table correctly OR calculating the ingredients for smoothies correctly for Lupe's class or Julio's class. No credit will be given for an incorrect answer.
TOTAL	11	

	3 Servings	6 Servings	12 Servings
Yogurt	3 cups	6 cups	12 cups
Strawberries	9	18	36
Apple Juice	6 ounces	12 ounces	24 ounces
Honey	1.5 Tablespoons	3 Tablespoons	6 Tablespoons

Performance Task Rubrics

Page PT3 Class Elections

CCSS Content Standard(s)	6.RP.3, *Preparation for* 6.RP.3c
Mathematical Practices	MP1, MP2, MP4, MP7
Depth of Knowledge	DOK2, DOK3

<table>
<tr><th>Part</th><th>Max Points</th><th>Scoring Rubric</th></tr>
<tr><td>A</td><td>1</td><td>Full Credit:
Tia received 45% of the vote. $160 \times 0.45 = 72$;
Tia received 72 votes.
Francisco received $\frac{2}{5}$ of the votes which is 40%. Francisco received
$160 \times 0.40 = 64$ votes.
Tia received 8 more votes than Francisco.
No credit will be given for an incorrect answer.</td></tr>
<tr><td>B</td><td>2</td><td>Full Credit:
Tia = 45% of the grid
Francisco = 40% of the grid
Morgan = 15% of the grid
Sample grid is shown.
<table>
<tr><td>T</td><td>T</td><td>T</td><td>T</td><td>T</td><td>M</td><td>F</td><td>F</td><td>F</td><td>F</td></tr>
<tr><td>T</td><td>T</td><td>T</td><td>T</td><td>T</td><td>M</td><td>F</td><td>F</td><td>F</td><td>F</td></tr>
<tr><td>T</td><td>T</td><td>T</td><td>T</td><td>T</td><td>M</td><td>F</td><td>F</td><td>F</td><td>F</td></tr>
<tr><td>T</td><td>T</td><td>T</td><td>T</td><td>T</td><td>M</td><td>F</td><td>F</td><td>F</td><td>F</td></tr>
<tr><td>T</td><td>T</td><td>T</td><td>T</td><td>T</td><td>M</td><td>F</td><td>F</td><td>F</td><td>F</td></tr>
<tr><td>T</td><td>T</td><td>T</td><td>T</td><td>M</td><td>M</td><td>F</td><td>F</td><td>F</td><td>F</td></tr>
<tr><td>T</td><td>T</td><td>T</td><td>T</td><td>M</td><td>M</td><td>F</td><td>F</td><td>F</td><td>F</td></tr>
<tr><td>T</td><td>T</td><td>T</td><td>T</td><td>M</td><td>M</td><td>F</td><td>F</td><td>F</td><td>F</td></tr>
<tr><td>T</td><td>T</td><td>T</td><td>T</td><td>M</td><td>M</td><td>F</td><td>F</td><td>F</td><td>F</td></tr>
<tr><td>T</td><td>T</td><td>T</td><td>T</td><td>M</td><td>M</td><td>F</td><td>F</td><td>F</td><td>F</td></tr>
</table>
Partial Credit will be given for correct percents given for each candidate OR a correct model.
No credit will be given for an incorrect answer.</td></tr>
<tr><td>C</td><td>2</td><td>Full Credit:
Morgan would be elected vice president.
She receives 56 votes to Francisco's 32 votes.
Or
She receives 35% of the vote to Francisco's 20%
Or
She receives $\frac{7}{20}$ of the vote to Francisco's $\frac{1}{5}$.
Partial Credit will be given for identifying the correct student as vice president with no explanation provided.
No credit will be given for an incorrect answer.</td></tr>
<tr><td>D</td><td>1</td><td>Full Credit:
$\frac{1}{3}$ of Morgan's votes can be expressed as either 8 votes or 5% of the vote.
Francisco is incorrect, because getting $\frac{1}{3}$ of Morgan's votes would give him 72 votes or 45% of the vote, which ties him with Tia.
No credit will be given for an incorrect answer.</td></tr>
<tr><td>E</td><td>1</td><td>Full Credit:
Sample answers:
Morgan would win the election if she receives 21 of Tia's votes and 15 of Francisco's votes. That would give her 60 votes, Tia 51 votes and Francisco 49 votes.
Or
Morgan would win the election if she receives 18 of Tia's votes and 14 of Francisco's votes. That would give Morgan 56 votes, Tia 54 votes and Francisco 50 votes.
No credit will be given for an incorrect answer.</td></tr>
<tr><td>F</td><td>2</td><td>Full Credit:
Morgan received 24 votes in the original election. Twice as many votes is 48 votes. 48 votes of 160 votes is 30%.
To convert 30% into an equivalent decimal, divide by 100 and remove the % sign or move the decimal point two places to the left, 0.30.
Partial Credit will be given for a correct response with no explanation.
No credit will be given for an incorrect answer.</td></tr>
<tr><td>TOTAL</td><td>9</td><td></td></tr>
</table>

Page PT5 *Stuffed Animals*

CCSS Content Standard(s)	6.NS.2, 6.NS.3, 6.RP.3b
Mathematical Practices	MP1, MP2, MP3, MP6, MP8
Depth of Knowledge	DOK2, DOK3

Part	Max Points	Scoring Rubric
A	4	Full Credit:
		(see table below)

Animal	January	February	March
Lance Lion	12,375	11,175	10,575
Olivia Owl	15,675	14,155	13,395

January: (12,375 · 0.95) + (15,675 · 1.15) = 29,782.5 kg
February: (11,175 · 0.95) + (14,155 · 1.15) = 26,894.5 kg
March: (10,575 · 0.95) + (13,395 · 1.15) = 25,450.5 kg

Partial Credit (2 points) will be given for a correctly completed table OR amount of stuffing needed calculated correctly.

No credit will be given for an incorrect answer.

Part	Max Points	Scoring Rubric
B	2	Full Credit: Stuffing needed: 29,782.5 + 26,894.5 + 25,450.5 = 82,127.5 kg Number of cases needed: 82,127.5 ÷ 50 = 1,642.55 Happy Stuffs will need 1,643 cases of stuffing. Accounting for the change in stuffing use: Stuffing needed: 27,596.25 + 24,920.25 + 23,582.25 = 76,098.75 kg Number of cases needed: 76,098.75 ÷ 50 = 1,521.975 Happy Stuffs would need 1,522 cases of stuffing. The production manager's estimate is reasonable. The company will need 121 fewer cases of stuffing, which is close to the estimate. Partial credit will be given for a correct calculation of how many fewer cases are needed but does not give a reasonable explanation. No credit will be given for an incorrect answer.
C	1	Full Credit: 1 Point Company A: 39 ÷ 50 = 0.78; $0.78 per kilogram Company Z: 56.25 ÷ 75 = 0.75; $0.75 per kilogram Company Z has the better price. No credit will be given for an incorrect answer.
D	2	Full Credit: Company A — Company Z January: $23,230.35 — $22,336.88 February: $20,977.71 — $20,170.88 March: $19,851.39 — $19,087.88 Savings January: 23,230.35 − 22,336.88 = 893.47; $893.47 February: 20,977.71 − 20,170.88 = 806.83; $806.83 March: 19,851.39 − 19,087.88 = 763.51; $763.51 Partial Credit will be given if monthly savings are calculated for 2 months. No credit will be given for an incorrect answer.
E	1	Full Credit: A correct response will state that the current supplier must lower the unit cost below $0.75 per kilogram of stuffing. Sample answer: Company A could offer to sell the 50-kilogram cases of stuffing for $37 which would be a unit cost of $0.74 per kilogram. No credit will be given for an incorrect answer.
TOTAL	10	

Page PT7 *Banana Nut Muffins*

CCSS Content Standard(s)	6.NS.1, 6.RP.3, 6.RP.3d
Mathematical Practices	MP1, MP2, MP4, MP7
Depth of Knowledge	DOK2, DOK3

Part	Max Points	Scoring Rubric
A	2	**Full Credit:** Sample answer: In order to make 27 muffins James needs to make $1\frac{1}{2}$ batches. Amount of baking soda: $1\frac{1}{2} \times 1\frac{1}{2} = 2\frac{1}{4}$; $2\frac{1}{4}$ tsp Amount of bananas: $1\frac{1}{2} \times 2\frac{1}{4} = 3\frac{3}{8}$; $3\frac{3}{8}$ c Amount of butter: $1\frac{1}{2} \times \frac{3}{4} = 1\frac{1}{8}$; $1\frac{1}{8}$ c Partial Credit will be given for two correct multiplication equations and solutions OR an explanation of the number of batches the recipe needed. No credit will be given for an incorrect answer.
B	2	**Full Credit:** 162 muffins; Sample answer: To find the number of muffins James can make, first determine the number of batches he can make. To find the number of batches he can make, divide the number of cups of pecans, 6, by $\frac{2}{3}$ since it takes $\frac{2}{3}$ c per batch. $6 \div \frac{2}{3} = 9$ Then multiply that number by 18, since each batch makes 18 muffins. $9 \times 18 = 162$ Partial Credit will be given for the correct number of muffins OR for an explanation of the process to determine the number of muffins No credit will be given for an incorrect answer.
C	1	**Full Credit:** 4 batches; Sample drawing: 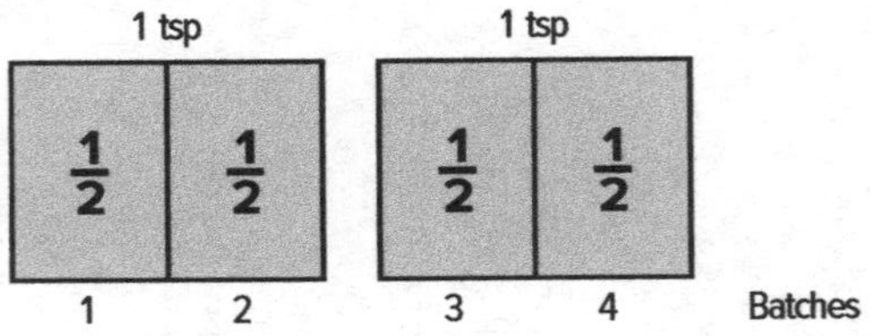 No credit will be given for an incorrect answer.
D	2	**Full Credit:** James has enough butter to make 12 batches of muffins. Number of batches using available butter: $9 \div \frac{3}{4} = 12$; 12 batches Partial Credit of 1 point each will be given for a correct equation OR the correct number of batches. No credit will be given for an incorrect answer.
E	1	**Full Credit:** No, James is not correct. Sample answer: One-half dozen muffins, or 6 muffins, is $\frac{1}{3}$ of 18. James needs $\frac{1}{3}$ of the amount of each ingredient to make $\frac{1}{2}$ dozen muffins. No credit will be given for an incorrect answer.
F	1	**Full Credit:** 32; James needs a 2-quart or 1-gallon size container for the tea. $2 \times \frac{1}{2}$ cup = 1 cup; 4 cups = 1 quart $8 \times \frac{1}{2} = 1$ quart; $16 \times \frac{1}{2} = 2$ quarts $2 \times 16 \times \frac{1}{2} = 4$ quarts = 1 gallon $32\left(\frac{1}{2}\text{ cups}\right) = 1$ gallon No credit will be given for an incorrect answer.
TOTAL	9	

Page PT9 Tracking Temperatures

CCSS Content Standard(s)	**6.NS.5, 6.NS.6a, 6.NS.6c, 6.NS.7, 6.NS.7a, 6.NS.7b, 6.NS.7c**
Mathematical Practices	**MP1, MP2, MP3, MP4, MP7**
Depth of Knowledge	**DOK2, DOK3**

Part	Max Points	Scoring Rubric
A	2	**Full Credit:** • Willport's average temperature is the only value that must be negative, because it is the opposite of an absolute value. • Either Mayes's or Norwick's average temperature is negative, but we do not know which one without more information. • Toniville's average temperature is an absolute value, so it will be positive. • Boone's and Hughes's average temperatures could be positive. Partial Credit will be given for a correct answer OR at least two valid reasons why the other cities may or may not have negative temperatures. No credit will be given for an incorrect answer.
B	1	**Full Credit:** Yes, it is possible for the average temperatures of Toniville and Boone to be the same if Boone's average temperature is positive. The absolute value of a positive number is an equivalent positive number. No credit will be given for an incorrect answer.
C	1	**Full Credit:** Either Mayes's or Norwick's average temperature must be positive. If Norwick's temperature is the one that is positive, then it will be greater than Toniville's temperature because the absolute value of Mayes's and Norwick's temperature is greater than that of Boone's temperature. No credit will be given for an incorrect answer.
D	2	**Full Credit:** The following conditions must be met: • Mayes's average temperature must be positive and greater than Boone's average temperature. • Norwick's average temperature must be the opposite of Mayes's average temperature. • Toniville's average temperature and Boone's average temperature must be equal. • Willport's average temperature must be negative and the opposite of Hughes's average temperature. Note: Hughes's and Mayes's average temperatures are not compared. Sample answer: Hughes: 13°C, Mayes: 9°C, Toniville and Boone: 6°C, Norwick: −9°C, Willport: −13°C Partial Credit will be given for correct scenario meeting all conditions, but not correctly ordered from greatest to least. No credit will be given for an incorrect answer.
E	1	**Full Credit:** Answers will depend on the response from Part D. Check student's graph for points for 6 cities identified based on response in Part D. No credit will be given for an incorrect answer.
F	1	**Full Credit:** Each point would move up 7 units on the graph. The average temperatures for Mayes and Norwick would no longer be opposites. For example, in January, the average temperatures were opposites such as 9°C for Mayes and −9°C for Norwick. If the average temperature increased by 7°C in February, the average temperature would be 16°C for Mayes and −2°C for Norwick. No credit will be given for an incorrect answer.
TOTAL	**8**	

Page PT11 *Matt's Family*

CCSS Content Standard(s)	6.EE.1, 6.EE.2, 6.EE.2a, 6.EE.2c, 6.EE.3, 6.EE.4, 6.EE.6
Mathematical Practices	MP1, MP2, MP3, MP4, MP7
Depth of Knowledge	DOK2, DOK3

Part	Max Points	Scoring Rubric
A	2	**Full Credit:** Sample expressions shown. Yolanda's age: $m^2 - 1$ Benito's age: $3m + 2$ **Partial Credit** will be given for one correct expression. **No credit** will be given for an incorrect answer.
B	1	**Full Credit:** Sample expression shown. Luke's age is the sum of Xavier and Benito's ages. Xavier's age: $8(m - 1)$ Benito's age: $3m + 2$ So Luke's age can be written $8(m - 1) + 3m + 2$. Simplifying $8(m - 1) + 3m + 2$ $8m - 8 + 3m + 2$ $11m - 6$ Luke's age: $11m - 6$ **No credit** will be given for an incorrect answer.
C	2	**Full Credit:** 27; Julio is 12 years old and Nicki is 15 years old. Nicki's age is the difference of Xavier and Benito's age. Matt is 5 years old. Xavier's age: $8(5 - 1) = 8(4) = 32$ years old. Benito's age: $3(5) + 2 = 15 + 2 = 17$ years old. So Nicki is $32 - 17$ or 15 years old. Julio is one-half of Yolanda's age. Yolanda's age: $m^2 - 1$, so Julio's age is $\frac{m^2 - 1}{2}$; $\frac{(5)^2 - 1}{2} = \frac{25 - 1}{2} = 12$ **Partial Credit** will be given for a correct answer with no explanation. **No credit** will be given for an incorrect answer.
D	1	**Full Credit:** Currently Yolanda's age is $m^2 - 1$. In four years her age will be $m^2 + 3$. Using $m = 5$; $5^2 + 3$ $5^2 + 3 = 25 + 3 = 28$ Yolanda will be 28 years old. **No credit** will be given for an incorrect answer.
E	2	**Full Credit:** Sample expression shown. Kathy's current age: $m^3 - 43 = 125 - 43 = 82$ years old Kathy's age in four years: $m^3 - 43 + 4 = m^3 - 39 = 86$ years old In order to calculate Kathy's age in four years you need to first substitute 5 for the variable *m*. Then you would need to evaluate 5^3, which means you would multiply $5 \cdot 5 \cdot 5 = 125$. Finally you would subtract $125 - 39 = 86$. **Partial Credit** will be given for a correct answer with no explanation. **No credit** will be given for an incorrect answer.
F	2	**Full Credit:** Matt's current age is 5. Anthony's age in 4 years will be 21. Sample expressions shown. $4m + 1$; $m^2 - 4$; $6m - 9$; $3(m + 2)$; $\frac{105}{m}$ **Partial Credit** will be given for three or four correct expressions. **No credit** will be given for an incorrect answer.
TOTAL	10	

Page PT13 Party Favors

CCSS Content Standard(s)	**6.EE.5, 6.EE.7, 6.RP.3**
Mathematical Practices	**MP1, MP2, MP3, MP6, MP7, MP8**
Depth of Knowledge	**DOK2, DOK3**

<table>
<tr><th>Part</th><th>Max Points</th><th>Scoring Rubric</th></tr>
<tr><td>A</td><td>1</td><td>Full Credit:
Sample answer: To find Kyra's cost you first have to find the cost of each party favor. Use the information from Austin's offer to write an equation: $p \cdot 34 = 204$. Divide each side of the equation by 34 to solve the equation: $204 \div 34 = 6$. Since each party favor costs $6, Kyra's cost would be $6 \cdot 42 = 252$. Kyra's cost would be $252.

No credit will be given for an incorrect answer.</td></tr>
<tr><td>B</td><td>1</td><td>Full Credit:
Sample answer: $(6 \cdot 20) + (5 \cdot 14) = t$
$t = 190$; $190
Austin's cost for favors at Great Party Suppliers will be $190.

No credit will be given for an incorrect answer.</td></tr>
<tr><td>C</td><td>1</td><td>Full Credit:
$\frac{c}{28} = \frac{204}{34}$
$c = 168$; $168
The cost of 28 party favors will be $168.

No credit will be given for an incorrect answer.</td></tr>
<tr><td>D</td><td>2</td><td>Full Credit:
Austin is not correct. The difference between Favors for You and Part Central is $57 and the difference between Favors for You and Great Party Suppliers is only $1.

If they purchase their party favors together from Party Central, they will spend:
$6(34 + 42) = 456$; $456

If they purchase their party favors together from Great Party Suppliers, they will spend:
$(6 \cdot 20) + (5 \cdot 56) = 120 + 280 = 400$, $400

If they purchase their party favors together at Favors for You, they will spend:
$7(34 + 42) = 532$
$532 \div 4 = 133$
$532 - 133 = 399$; $399

Difference between Party Central and Favors for You:
$456 - 399 = 57$; $57
The savings will be $57.

Difference between Great Party Suppliers and Favors for You:
$400 - 399 = 1$; $1
The savings will be $1.

Partial Credit will be given for a correct answer with no explanation.
No credit will be given for an incorrect answer.</td></tr>
<tr><td>E</td><td>3</td><td>Full Credit:
Sample equations shown.
<table>
<tr><th>Store</th><th>Equation</th><th>Total Cost</th><th>Cost per Favor</th></tr>
<tr><td>Party Central</td><td>$6 \cdot 100 = p$</td><td>$6 \cdot 100 =$ $600</td><td>$6.00</td></tr>
<tr><td>Great Party Suppliers</td><td>$(6 \cdot 20) + (5 \cdot 80) = g$</td><td>$120 + 400 =$ $520</td><td>$5.20</td></tr>
<tr><td>Favors for You</td><td>$(7 \cdot 100) - \frac{(7 \cdot 100)}{4} = f$</td><td>$525</td><td>$5.25</td></tr>
</table>
Partial Credit (1 point) will be given for 2 correct equations OR 2 correct costs per favor.

No credit will be given for an incorrect answer.</td></tr>
<tr><td>TOTAL</td><td>8</td><td></td></tr>
</table>

Page PT15 Pizza Party

CCSS Content Standard(s)	6.EE.2, 6.EE.2c, 6.EE.5, 6.EE.6, 6.EE.8, 6.EE.9
Mathematical Practices	MP1, MP2, MP3, MP4, MP6, MP7
Depth of Knowledge	DOK2, DOK3

Part	Max Points	Scoring Rubric
A	2	**Full Credit:** Sample function table shown. (see table below) Graphs must contain these points: (0, 8), (1, 9.25), (2, 10.5), (3, 11.75), (4, 13), and (5, 14.25). Sample graph shown. (see graph below) Partial Credit will be given for a correct table OR a correct graph. No credit will be given for an incorrect answer.
B	1	**Full Credit:** Sample answer: The student assistant multiplied the base price for a pizza by the cost of two toppings instead of adding it. Also, the student assistant incorrectly used \$2.25 for the cost of two toppings instead of \$2.50. The correct function should be: $p(8.00 + 2.50)$. No credit will be given for an incorrect answer.
C	1	**Full Credit:** $350 \times 0.56 = 196$; 196 students. $196 \div 4 = 49$; For 196 students to each receive $\frac{1}{4}$ of a pizza, Mrs. Alabi will need to order 49 pizzas. No credit will be given for an incorrect answer.
D	2	**Full Credit:** All 1-topping pizzas: $9.25p \leq 475$; $p \leq 51.3$. Mrs. Alabi will be able to order all 1-topping pizzas because she needs only 49 pizzas. All 2-topping pizzas: $10.50p \leq 475$; $p \leq 45.2$. Mrs. Alabi will not be able to order enough 2-topping pizzas because she needs 49 pizzas. Partial Credit will be given for a correct inequality and solution for 1-topping pizzas OR 2-topping pizzas. No credit will be given for an incorrect answer.
E	2	**Full Credit:** Answer should include the following elements: a total cost of not more than \$475; at least one 1-topping pizza; at least one 2-topping pizza; a total of at least 49 pizzas. Sample answer: Mrs. Alabi has enough money to order: 17 2–topping pizzas @ 10.50 = \$178.50; 32 1–topping pizzas @ 9.25 = \$296.00; 49 total pizzas = \$474.50. No credit will be given for an incorrect answer.
TOTAL	8	

Part A sample function table:

Input (x)	x(\$1.25) + \$8.00	Output
0	0(1.25) + 8.00	8.00
1	1(1.25) + 8.00	9.25
2	2(1.25) + 8.00	10.50
3	3(1.25) + 8.00	11.75
4	4(1.25) + 8.00	13.00
5	5(1.25) + 8.00	14.25

Part A sample graph:

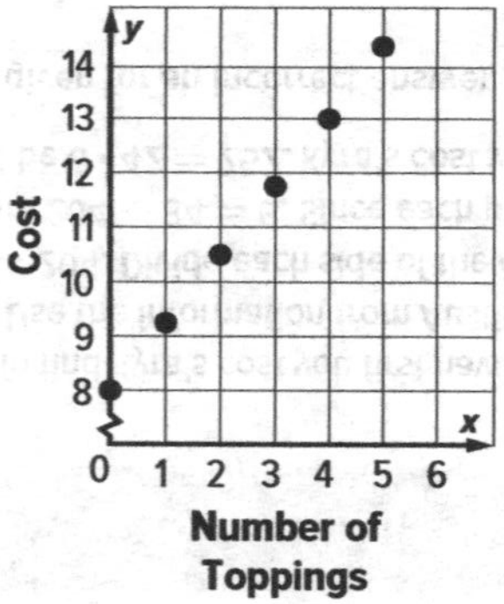

Page PT17 Home Addition

CCSS Content Standard(s)	6.G.1, 6.G.3
Mathematical Practices	MP1, MP2, MP4, MP7
Depth of Knowledge	DOK2, DOK3

Part	Max Points	Scoring Rubric
A	2	**Full Credit:** Sample diagram as shown: 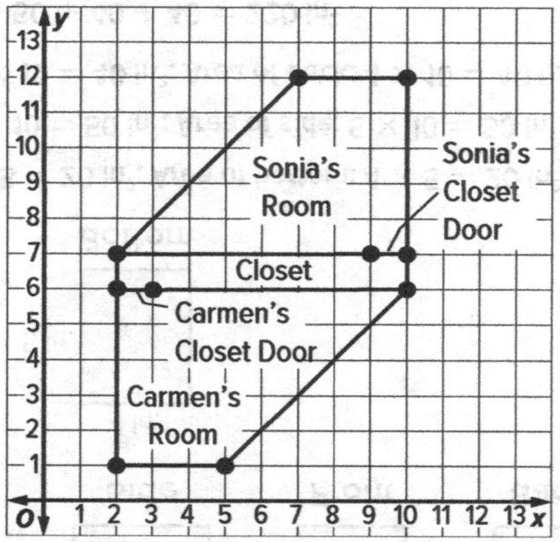 Sonia's closet door is located from coordinates (9, 7), (10, 7). Diagram should include coordinates of each girl's room and closet doors with room, closet and doors labeled. Partial Credit will be given for a correct diagram of the girls' bedrooms without the closet doors or closet located. No credit will be given for an incorrect answer.
B	2	**Full Credit:** 64 tiles; Each square is 2 feet × 2 feet or 4 square feet. The closet is 8 squares, or 32 square feet. Each tile is 0.5 square foot; $0.5t = 32$; $t = 64$ The girls will need 64 tiles for the floor of the closet. Partial Credit will be given for the correct area of the closet OR the correct number of tiles needed for the floor of the closet. No credit will be given for an incorrect answer.
C	2	**Full Credit:** Happy Flooring is the better deal. The bedrooms are congruent, so they have the same area. To find the area of each bedroom, use the formula for the area of a trapezoid. $A = \frac{1}{2}h(b_1 + b_2)$; $A = \frac{1}{2}(10)(6 + 16)$; $A = 110$ square feet The area of one bedroom is 110 square feet, so the area of both bedrooms is 220 square feet. The area of the closet is 32 square feet. **Mama's Flooring** 3.50 × 32 = 112 2.95 × 220 = 649 Installation = 400 112 + 649 + 400 = $1,161 **Happy Flooring** 4.25 × 32 = 136 3.10 × 220 = 682 Installation for carpet: 0.5 × 20 = 10 Installation for tile: 4 × 32 = 128 136 + 682 + 10 + 128 = $956 Partial Credit will be given for the correct area of the bedrooms OR correctly determining the better deal for the tile and carpet. No credit will be given for an incorrect answer.
D	2	**Full Credit:** Sample answer: Coordinates for bathroom: (8, 4), (10, 6), (10, 8), (13, 8), (13, 4) Answer must include coordinates that make a figure that has an area of 56 square feet. Cost for tile and installation for bathroom: 4.25 × 56 = 238 Installation: 4 × 56 = 224 238 + 224 = 462; $462 Partial Credit will be given for correct coordinates for bathroom with area of 56 square feet OR correct cost for tile and installation for bathroom. No credit will be given for an incorrect answer.
TOTAL	8	

Page PT19 Trophies

CCSS Content Standard(s)	6.G.2, 6.G.4
Mathematical Practices	MP1, MP2, MP3, MP4, MP5, MP6
Depth of Knowledge	DOK2, DOK3

Part	Max Points	Scoring Rubric
A	2	**Full Credit:** 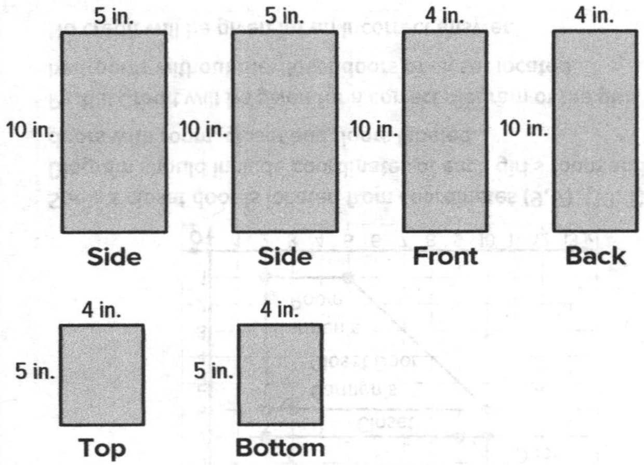 Area of top: $4 \times 5 = 20$ in^2; Area of bottom: $4 \times 5 = 20$ in^2 Area of side: $5 \times 10 = 50$ in^2; Area of side: $5 \times 10 = 50$ in^2 Area of front: $4 \times 10 = 40$ in^2; Area of back: $4 \times 10 = 40$ in^2 $20 + 20 + 50 + 50 + 40 + 40 = 220$ in^2 Partial Credit will be given for an accurate drawing of the trophy OR the correct surface area of the trophy. No credit will be given for an incorrect answer.
B	1	**Full Credit:** Sample answer: The volume of the trophy is $5 \times 4 \times 10 = 200$ cubic inches. If the trophy's dimensions were changed to 5 in., 5 in., and 8 in., the volume would still be 200 cubic inches. No credit will be given for an incorrect answer.
C	1	**Full Credit:** Top and bottom: $2(8 \times 9) = 144$ in^2 Two sides: $2(9 \times 14) = 252$ in^2 Front and back: $2(8 \times 14) = 224$ in^2 $144 + 252 + 224 = 620$ in^2 $620 - 220 = 400$; The difference between the surface area of the wrapped trophy and the trophy itself is 400 square inches. No credit will be given for an incorrect answer.
D	2	**Full Credit:** Sample drawing of net: 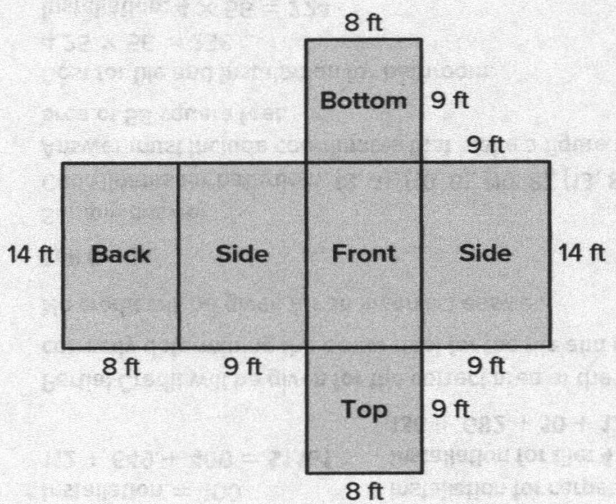 Net must show a box in which the trophy and bubble wrap can fit securely. Approximate measures of box could be: $14 \times 9 \times 8$. Volume would be approximately $14 \times 9 \times 8 = 1{,}008$ cubic inches. Partial Credit will be given for a drawing of a net that would fit the wrapped trophy OR for calculating the volume of the box. No credit will be given for an incorrect answer.
TOTAL	6	

Page PT21 Drink Study

Note to teacher: For this Performance Task, students will need to do some research either as class work or homework and record the number of grams of sugar in 12 ounces of different types of drinks.

CCSS Content Standard(s)	**6.SP.3, 6.SP.5, 6.SP.5b, 6.SP.5c, 6.SP.5d**
Mathematical Practices	**MP1, MP2, MP3, MP5, MP6**
Depth of Knowledge	**DOK2, DOK3**

Part	Max Points	Scoring Rubric
A	**2**	**Full Credit:** **Sample data: 0, 27, 39, 23, 25, 0, 0, 27, 32, 42** **Sample answer:** **Sugar in Drinks (g)** 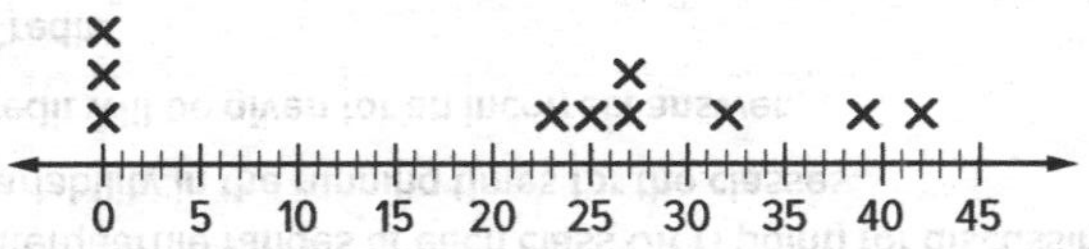 **Mean: 21.5; Median: 26; Mode: 0** **Partial Credit will be given for a correct line plot OR correct mean, median, and mode values.** **No credit will be given for an incorrect answer.**
B	**1**	**Full Credit:** **Sample answer: Interquartile range: 32 − 0 = 32** **To find the interquartile range, subtract the first quartile from the third quartile.** **No credit will be given for an incorrect answer.**
C	**1**	**Full Credit:** **Sample answer: A drink with a sugar amount of 80 grams would be an outlier. It is much greater than the other drinks' sugar amounts. The mode would not be affected by this outlier. If I included the drink with 80 grams of sugar with the other data, the median and mean would increase. The interquartile range would also increase.** **No credit will be given for an incorrect answer.**
D	**1**	**Full Credit:** **Sample answer: To find the mean absolute deviation, first find the mean, which is 21.5. Then find the absolute value of the difference between each drink's sugar amount and the mean: 21.5, 21.5, 21.5, 1.5, 3.5, 5.5, 5.5, 10.5, 17.5, 20.5. Finally, add those differences and divide by the number of drinks, 10. Mean absolute deviation: 12.9** **The mean absolute deviation tells you the average distance each drink's sugar amount is from the mean. The greater the value, the more varied the data.** **No credit will be given for an incorrect answer.**
E	**2**	**Full Credit:** **Sample answer: We both have drinks with a variety of sugar amounts. My mode was less than my classmate's mode.** **Mean: 27; Median: 24.95; Mode: 0** **If my classmate had all drinks with 0 grams of sugar, it would change the statistical measures of the data quite a bit. The median of the combined data would be 0, and the mean would drop to 10.75. On a data display it would appear that the majority of the drinks available have 0 grams of sugar.** **Partial Credit will be given for information on a combined data set OR analysis of how data with all 0 grams of sugar impact the data.** **No credit will be given for an incorrect answer.**
F	**2**	**Full Credit:** **Answers will vary. Students should compare the given value to the measures of central tendency for the drinks they researched.** **Sample answer: The total sugar in 12 ounces of this drink is lower than the average of the top 10 competitors. I would also highlight other benefits based on information on nutritional fact labels.** **Partial Credit will be given for one valid point for sales of the new product based on data collected for other drinks.** **No credit will be given for an incorrect answer.**
TOTAL	**9**	

Chapter 12 Performance Task Rubric

Page PT23 Track Results

CCSS Content Standard(s)	**6.SP.4, 6.SP.5, 6.SP.5b, 6.SP.5c, 6.SP.5d**
Mathematical Practices	**MP1, MP2, MP3, MP4, MP7**
Depth of Knowledge	**DOK2, DOK3**

Part	Max Points	Scoring Rubric
A	3	**Full Credit:** 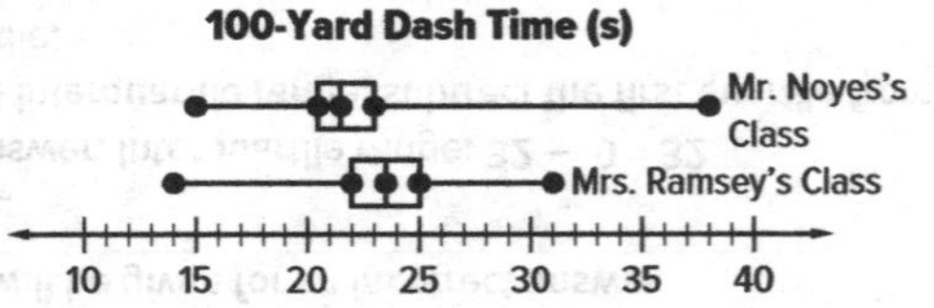 **The interquartile range is 2.5 for Mr. Noyes's class and 3.0 for Mrs. Ramsey's class. Therefore, there is a greater variability in the times for Mrs. Ramsey's class.** **Partial Credit (2 points) will be given for correct box plots and the interquartile ranges of each class OR (1 point) for discussing the variability in the running times for the classes.** **No credit will be given for an incorrect answer.**
B	1	**Full Credit:** **Sample answer: Their times could have been 20, 22, and 25 seconds. Any three times whose mean is greater than 21.5, the original median, will increase the median of the data set.** **No credit will be given for an incorrect answer.**
C	2	**Full Credit:** **Answers will vary.** **A correct response must contain at least two of the following:** • **The median time for Mr. Noyes's class times is faster than the median time for Mrs. Ramsey's class times.** • **The mode time of Mr. Noyes's class times is faster than the mode time for Mrs. Ramsey's class times.** • **Since the Q3 is a faster time for Mr. Noyes's class, the top 75% of the students in his class ran faster than the top 75% in Mrs. Ramsey's class.** **Partial Credit will be given for correctly stating one accurate reason to support the argument.** **No credit will be given for an incorrect answer.**
D	2	**Full Credit:** **14.05 seconds; Sample line plot shown:** 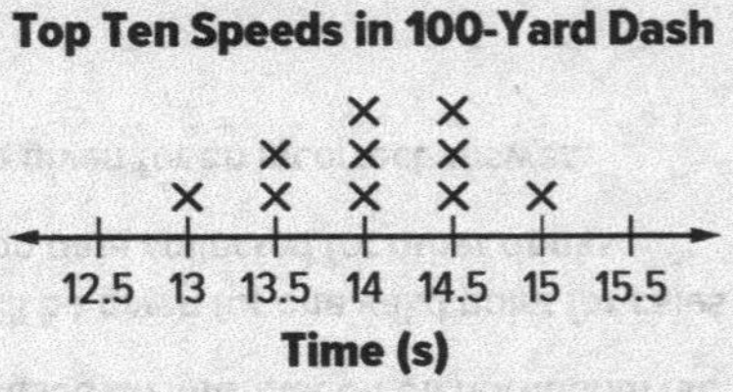 **The mean time of the top ten runners for the city-wide competition is (13 + 13.5 + 13.5 + 14 + 14 + 14 + 14.5 + 14.5 + 14.5 + 15) ÷ 10 = 14.05 seconds.** **Partial Credit will be given for a correct line plot OR for correctly calculating the mean of the top ten speeds.** **No credit will be given for an incorrect answer.**
E	2	**Full Credit:** **Sample answer: The top ten times for the two classes combined are 14, 15, 15, 18, 18, 18, 20, 21, 21, and 21 seconds. The median of the combined class time is 4 seconds slower than the top ten times from last year. The interquartile range of the combined class is 5 seconds greater than the top ten times last year, which means that the top ten combined class times are more varied than the top ten runners from last year.** **I would expect the top 25% of the sixth-graders from the combined classes to do well this year at the city-wide competition.** 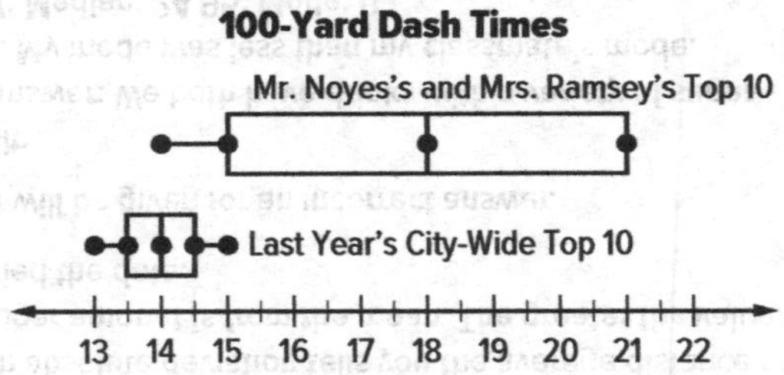 **Partial Credit will be given for a correct explanation OR box plot.** **No credit will be given for an incorrect answer.**
TOTAL	**10**	

GLENCOE MATH

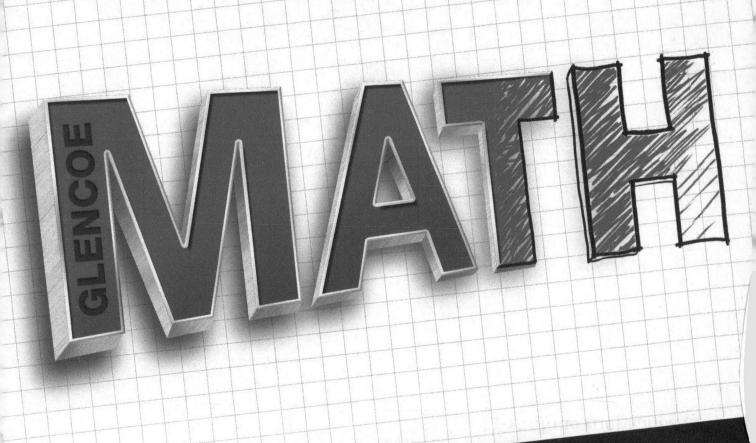

PRACTICE MASTERS

AUTHORS
Carter • Cuevas • Day • Malloy
Kersaint • Reynosa • Silbey • Vielhaber

Mc Graw Hill Education

Bothell, WA • Chicago, IL • Columbus, OH • New York, NY

mheducation.com/prek-12

McGraw Hill Education

Send all inquiries to:
McGraw-Hill Education
8787 Orion Place
Columbus, OH 43240

ISBN: 978-0-07-898944-5
MHID: 0-07-898944-2

Printed in the United States of America.

1 2 3 4 5 6 QVS 21 20 19 18 17

Contents

NAME ______________________ DATE ______________ PERIOD ________

Lesson 1 **Multi-Step** Problem Solving

Multi-Step Example

The table shows the school supplies Jin has. He wants to put the pencils, erasers, and notepads in bags, with each item distributed evenly in the bags. If each bag is $2, what is the greatest amount of money he would spend? **6.NS.4,** MP 1

Item	Number
Pencils	48
Pens	32
Erasers	60
Notepads	36

Ⓐ $8 Ⓒ $24

Ⓑ $16 Ⓓ $120

Use a problem-solving model to solve this problem.

1 Understand

Read the problem. Circle the information you know.
Underline what the problem is asking you to find.

2 Plan

What will you need to do to solve the problem? Write your plan in steps.

Step 1 Find the common factors of the number of pencils, the number of erasers, and the number of notepads.

Step 2 Find the greatest common factor of these factors.

Step 3 Determine the greatest amount of money he would spend.

Read to Succeed!
Jin is not putting pens in the bags. So, the number of pens is not needed to solve the problem.

3 Solve

Use your plan to solve the problem. Show your steps.

factors of 48: ______________________________

factors of 60: ______________________________

factors of 36: ______________________________

The common factors are ______________________________.

The greatest common factor is ______. The greatest number of bags he can fill is ______. The most amount of money he would spend is ______.

So, the correct answer is ______. Fill in that answer choice.

4 Check

How do you know your solution is reasonable?

____

____

____

NAME ______________________ DATE ______________ PERIOD ________

Lesson 1 *(continued)*

Use a problem-solving model to solve each problem.

1 The table shows the number of muffins Ana baked. She wants to put the oatmeal, banana, and blueberry muffins in rows with the same number of each type of muffin. What is the greatest number of muffins in each row? **6.NS.4, MP 2**

Muffins	
Type	**Number**
Oatmeal	16
Banana	12
Raisin	10
Blueberry	20

Ⓐ 2 muffins

Ⓑ 4 muffins

Ⓒ 12 muffins

Ⓓ 20 muffins

2 Maria will use all the glass, wooden, and pottery beads she buys to make a necklace. The table shows the number of beads that come in each pack. Maria wants to use the same number of each type of bead. What is the least number of each type of bead she will buy? The cost per pack is $1. How much will she spend? **6.NS.4, MP 4**

Beads	
Type	**Number per Pack**
Glass	6
Wooden	3
Pottery	5

3 Carlos has three pieces of fabric. The yellow fabric is 45 inches wide, the blue fabric is 36 inches wide, and the red fabric is 27 inches wide. Carlos wants to cut all three pieces into equal width strips that are as wide as possible. How many strips of each color will he have? **6.NS.4, MP 4**

____

____

____

4 **H.O.T. Problem** Suppose the LCM of three numbers is 20, the GCF is 2, and the sum of the three numbers is 16. What are the three numbers? **6.NS.4, MP 7**

____

Lesson 2 **Multi-Step** Problem Solving

Multi-Step Example

The table shows the types of sandwiches sold on Friday. What is the denominator when the ratio of veggie sandwiches to the total number of sandwiches is written as a fraction in simplest form? 6.RP.1, MP 1

Sandwich	Number Sold
Turkey	9
Tuna	11
Veggie	6
Chicken	14

Use a problem-solving model to solve this problem.

1 Understand

Read the problem. Circle the information you know.
Underline what the problem is asking you to find.

2 Plan

What will you need to do to solve the problem?
Write your plan in steps.

Step 1 Determine the total number of sandwiches sold.

Step 2 Express the ratio of veggie sandwiches to total sandwiches as a fraction in simplest form.

Read to Succeed!
The order of the words of the ratio gives the order of the values to use as the numerator and denominator.

3 Solve

Use your plan to solve the problem. Show your steps.

The total number of sandwiches sold is ☐ + ☐ + ☐ + ☐ or ☐ sandwiches.

The ratio of veggie to total sandwiches is $\frac{\square}{\square}$ or $\frac{\square}{\square}$.

So, the simplified denominator is ☐.

4 Check

How do you know your solution is accurate?

____

____

____

Lesson 2 *(continued)*

Use a problem-solving model to solve each problem.

1 The table shows the types of breakfasts sold on Thursday. What is the denominator of the simplified ratio of oatmeal orders to total orders? **6.RP.1, MP 1**

Breakfast	Number Sold
Omelets	14
Pancakes	17
Waffles	11
Oatmeal	8

____

2 How many blue counters must be added so that the ratio of yellow counters to total counters is 1:6? **6.RP.1, MP 4**

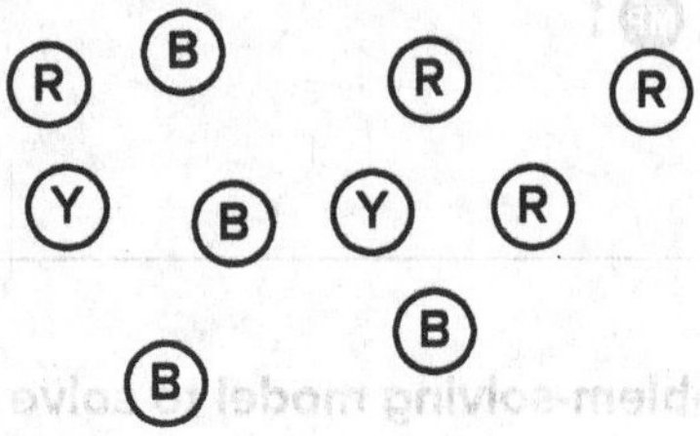

____

3 Cantrise surveyed 100 students about their favorite type of music. After she makes the graph, she receives two more votes for rap. What is the new ratio of rap to total types of music? **6.RP.1, MP 1**

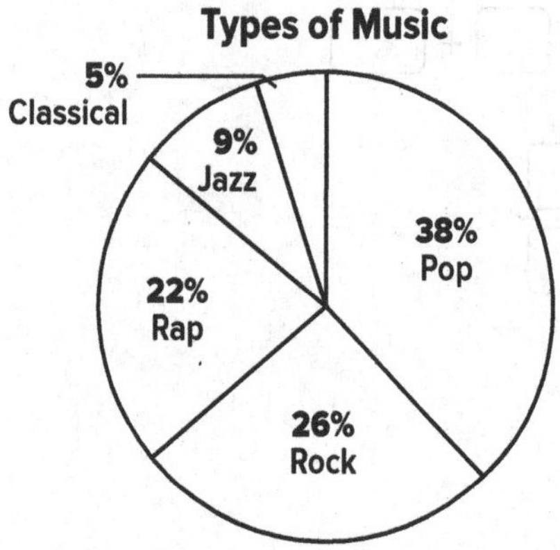

____

4 **H.O.T. Problem** The ratio of blue circles to total circles is 4 to 5. There are more than 10 circles. Describe what this group of circles might look like. Explain. **6.RP.1, MP 3**

____

____

____

____

Lesson 3 **Multi-Step** Problem Solving

Multi-Step Example

Zariah is training for a 5-kilometer run, which is about 3 miles. She begins her training by running 1 mile each day for 5 days. She records the number of minutes it takes her to run a mile, as seen in the table. What is her average time in feet per second? **6.RP.2, MP 1**

Day	Time (min)
1	15
2	13
3	16
4	14
5	12

Ⓐ 0.16 ft/sec

Ⓑ 5.87 ft/sec

Ⓒ 6.29 ft/sec

Ⓓ 18.86 ft/sec

Use a problem-solving model to solve this problem.

1 Understand

Read the problem. Circle the information you know.
Underline what the problem is asking you to find.

2 Plan

What will you need to do to solve the problem? Write your plan in steps.

Step 1 Determine the average time in minutes. Convert to seconds.

Step 2 Divide the number of feet in a mile by the seconds.

Read to Succeed!
There are 5,280 feet in a mile.

3 Solve

Use your plan to solve the problem. Show your steps.

The average time is $\frac{15 + 13 + 16 + 14 + 12}{5}$ or ☐ minutes.

14 minutes × 60 = ☐ seconds

So, Zariah runs an average of 5,280 ÷ 840 or about ☐ feet per second.

Choice ______ is correct. Fill in that answer choice.

4 Check

How do you know your solution is accurate?

____

____

____

Lesson 3 *(continued)*

Use a problem-solving model to solve each problem.

1 Adam records his biking speed for five consecutive days in a table. There are 5,280 feet in a mile. Determine his average rate in feet per minute. Round to the nearest foot per minute. **6.RP.2, MP 1**

Day	Speed (miles per hour)
1	12
2	15
3	13
4	16
5	17

Ⓐ 1,232 ft/min
Ⓑ 1,276 ft/min
Ⓒ 1,285 ft/min
Ⓓ 1,320 ft/min

2 At a grocery store, a 24-pack of 16.9-ounce water bottles is sold for $4.99. There are 128 ounces in a gallon. Determine the price per gallon. Round to the nearest penny. **6.RP.3b, MP 1**

______

3 Carmine is in charge of buying shirts for the senior class. She contacted three companies and recorded their pricing information in the table below. What is the best price per shirt? **6.RP.3b, MP 1**

Company	Cost
A	20 shirts for $38.40
B	25 shirts for $48.75
C	30 shirts for $57.00

______

4 **H.O.T. Problem** Using the rate $\frac{y \text{ yards}}{m \text{ minutes}}$, predict what will happen to the value of the ratio for each scenario in the table below. Explain your reasoning. **6.RP.3, MP 7**

y	*m*	value of ratio
increases	unchanged	
unchanged	increases	
decreases	unchanged	
unchanged	decreases	

______

______

______

______

______

Lesson 4 Multi-Step Problem Solving

Multi-Step Example

Marybeth makes money babysitting. She charges a flat rate of \$5 plus \$5 per hour for each child. The rule $5(n) + 5$ can be used to calculate her fee per hour, where n is the number of children. Complete the table to find the ratios of the number of children to the fee. Write each ratio in simplest form. **6.RP.3, MP 1**

Number of children	1	2	3	4	5
Babysitting fee per hour (\$)					

Ⓐ $\frac{1}{10}, \frac{2}{15}, \frac{3}{20}, \frac{4}{25}, \frac{1}{6}$

Ⓑ $\frac{1}{11}, \frac{1}{6}, \frac{3}{13}, \frac{2}{7}, \frac{1}{3}$

Ⓒ $\frac{10}{1}, \frac{15}{2}, \frac{20}{3}, \frac{25}{4}, \frac{6}{1}$

Ⓓ $\frac{1}{10}, \frac{2}{15}, \frac{3}{20}, \frac{4}{25}, \frac{1}{30}$

Use a problem-solving model to solve this problem.

1 Understand

Read the problem. Circle the information you know.
Underline what the problem is asking you to find.

2 Plan

What will you need to do to solve the problem? Write your plan in steps.

Step 1 Use the rule to determine each fee.

Step 2 Express each ratio in simplest form.

Read to Succeed!
Remember to multiply first then add 5.

3 Solve

Use your plan to solve the problem. Show your steps.

Complete the table.

Number of children	1	2	3	4	5
Babysitting fee per hour (\$)					

The number of children is the numerator and the fee is the denominator. Simplify if necessary.

The ratios are $\frac{\square}{\square}, \frac{\square}{\square}, \frac{\square}{\square}, \frac{\square}{\square}, \frac{\square}{\square}$.

Choice ______ is correct. Fill in that answer choice.

4 Check

How do you know your solution is accurate?

____

____

NAME ______________________ DATE ______________ PERIOD ________

Lesson 4 *(continued)*

Use a problem-solving model to solve each problem.

1 A rental car company charges $0.25 a mile. Use a table like the one below to find the ratios of the cost to the number of miles in simplest form. **6.RP.3, MP 1**

Cost ($)				
Number of Miles	5	10	15	20

Ⓐ $\frac{5}{4}, \frac{5}{2}, \frac{15}{4}, \frac{5}{1}$

Ⓑ $\frac{4}{1}, \frac{4}{1}, \frac{4}{1}, \frac{4}{1}$

Ⓒ $\frac{1}{4}, \frac{1}{4}, \frac{1}{4}, \frac{1}{4}$

Ⓓ $\frac{1}{25}, \frac{2}{5}, \frac{3}{75}, \frac{5}{1}$

2 The tables below show the swimming speeds of the King penguin and the Emperor penguin. Determine which table shows an equivalent ratio. Write the equivalent ratio in miles per hour rounded to the hundredths place. **6.RP.3, MP 6**

King Penguin				
Distance (ft)	440	860	1,300	1,740
Time (min)	1	2	3	4

Emperor Penguin				
Distance (ft)	412	824	1,236	1,648
Time (min)	1	2	3	4

____

3 Joshua is trying to determine the number of pizzas to order. The table below shows the number of people pizzas will feed. If each pizza costs $7, how much will it cost to feed 36 people? **6.RP.3, MP 1**

Number of Pizzas	5	7	9
Number of People	15	21	27

____

4 **H.O.T. Problem** On the blueprint below, 1 inch represents 2.5 feet of the house. Complete a table that shows the corresponding number of feet of the house for each length on the blueprint. **6.RP.3a, MP 2**

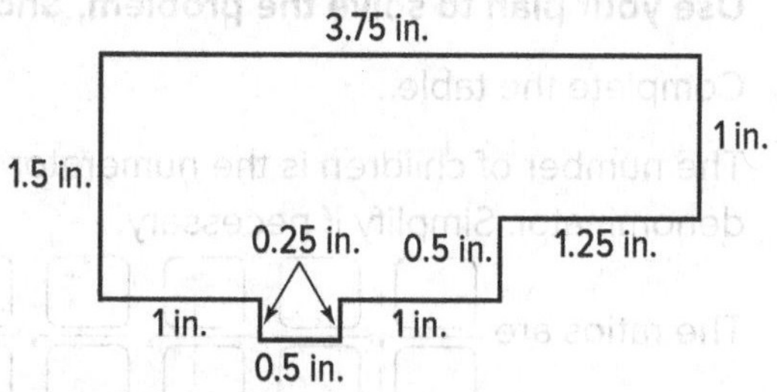

NAME ______________________________ DATE ______________ PERIOD ________

Lesson 5 Multi-Step Problem Solving

Chapter 1

Multi-Step Example

The height of a tree over time is an equivalent ratio. Which statement best describes the graph of the ratio table? **6.RP.3a, MP 1**

Time (yr)	Height (ft)
3	4.2
5	7
8	11.2

Ⓐ For every one unit right, the line goes up three units.

Ⓑ The line appears to pass through (1, 2).

Ⓒ The line decreases from left to right.

Ⓓ The line appears to pass through the origin.

Use a problem-solving model to solve this problem.

1 Understand

Read the problem. Circle the information you know.
Underline what the problem is asking you to find.

2 Plan

What will you need to do to solve the problem? Write your plan in steps.

Step 1 Graph the ordered pairs.

Step 2 Determine which statement is true based on the graph.

Read to Succeed!
When graphing ordered pairs, the x-coordinate tells how far to the right or left. The y-coordinate tells how far up or down.

3 Solve

Use your plan to solve the problem. Show your steps.

Graph the ordered pairs.

The line appears to increase from left to right, so C is incorrect.
The line also does not pass through (1, 2), so B is incorrect.
For every unit right, the line goes up 1.4 units, so A is incorrect.

Since the line appears to pass through the origin, choice ______ is correct. Fill in that answer choice.

4 Check

How do you know your solution is accurate?

____

____

Lesson 5 *(continued)*

Use a problem-solving model to solve each problem.

1 The number of bricks is an equivalent ratio to the height of a wall. Which statement describes the pattern in the graph of the ordered pairs? **6.RP.3a, MP 1**

Number of Bricks	Height of Wall (ft)
48	4
84	7
108	9

Ⓐ As the height of the wall increases by 1 foot, there are 12 more bricks in the wall.

Ⓑ For every 48 bricks, the wall increases by 3 feet.

Ⓒ For every 36 bricks, the wall increases by 4 feet.

Ⓓ As the height of the wall increases by 2 feet, there are 48 more bricks in the wall.

2 The perimeter of an equilateral triangle is an equivalent ratio to the length of the sides. An equilateral triangle with a side length of 3 cm has a perimeter of 9 cm, and an equilateral triangle with a side length of 5 cm has a perimeter of 15 cm. What is the perimeter of a triangle with a side length of 7 cm? **6.RP.3, MP 8**

____

3 Five yards of material is needed to make 2 curtains. This represents an equivalent ratio. If the equivalent ratio (curtains, material) were graphed, what would the *y*-coordinate be if the *x*-coordinate was 7? **6.RP.3a, MP 4**

____

4 **H.O.T. Problem** The table below shows the hours and wages for two friends who babysit. Use a graph to determine if they make the same hourly rate. Explain why or why not. **6.RP.3a, MP 3**

	Zoe	Felix
Hours	3	5
Wages	$21	$30

____

____

____

Lesson 6 Multi-Step Problem Solving

Multi-Step Example

Luna wants to burn as many Calories as possible per minute of exercise. Which exercise should Luna choose? **6.RP.3, MP 1**

Exercise	Calories	Minutes
Walking	300	60
Jump rope	110	10
Biking	270	30
Aerobics	160	20

Ⓐ walking

Ⓑ jump rope

Ⓒ biking

Ⓓ aerobics

Use a problem-solving model to solve this problem.

1 Understand

Read the problem. Circle the information you know.
Underline what the problem is asking you to find.

2 Plan

What will you need to do to solve the problem? Write your plan in steps.

Step 1 Determine the unit rate for each exercise.

Step 2 Compare the unit rates.

> **Read to Succeed!**
> Unit rates have a denominator of 1 when simplified.

3 Solve

Use your plan to solve the problem. Show your steps.

Walking: $\frac{300}{60} = \frac{\square}{\square}$ Jump rope: $\frac{110}{10} = \frac{\square}{\square}$

Biking: $\frac{270}{30} = \frac{\square}{\square}$ Aerobics: $\frac{160}{20} = \frac{\square}{\square}$

The unit rate ☐ Calories per minute is the greatest.

So, Luna should __________. Choice ______ is correct. Fill in that answer choice.

4 Check

How do you know your solution is accurate?

____

____

Lesson 6 *(continued)*

Use a problem-solving model to solve each problem.

1 Santiago needs to buy apples to make applesauce. He is looking for a better price than $1.29 per pound. Which store has a better price per pound? **6.RP.3, MP 4**

Store	Price ($)	Weight (lb)
Store A	4.08	3
Store B	5.04	4
Store C	12.88	7
Store D	7.35	5

Ⓐ Store A
Ⓑ Store B
Ⓒ Store C
Ⓓ Store D

2 Josh is making a scale model of his grandfather's airplane. Given the information in the diagram, how many inches is the wingspan of the model? **6.RP.3, MP 2**

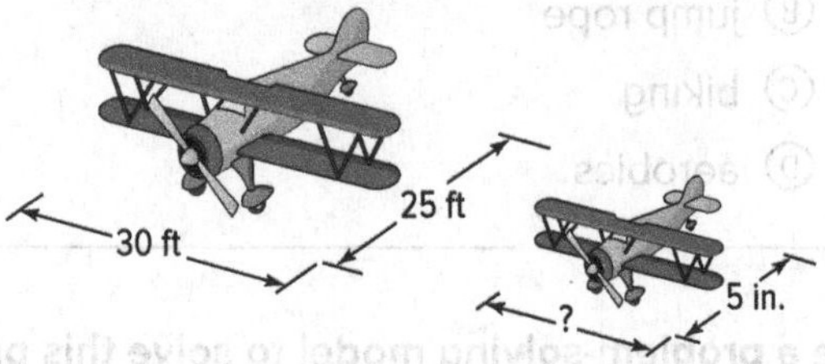

____

3 Riley and Magdalena bought beads to make a necklace. The beads that Riley bought cost 48¢ for 12. Magdalena bought 20 beads for $1.00. How much less do 15 of Riley's beads cost than 15 of Magdalena's beads? **6.RP.3b, MP 4**

____

4 **H.O.T. Problem** Jacob bought 5 pencils for 80¢. At this same rate, how many pencils can he buy for 40¢? Explain why this answer does not make sense. **6.RP.3b, MP 3**

____

____

NAME______________________________ DATE ______________ PERIOD __________

Lesson 7 Multi-Step Problem Solving

Multi-Step Example

The table shows the results of a survey of a group of people about their favorite animal. If 600 people were asked about their favorite animal, predict how many more people would prefer dogs than cats. **6.RP.3, MP 1**

Favorite Animal	
Animal	**Number of Responses**
Bird	5
Cat	13
Dog	19
Iguana	3

Ⓐ 6
Ⓑ 285
Ⓒ 195
Ⓓ 90

Use a problem-solving model to solve this problem.

1 Understand

Read the problem. Circle the information you know.
Underline what the problem is asking you to find.

2 Plan

What will you need to do to solve the problem? Write your plan in steps.

Step 1 Determine the total number of people surveyed.

Step 2 Set up equivalent ratios.

Step 3 Subtract to determine how much more.

Read to Succeed!
When setting up equivalent ratios, remember to keep the same attribute in the numerators and the same in the denominators.

3 Solve

Use your plan to solve the problem. Show your steps.

There were 5 + 13 + 19 + 3 or 40 people surveyed.

Dogs: $\frac{19}{40} = \frac{?}{600}$ $= \square$

Cats: $\frac{13}{40} = \frac{?}{600}$ $= \square$

So, 285 − 195 or $\square$ more people chose dogs than cats. Choice ______ is correct.

Fill in that answer choice.

4 Check

How do you know your solution is accurate?

____

____

Lesson 7 *(continued)*

Use a problem-solving model to solve each problem.

1 The table shows the results of a survey of a group of people about their favorite sport. If 100 people were asked about their favorite sport, predict how many more people would prefer hockey than volleyball. **6.RP.3, MP 1**

Favorite Sport	
Sport	**Number of Responses**
Baseball	7
Soccer	10
Volleyball	5
Hockey	8

Ⓐ 3
Ⓑ 10
Ⓒ 30
Ⓓ 70

2 Keshia rides her bike 10 miles per hour. At this rate, how many more minutes will it take her to ride 30 miles than 25 miles? **6.RP.3b, MP 2**

____

3 Marisol pays $12 for 4 notebooks. How many notebooks could she buy with $62? **6.RP.3b, MP 2**

____

4 **H.O.T. Problem** Jamal is helping a friend with her homework. Look at his friend's work to determine and explain the error she made finding the answer to the problem. **6.RP.3, MP 3**

If 3 dogs eat 4 pounds of food per day, how many dogs eat 15 pounds of food?

$\frac{3}{4} = \frac{15}{?}$

Multiply numerator and denominator by 5. Therefore, 20 dogs eat 15 pounds of food.

____
____
____
____
____

Lesson 1 Multi-Step Problem Solving

Multi-Step Example

The frequency table shows the favorite lunch of some sixth graders. What decimal represents the part of these students that chose pizza or burgers?
Preparation for **6.RP.3c, MP 1**

Food	Tally	Frequency
Tacos	|||	3
Burgers	𝍸 |	6
Chicken	𝍸	5
Pizza	𝍸 𝍸 |	11

Use a problem-solving model to solve this problem.

1 Understand

Read the problem. Circle the information you know.
Underline what the problem is asking you to find.

Read to Succeed!
Use a pencil to write your answer in the boxes at the top of the grid and to fill in the correct bubble(s) of each digit in your answer below.

2 Plan

What will you need to do to solve the problem? Write your plan in steps.

Step 1 Determine the total number of students surveyed.

Step 2 Determine the number of students that chose ________ or ________.

Step 3 ________ to find the decimal.

3 Solve

Use your plan to solve the problem. Show your steps.

There were 3 + 6 + 5 + 11 or ______ students surveyed.

There were 6 + 11 or ______ students that chose pizza or burgers.

______ ÷ ______ = ______

So, ______ of the students surveyed chose pizza or burgers.

4 Check

How do you know your solution is accurate?

____

____

Lesson 1 *(continued)*

Use a problem-solving model to solve each problem.

1 The graph shows how Bianca spends her weekly allowance. What decimal represents the part of Bianca's allowance that she spends on entertainment and clothes? *Preparation for* **6.RP.3c, MP 1**

Allowance

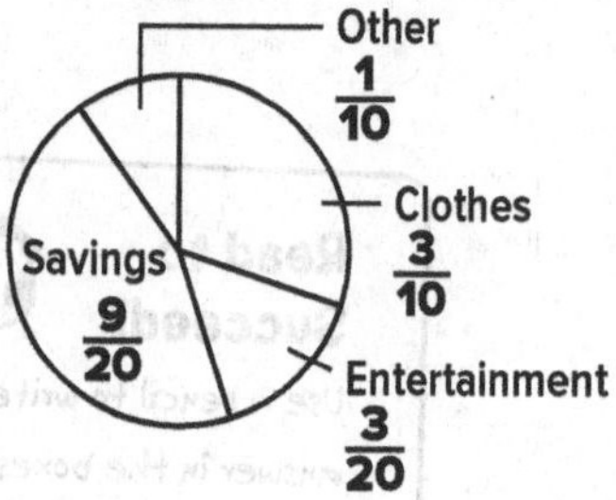

____

2 Renee's goal is to run for at least $\frac{1}{4}$ mile more than she ran the previous day. On Day 1, she ran 0.75 mile. The table shows the distances she ran for the next five days. On which of these days did she NOT reach her goal? *Preparation for* **6.RP.3c, MP 2**

Day	Distance (mi)
2	$1\frac{1}{10}$
3	1.35
4	$1\frac{1}{2}$
5	$1\frac{4}{5}$
6	2.05

____

3 The table shows the number of free throws that Raj successfully made and the number of his attempts over four days.

Day	Shots Made	Attempts
Monday	21	30
Tuesday	18	25
Wednesday	20	32
Thursday	18	24

On which day was the fraction of shots made to attempts the greatest? Express the fraction as a decimal.
Preparation for **6.RP.3c, MP 2**

____

4 H.O.T. Problem Write a fraction that is between 0.1 and $\frac{1}{5}$, has a whole-number numerator, and has a denominator of 100. Then express an equivalent decimal form for this fraction. *Preparation for* **6.RP.3c, MP 6**

____

NAME ______________________ DATE ______________ PERIOD __________

Lesson 2 Multi-Step Problem Solving

Multi-Step Example

The table shows the percent of time Allison spent studying each of her school subjects last week. What fraction of the subjects studied were math or history?

Preparation for **6.RP.3c, MP 1**

Ⓐ $\frac{1}{10}$ Ⓒ $\frac{3}{10}$

Ⓑ $\frac{2}{5}$ Ⓓ $\frac{4}{5}$

Subject	Time Spent Studying (% of week)
Math	30
Science	10
Language Arts	15
History	10
Reading	20
Music	15

Use a problem-solving model to solve this problem.

1 Understand

Read the problem. Circle the information you know.
Underline what the problem is asking you to find.

2 Plan

What will you need to do to solve the problem? Write your plan in steps.

Step 1 Determine the total percent for both math and history.

Step 2 Express the percent as a fraction in simplest form.

Read to Succeed!
Including Math or History means both subjects need to be included in the studying time.

3 Solve

Use your plan to solve the problem. Show your steps.

Math: ☐ %

History: ☐ %

Total percentage spent on Math and History: ☐ %

So, the simplified fraction is ☐ % or $\frac{☐}{☐}$. So, choice ☐ is correct.

Fill in that answer choice.

4 Check

How do you know your solution is accurate?

____

____

Lesson 2 *(continued)*

Use a problem-solving model to solve each problem.

1 The table shows the percent of each type of car rented last month. What fraction of the rentals were for a sedan or a truck?
Preparation for **6.RP.3c, MP 1**

Type of Car	Percent Rented
Minivan	13
Sport utility	37
Sedan	9
Convertible	4
Sports car	6
Truck	31

Ⓐ $\frac{9}{100}$

Ⓑ $\frac{1}{10}$

Ⓒ $\frac{3}{20}$

Ⓓ $\frac{2}{5}$

2 Benito had 10 days of vacation. He spent $\frac{1}{5}$ of his vacation fishing. He spent 30% of his vacation at soccer camp. He spent the rest of the time at the beach. What percent of his vacation did Benito spend at the beach?
Preparation for **6.RP.3c, MP 2**

____

3 Patricia spent 10 hours at the pool last week. She practiced the butterfly for 2 hours, the breaststroke for 5 hours, and the backstroke for 3 hours. This week she only has 5 hours at the pool. She wants to keep the same percentage of time spent on each stroke. How many more hours will she practice the breaststroke than the backstroke?
Preparation for **6.RP.3c, MP 2**

____

4 **H.O.T. Problem** Ramiro's garden is shown below. What percent of the total area of the garden do the cucumbers cover? Explain.
Preparation for **6.RP.3c, MP 1**

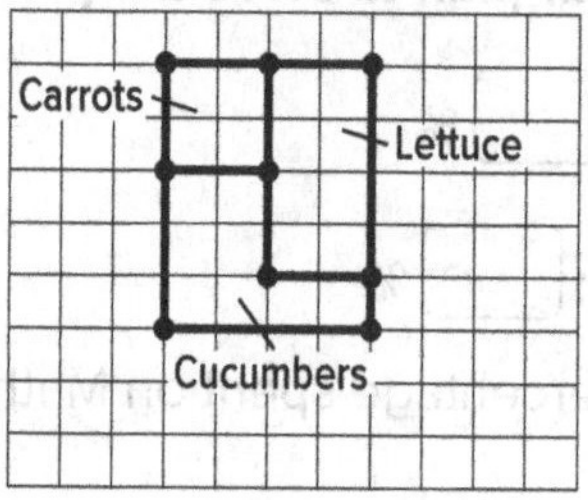

____

____

____

____

____

____

Lesson 3 **Multi-Step** Problem Solving

Multi-Step Example

A yogurt company decreased the amount of yogurt in each container it sells. The new containers contain 15% less yogurt than the original containers. The number line shows the original amount in each container. Which point represents the new amount? *Preparation for* **6.RP.3c,** **MP** **1**

Ⓐ Point *A*

Ⓑ Point *B*

Ⓒ Point *C*

Ⓓ Point *D*

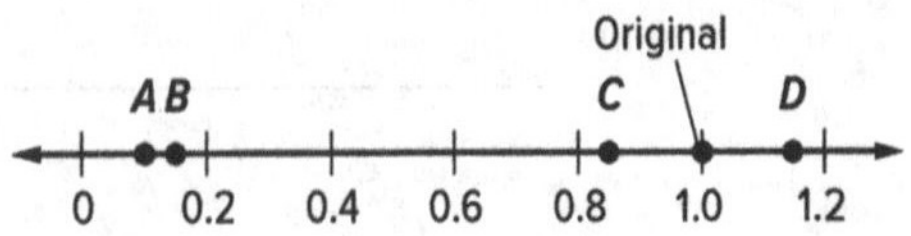

Use a problem-solving model to solve this problem.

1 Understand

Read the problem. Circle the information you know.
Underline what the problem is asking you to find.

2 Plan

What will you need to do to solve the problem? Write your plan in steps.

Step 1 Determine the new percentage of the amount of yogurt.

Step 2 Locate the percent as a decimal on the number line.

3 Solve

Use your plan to solve the problem. Show your steps.

The new percentage of yogurt is 100% − 15% or ____%.

As a decimal, 85% is ____.

Point ____ is located at 0.85. So, choice ____ is correct. Fill in that answer choice.

Read to Succeed!
To express a percent as a decimal, move the decimal point two places to the left and remove the percent symbol.

4 Check

How do you know your solution is accurate?

____

____

____

NAME ______________________ DATE ______________ PERIOD ________

Lesson 3 *(continued)*

Use a problem-solving model to solve each problem.

1 Dexter is tracking his progress in completing math assignments. He has completed 30% of his assignments. What decimal represents the part he has not completed? *Preparation for* **6.RP.3c, MP 2**

Ⓐ 0.03

Ⓑ 0.07

Ⓒ 0.3

Ⓓ 0.7

2 Trista deposited money into a savings account and left it there for several years. She earned a total of 12% interest on her deposit. How much did she earn per dollar? *Preparation for* **6.RP.3c, MP 2**

____

3 The table shows the number of students who earned various grades in English. What percent of the students earned an A or a B? *Preparation for* **6.RP.3c, MP 2**

Grade	Tally	Frequency
A	𝍸 IIII	9
B	𝍸 II	7
C	III	3
D	I	1

____

4 **H.O.T. Problem** Autumn used 0.675 of the battery life on her MP3 player. Her brother borrowed it to play a game and used 0.2 of the total battery life. If her MP3 player shows the battery life that remains as a percentage, what does it show after her brother is finished? *Preparation for* **6.RP.3c, MP 6**

____

Lesson 4 Multi-Step Problem Solving

Multi-Step Example

The table shows a salesperson's commissions during several consecutive years. If the 2010 commission is considered 100% of expected commissions, what percent of expected commissions is the 2011 value?
Preparation for **6.RP.3c, MP 1**

Use a problem-solving model to solve this problem.

1 Understand

Read the problem. Circle the information you know.
Underline what the problem is asking you to find.

2 Plan

What will you need to do to solve the problem? Write your plan in steps.

Step 1 Locate the amounts for 2010 and 2011.

Step 2 Express the amount of 2011 in relation to 2010 as a fraction and convert to a percent.

Read to Succeed!
The amount for 2010 is 100%. So, $10,000 is 100%.

3 Solve

Use your plan to solve the problem. Show your steps.

2011 ⟶ $12,150
2010 ⟶ $10,000

$\frac{\$12{,}150}{\$10{,}000} = \square = \square\%$

So, the amount for 2011 is ☐% of 2010.

4 Check

How do you know your solution is accurate?

____

____

____

Lesson 4 *(continued)*

Use a problem-solving model to solve each problem.

1 It is recommended that 13-year-old girls get 45 mg of Vitamin C each day. The table shows the Vitamin C content of various foods. What percentage of the recommended daily amount will Adelina receive if she eats all of these foods in one day? *Preparation for* **6.RP.3c, MP 2**

Food	Approx. Vitamin C Content (mg)
1 orange	70
1 green pepper	100
1 cup cooked broccoli	100

____

2 Adam has read $\frac{3}{20}$ of a novel in one week. The table shows the fraction of his reading that he completed each day. What percent of the novel did he read on Wednesday? *Preparation for* **6.RP.3c, MP 6**

Day	Fraction of Reading Completed
Monday	$\frac{1}{6}$
Tuesday	$\frac{2}{9}$
Wednesday	$\frac{1}{30}$
Thursday	$\frac{4}{9}$
Friday	$\frac{2}{15}$

____

3 If the volume of Rectangular Prism *A* is 0.01% of the volume of Rectangular Prism *B*, what is the ratio of the number of unit cubes it takes to fill Prism *B* to the number it takes to fill Prism *A*? *Preparation for* **6.RP.3c, MP 7**

____

4 **H.O.T. Problem** Shina wants to plot the values below as decimals on a histogram. Arrange the parts of the same whole shown in increasing order. *Preparation for* **6.RP.3c, MP 7**

0.01% 0.05 0.5% 0.500 50 500% 1%

____

Lesson 5 Multi-Step Problem Solving

Multi-Step Example

The table shows the portion of an exercise class that each student spent jogging. Who spent the most time jogging?
Preparation for **6.RP.3c, MP 4**

Ⓐ Kyle
Ⓑ Lin
Ⓒ Rabi
Ⓓ Sofia

Exercise Class	
Student	Portion Spent Jogging
Kyle	$\frac{3}{8}$
Lin	0.43
Rabi	35%
Sofia	$\frac{2}{5}$

Use a problem-solving model to solve this problem.

Read to Succeed!
It is usually easier to compare fractions, decimals, and percents when all the numbers are written as decimals.

1 Understand

Read the problem. Circle the information you know.
Underline what the problem is asking you to find.

2 Plan

What will you need to do to solve the problem? Write your plan in steps.

Step 1 Express each number as a decimal with the same number of places.

Step 2 Locate and compare the numbers on a number line.

3 Solve

Use your plan to solve the problem. Show your steps.

Write each number as a decimal.

$\frac{3}{8}$ = ______ 0.43 = ______ 35% = ______ $\frac{2}{5}$ = ______

Locate the numbers on a number line.

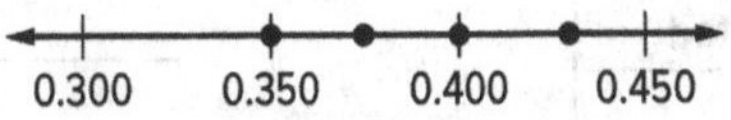

From least to greatest, the numbers are 35%, $\frac{3}{8}$, $\frac{2}{5}$, and 0.43. Since ______ is the greatest number, Lin spent the most time jogging.

So, choice ______ is the correct answer. Fill in that answer choice.

4 Check

How do you know your solution is reasonable?

____

____

____

Lesson 5 *(continued)*

Use a problem-solving model to solve each problem.

1 The table shows the middle school band concert attendance by class. Which class has the greatest part of the class attending the concert? *Preparation for* **6.RP.3c, MP 1**

Middle School Band Concert	
Class	**Attendance**
Class A	$\frac{4}{5}$
Class B	0.75
Class C	85%
Class D	$\frac{5}{6}$

Ⓐ Class A
Ⓑ Class B
Ⓒ Class C
Ⓓ Class D

2 The after-school activities of the students in Caleb's class are shown in the table. Order the activities from least to greatest. *Preparation for* **6.RP.3c, MP 2**

After-School Activities	
Activity	**Portion of Students**
Basketball	0.1
Computers	$\frac{3}{10}$
Gymnastics	$\frac{1}{5}$
Soccer	25%
Other	$\frac{3}{20}$

____

____

____

3 The table shows the portion of math homework that students have completed. Which students have completed more than 75% of their math homework? *Preparation for* **6.RP.3c, MP 2**

Math Homework	
Student	**Portion Completed**
Clara	$\frac{5}{8}$
Kyra	68%
Levon	0.85
Noah	$\frac{8}{10}$
Soto	$\frac{3}{4}$

____

____

4 **H.O.T. Problem** Write a fraction that is between 0.4 and 60%. Then write a fraction that is between 0.4 and the fraction you wrote. Write all four numbers in order from least to greatest. Explain how you know your answer is correct. *Preparation for* **6.RP.3c, MP 3**

____

____

____

____

____

NAME _______________ DATE _______________ PERIOD _______________

Lesson 6 Multi-Step Problem Solving

Chapter 2

Multi-Step Example

Sabrina takes her car to the car wash and gets the Gold Star service that includes a wash, wax, and interior cleaning. This service costs $51.99, but she must also pay a 6% sales tax. Estimate the total amount Sabrina paid at the car wash. **6.RP.3, MP 1**

Ⓐ $3.00

Ⓑ $47.50

Ⓒ $53.00

Ⓓ $75.00

Use a problem-solving model to solve this problem.

1 Understand

Read the problem. Circle the information you know.
Underline what the problem is asking you to find.

2 Plan

What will you need to do to solve the problem? Write your plan in steps.

Step 1 Determine the benchmark percent and use it to determine the tax.

Step 2 Add to determine the total spent.

3 Solve

Use your plan to solve the problem. Show your steps.

6% is a multiple of ☐%. $51.99 is about $☐.

Using the benchmark percent, 1% of $50 is $☐.

The total tax is 6 × $0.50 or $☐.

The total spent is about $☐ + $☐ or $☐.

So, Sabrina will spend about $☐. Choice ☐ is correct.
Fill in that answer choice.

Read to Succeed!
By using a benchmark percent, you can mentally estimate the total.

4 Check

How do you know your solution is accurate?

____

____

NAME________________________ DATE ______________ PERIOD ________

Lesson 6 *(continued)*

Use a problem-solving model to solve each problem.

1 A sporting goods store purchases a skateboard for $100 and marks the price up by 40%. The store is having a sale where everything is 15% off the sticker price. Estimate the final price of a skateboard. **6.RP.3, MP 1**

Ⓐ $119
Ⓑ $125
Ⓒ $140
Ⓓ $155

2 Emilio buys 3 pizzas, 4 subs, and 8 sodas for a party. The sales tax is 7.5%. What is the minimum number of $20 bills Emilio should pay with? **6.RP.3, MP 6**

Item	Cost ($)
Pizza	10
Sub	5
Soda	1

3 There were 485 people who went to an amusement park on Monday. Sixty percent of the people wanted to ride the new roller coaster. Twenty-three percent of those people decided not to ride the coaster because the line was too long. About how many people waited in line? **6.RP.3, MP 1**

4 **H.O.T. Problem** Suppose the area of the rectangle below was increased by 20%. What would be the perimeter of the larger rectangle if the length of 15 feet stayed the same? Explain the steps you used to solve this problem. **6.RP.3, MP 1**

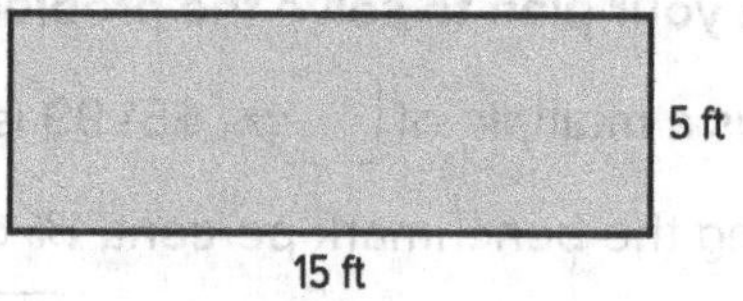

Lesson 7 Multi-Step Problem Solving

Multi-Step Example

Students were asked which night they planned on attending the book fair. The results of a survey are shown in the table. If 25% of the people who responded with Thursday did not go to the book fair that night, how many people did go on Thursday? **6.RP.3, MP 1**

Day	Number of People
Monday	55
Tuesday	80
Wednesday	70
Thursday	112
Friday	65

Ⓐ 112

Ⓑ 100

Ⓒ 84

Ⓓ 28

Use a problem-solving model to solve this problem.

1 Understand

Read the problem. Circle the information you know.
Underline what the problem is asking you to find.

2 Plan

What will you need to do to solve the problem? Write your plan in steps.

Step 1 Determine the number of people that did not go to the book fair on Thursday.

Step 2 Subtract to determine the number of people that did go on Thursday.

3 Solve

Use your plan to solve the problem. Show your steps.

The number of people that did not go to the book fair on Thursday is 25% of 112 or ☐.

The number of people that did go on Thursday is 112 − ☐ or ☐.

So, choice ☐ is correct. Fill in that answer choice.

Read to Succeed!
When finding the percent of a number, rename the percent as a decimal and multiply.

4 Check

How do you know your solution is accurate?

____

____

____

Lesson 7 *(continued)*

Use a problem-solving model to solve each problem.

1 Students were surveyed about their summer plans. Of the people that stated they were traveling abroad, 30% did not actually travel abroad. Of the people that stated they were going to summer camp, 25% did not actually go. How many more students went to summer camp than traveled abroad? **6.RP.3, MP 1**

Summer Plans	Number of People
Summer camp	252
Traveling abroad	180
Visiting grandparent	327

Ⓐ 315
Ⓑ 189
Ⓒ 126
Ⓓ 63

2 Five hundred students were asked what color they prefer for the new school colors. The results are shown in the circle graph. How many students prefer red or black? **6.RP.3, MP 4**

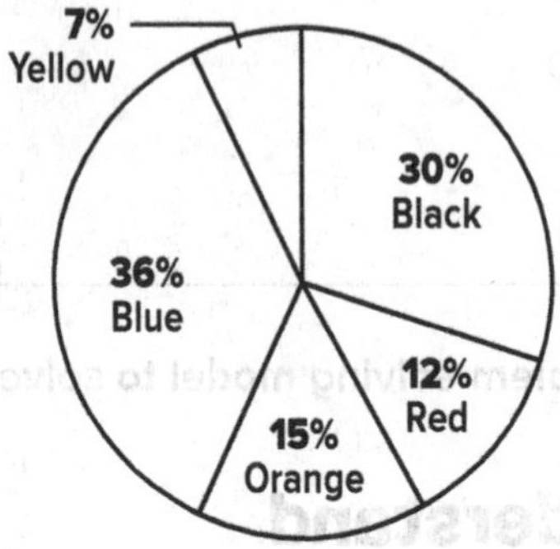

____

3 Carmen is going to buy a pair of sneakers that cost $63. The sales tax rate is 7.5%. What is the total cost of the sneakers to the nearest cent? **6.RP.3, MP 1**

____

4 **H.O.T. Problem** Aaron is estimating the growth of his puppy over time. If the puppy continues to grow at the same rate, how old will the puppy be when it is 250% of its 2-month weight? **6.RP.3, MP 8**

Age (months)	Weight (lb)
2	4
3	5.5

____

NAME ____________________ DATE ____________ PERIOD ________

Lesson 8 Multi-Step Problem Solving

Multi-Step Example

The table shows the percentage of each type of popcorn flavor at a specialty food store. A store clerk put all of the bags of cinnamon popcorn and cheese popcorn in a display in the front of the store. If the clerk put 60 bags up front, how many bags of popcorn does the store have in all? 6.RP.3, MP 1

Popcorn Flavor	
Kettle corn	60%
Cinnamon	15%
Caramel	10%
Cheese	15%

Ⓐ 18

Ⓑ 100

Ⓒ 200

Ⓓ 400

Use a problem-solving model to solve this problem.

1 Understand

Read the problem. Circle the information you know.
Underline what the problem is asking you to find.

2 Plan

What will you need to do to solve the problem? Write your plan in steps.

Step 1 Determine the total percentage of bags displayed.

Step 2 Use the percent proportion to determine the whole.

3 Solve

Use your plan to solve the problem. Show your steps.

The percentage of bags displayed is 15% + 15% or ☐%.

The situation can be represented by the proportion: $\frac{60}{■} = \frac{30}{100}$.

Since 30 × ☐ is 60, multiply 100 × ☐.

So, the total number of bags the store has is 100 × 2 or ☐.

Choice ☐ is correct. Fill in that answer choice.

Read to Succeed!
The percent proportion is $\frac{part}{whole} = \frac{percent}{100}$.

4 Check

How do you know your solution is accurate?

____

____

NAME________________________ DATE ______________ PERIOD ________

Lesson 8 *(continued)*

Use a problem-solving model to solve each problem.

1 The table shows the percentage of each type of puzzle in a toy store. During a sale, the store sold all of the 300-piece and 500-piece puzzles. If they sold 120 puzzles, how many puzzles did the store have before the sale? **6.RP.3, MP 2**

Jigsaw Puzzles	
300-piece	50%
500-piece	30%
750-piece	15%
1,000-piece	5%

Ⓐ 150
Ⓑ 240
Ⓒ 400
Ⓓ 600

2 Miguel surveyed 150 students about their favorite sport. The results are shown in the circle graph. How many more students chose basketball than tennis? **6.RP.3, MP 4**

Favorite Sport

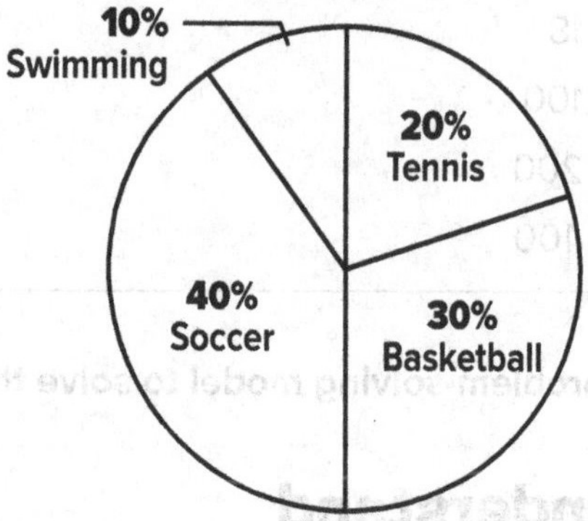

3 A tile wall is shown below. Each square tile has a side length of 4 inches. What percent of the wall is gray? **6.RP.3, MP 4**

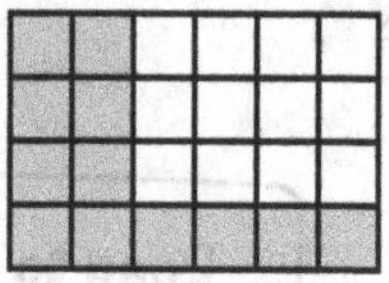

____

4 **H.O.T. Problem** The perimeter of a square is 30% of the perimeter of a larger square. If the perimeter of the smaller square is 4.8 inches, what is the side length of the larger square? Explain how you found your answer. **6.RP. 3, MP 7**

____

____

____

____

____

NAME______________________________ DATE ______________ PERIOD ________

Lesson 1 **Multi-Step** Problem Solving

Multi-Step Example

The table shows the amount of time Jessica spent practicing different strokes in swim class. If the class is 1 hour long, how many minutes did Jessica have left to practice freestyle?
6.NS.3, MP 1

Ⓐ 15.65 min
Ⓑ 24.35 min
Ⓒ 26.65 min
Ⓓ 44.35 min

Jessica's Swim Times	
Swimming Stroke	**Time (min)**
Backstroke	12.5
Breaststroke	13.75
Butterfly	18.1
Freestyle	?

Use a problem-solving model to solve this problem.

1 Understand

Read the problem. Circle the information you know. Underline what the problem is asking you to find.

Read to Succeed!
When adding and subtracting decimals, use zeros to write equivalent decimals. This way all the decimals have the same number of places and are aligned properly.

2 Plan

What will you need to do to solve the problem? Write your plan in steps.

Step 1 Estimate.

Step 2 Line up the decimal points and add.

Step 3 Subtract the sum from the number of minutes in an hour.

3 Solve

Use your plan to solve the problem. Show your steps.

Estimate: $12.5 + 13.75 + 18.1 \approx 13 + 14 + 18$ or ______ and $60 - 45 =$ ______

Line up the decimal points and add.

$$\begin{array}{r} 12.50 \\ 13.75 \\ +\ 18.10 \\ \hline \_\_\_\_ \end{array}$$

Subtract.

$$\begin{array}{r} 60.00 \\ -44.35 \\ \hline \_\_\_\_ \end{array}$$

So, ______ is the correct answer. Fill in that answer choice.

4 Check

How do you know your solution is reasonable?

____

____

Lesson 1 *(continued)*

Use a problem-solving model to solve each problem.

1 The table shows the weight of three different-size bags of peanuts sold at the store. How many more ounces of peanuts will you get if you buy 2 large bags than if you buy 2 small bags and 1 medium bag? **6.NS.3, MP 1**

Bags of Peanuts	
Bag Size	**Weight (oz)**
Small bag	12.5
Medium bag	18.25
Large bag	22.8

Ⓐ 2.25 oz
Ⓑ 2.35 oz
Ⓒ 12.35 oz
Ⓓ 18.85 oz

2 Brandon has a board that is 10 feet long. He wants to cut the board into 4 shelves. The lengths of 3 shelves are shown in the table. How long will the fourth shelf be? **6.NS.3, MP 6**

Brandon's Shelves	
Shelf	**Length (ft)**
Shelf A	1.5
Shelf B	2.35
Shelf C	3.85
Shelf D	?

3 Tia bought crackers for $3.68, two loaves of bread for $3.29 each, a bag of apples for $5.99, and three dozen eggs for $1.65 per dozen. How much more than $20 do these items cost? **6.NS.3, MP 1**

4 **H.O.T. Problem** Logan says that 41.75 − 24.689 = 17.079. What might he have done wrong? **6.NS.3, MP 3**

NAME______________________________ DATE ______________ PERIOD ________

Lesson 2 **Multi-Step** Problem Solving

Multi-Step Example

David, Nicole, and Tyrell each earn $19.15 an hour working at a science museum. The table shows how many hours they work each week. About how much more does Tyrell earn than Nicole in a week? Estimate to solve the problem. **6.NS.3,** MP 1

Employee	Hours Worked
David	18
Nicole	26.5
Tyrell	37.5

Use a problem-solving model to solve this problem.

Read to Succeed!
There are different ways to round numbers. It is easier to multiply numbers that have been rounded to the greatest place value.

1 Understand

Read the problem. Circle the information you know. Underline what the problem is asking you to find.

2 Plan

What will you need to do to solve the problem? Write your plan in steps.

Step 1 Determine the numbers needed to solve the problem. Round each number to the greatest place value to make it easier to compute.

Step 2 Multiply the rounded numbers to find about how much each employee earns.

Step 3 Subtract to find the difference.

3 Solve

Use your plan to solve the problem. Show your steps.

Round to the greatest place value. 19.15 ≈ 20 26.5 ≈ ______ 37.5 ≈ ______

Estimate what Nicole earns. Estimate what Tyrell earns. Find the difference.

20 × 30 = ______ 20 × 40 = ______ 800 − 600 = ______

Nicole makes about $600 a week. Tyrell makes about $800 a week. So Tyrell makes about $200 more than Nicole in a week.

4 Check

How do you know your solution is accurate?

____

____

____

Lesson 2 *(continued)*

Use a problem-solving model to solve each problem.

1 Irene, Pablo, and Kari each earn $8.85 an hour working at an animal shelter. The table shows how many hours they work each week. About how much less does Kari earn than Irene in a week? Estimate to solve the problem. **6.NS.3, MP 1**

Employee	Hours Worked
Irene	32.5
Pablo	15.75
Kari	21.5

2 The table shows the price per pound of some items at the deli in the grocery store. About how much would 3 pounds of Swiss cheese and 4 pounds of baked ham cost? **6.NS.3, MP 4**

Item	Cost per Pound
Baked ham	$6.25
Roast beef	$7.99
Swiss cheese	$4.95
Smoked turkey	$6.49

3 LaToya wants to buy a backpack for $15.39 and 5 notebooks for $4.25 each. She estimates that the items will cost $35. Is the actual cost more than or less than her estimate? Explain your reasoning. **6.NS.3, MP 3**

4 **H.O.T. Problem** Jason says that if both factors are rounded up when estimating the product of two factors, the estimated product will always be greater than the actual product. Is he correct? Explain your reasoning. **6.NS.3, MP 3**

Lesson 3 **Multi-Step** Problem Solving

Multi-Step Example

The table shows the ingredients in a fruit punch recipe. Luis has 3.75 quarts of cranberry juice. How many cups of cranberry juice will he have left after he makes 3 batches of the punch? **6.NS.3,** MP 1

Ⓐ 4.75 c
Ⓑ 5.25 c
Ⓒ 9.75 c
Ⓓ 15 c

Fruit Punch Recipe	
Ingredient	**Amount (c)**
Cranberry juice	3.25
Lemonade	1.75
Orange juice	2.25
Sparkling water	4.0

Use a problem-solving model to solve this problem.

Read to Succeed!
Hint: There are 4 cups in 1 quart.

1 Understand

Read the problem. Circle the information you know.
Underline what the problem is asking you to find.

2 Plan

What will you need to do to solve the problem? Write your plan in steps.

Step 1 Multiply to determine the number of cups in 3 batches of punch.

Step 2 Convert 3.75 quarts to cups.

Step 3 Subtract to determine the number of cups left.

3 Solve

Use your plan to solve the problem. Show your steps.

Multiply. $3.25 \times 3 = \square$

There are 4 cups in 1 quart. $3.75 \times 4 = \square$

So, there are 15 cups in 3.75 quarts.

Subtract. $15.00 - 9.75$

So, Luis will have 5.25 cups of cranberry juice left.

So, ______ is the correct answer. Fill in that answer choice.

4 Check

How do you know your solution is reasonable?

____

____

____

Lesson 3 *(continued)*

Use a problem-solving model to solve each problem.

1 The table shows the ingredients in 1 batch of snack mix. Kenji has 1.5 pounds of raisins. How many more ounces of raisins does he need to make 5 batches of snack mix? (*Hint*: There are 16 ounces in 1 pound.). **6.NS.3, MP 1**

Snack Mix Recipe	
Ingredient	**Weight (oz)**
Flaked coconut	3.5
Peanuts	6.8
Raisins	7.5
Toasted oats	4.25

Ⓐ 13.5 oz

Ⓑ 21.1 oz

Ⓒ 24 oz

Ⓓ 37.5 oz

2 Rosa had $35.20 in her checking account. She wrote 4 checks for $7.59 each. How much did she have left in her checking account after she wrote the 4 checks? **6.NS.3, MP 6**

____

3 The library is having a book sale. All hardcover books cost $3.75 and all paperback books cost $2.15. Shari has $20. How much more money does she need to buy 4 hardcover books and 4 paperback books? **6.NS.3, MP 5**

____

4 **H.O.T. Problem** Who solved the problem correctly? Explain your answer. **6.NS.3, MP 3**

Matthew	Kendra
175 × 0.05 8.75	175 × 0.05 0.875

____

____

____

____

Lesson 4 Multi-Step Problem Solving

Multi-Step Example

Cheyenne buys the amounts of fruit shown in the table. She uses 1.05 pounds of grapes in a salad. How many times more pounds of grapes does she have now than apples? Round to the nearest hundredth. **6.NS.3, MP 1**

Fruit	Pounds
Cherries	3.2
Grapes	4.8
Bananas	3.8
Apples	2.25

Ⓐ 1.47
Ⓑ 1.50
Ⓒ 1.67
Ⓓ 2.13

Use a problem-solving model to solve this problem.

1 Understand

Read the problem. Circle the information you know.
Underline what the problem is asking you to find.

2 Plan

What will you need to do to solve the problem? Write your plan in steps.

Step 1 ______ to find the number of pounds of grapes she has after making the salad.

Step 2 ______ to find how many times more pounds of grapes she has than apples.

Read to Succeed!
Divide to determine how many times greater one quantity is of another.

3 Solve

Use your plan to solve the problem. Show your steps.

She has 4.8 − 1.05 or ______ pounds of grapes left.

She has ______ ÷ 2.25 or ______ times more pounds of grapes than apples.

Since $1.6\overline{6}$ is about ______, the correct choice is ______. Fill in that answer choice.

4 Check

How do you know your solution is reasonable?

______

______

______

Lesson 4 *(continued)*

Use a problem-solving model to solve each problem.

1 The lengths of two wires are shown below. A project uses 32.6 centimeters of Wire B. How many times longer is Wire B than Wire A after part of Wire B is used? (Wires not drawn to scale.) **6.NS.3, MP 2**

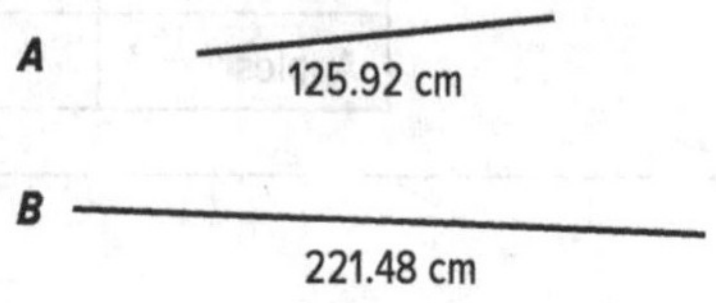

Ⓐ 1.5
Ⓑ 2.4
Ⓒ 3.9
Ⓓ 6.8

2 Mika has $9.50. She buys as many bumper stickers as she can afford. Then she earns $2.50 for her allowance. How much money, in dollars, does she have now? **6.NS.3, MP 1**

Item	Cost
Bumper sticker	$1.25
Hat	$6.00
Mug	$5.50

____

3 Elijah has a red ribbon that is 12.2 inches long. He has a green ribbon that is 3.4 inches longer than the red ribbon. He has a yellow ribbon that is 28.08 inches long. How many times longer is the yellow ribbon than the green ribbon? **6.NS.3, MP 4**

____

4 H.O.T. Problem Determine which quotient is greater, without performing the division. Explain your reasoning. **6.NS.3, MP 7**

$8.54 \div 5.23$ $8.54 \div 5.32$

____

____

____

NAME ______________________________ DATE ________________ PERIOD ________

Lesson 5 Multi-Step Problem Solving

Multi-Step Example

The table shows the number of cookies made for the bake sale. The cookies were put into bags with a dozen cookies in each bag. How many bags of a dozen cookies were there? **6.NS.2, MP 1**

Bake Sale Cookies	
Type	**Number**
Chocolate chip	125
Oatmeal	60
Peanut butter	245
Sugar	116

Ⓐ 60 Ⓒ 50

Ⓑ 55 Ⓓ 45

Use a problem-solving model to solve this problem.

Read to Succeed!
When you divide, use estimation to help you place the first digit in the quotient.

1 Understand

Read the problem. Circle the information you know.
Underline what the problem is asking you to find.

2 Plan

What will you need to do to solve the problem? Write your plan in steps.

Step 1 Add to determine the total number of cookies.

Step 2 Write the division problem. Estimate the quotient.

Step 3 Divide and interpret the quotient.

3 Solve

Use your plan to solve the problem. Show your steps.

Add
$$\begin{array}{r} 125 \\ 60 \\ 245 \\ +\,116 \\ \hline \square \end{array}$$

Divide $546 \div 12$

Estimate $500 \div 10 = \square$

Divide
$$\begin{array}{r} \square \text{ R} \square \\ 12\overline{)546} \\ -48 \\ \hline 66 \\ -60 \\ \hline 6 \end{array}$$

There are 45 bags with a dozen cookies in each bag and 6 cookies left over.

So, ______ is the correct answer. Fill in that answer choice.

4 Check

How do you know your solution is reasonable?

____

____

____

Lesson 5 *(continued)*

Use a problem-solving model to solve each problem.

1 There are 24 seats in each row of the middle school auditorium. The table shows the number of students from each grade who attended a concert in each grade of the middle school. If the students fill each row in the auditorium, how many rows are needed for all the students? **6.NS.2, MP 1**

Grade	Number of Students
Sixth	310
Seventh	256
Eighth	272

Ⓐ 25 rows

Ⓑ 32 rows

Ⓒ 35 rows

Ⓓ 38 rows

2 Ryland and Charlotte each drove their own cars on 4–day vacations. The table shows the number of miles each person drove each day. Ryland used 46 gallons of gas and Charlotte used 49 gallons of gas. Who got more miles per gallon on his or her trip? How much more? **6.NS.2, MP 6**

Day	Miles Ryland Drove	Miles Charlotte Drove
Thursday	425	157
Friday	312	253
Saturday	175	279
Sunday	330	536

____

3 A stadium holds 37,402 people. There are 18,682 empty seats in the stadium. Suppose, the stadium has 36 sections and each section has the same number of people. How many people are in each section of the stadium? **6.NS.2, MP 4**

____

4 **H.O.T. Problem** Write and solve a real-world problem that uses division. Make the solution to the problem the remainder in the division problem. **6.NS.2, MP 2**

____

____

____

____

____

NAME________________________________ DATE ____________ PERIOD ________

Lesson 6 Multi-Step Problem Solving

Multi-Step Example

The table shows the average monthly precipitation in Phoenix, Arizona and Atlanta, Georgia during the months of January, February, and March. About how many times more inches of precipitation does Atlanta have than Phoenix during the three months? 6.NS.2, MP 1

Average Precipitation (in.)		
Month	Phoenix, AZ	Atlanta, GA
January	0.8	5.0
February	0.8	4.7
March	1.1	5.4

Ⓐ about 3 times

Ⓑ about 4 times

Ⓒ about 5 times

Ⓓ about 7 times

Use a problem-solving model to solve this problem.

Read to Succeed!
You can use rounding and compatible numbers to help you estimate quotients.

1 Understand

Read the problem. Circle the information you know.
Underline what the problem is asking you to find.

2 Plan

What will you need to do to solve the problem? Write your plan in steps.

Step 1 Add to determine the precipitation for each city.

Step 2 Write the division problem.

Step 3 Use rounding to estimate the quotient.

3 Solve

Use your plan to solve the problem. Show your steps.

Add 0.8 + 0.8 + 1.1 = ☐ 5.0 + 4.7 + 5.4 = ☐

$2.7\overline{)15.1} \rightarrow 3\overline{)15}$ = 5

Round 2.7 to 3 and 15.1 to 15 to write compatible numbers.

So, _____ is the correct answer. Fill in that answer choice.

4 Check

How do you know your solution is reasonable?

____

____

____

Lesson 6 *(continued)*

Use a problem-solving model to solve each problem.

1 The table shows the average monthly snowfall on the north rim of the Grand Canyon and Zion National Park during the months of January, February, and March. About how many times more inches of snow fall on the north rim of the Grand Canyon than in Zion National Park during the three months? **6.NS.2, MP 1**

Average Snowfall (in.)		
Month	**Grand Canyon**	**Zion National Park**
January	34.8	3.3
February	25.2	1.7
March	28.2	1.1

Ⓐ about 6
Ⓑ about 10
Ⓒ about 12
Ⓓ about 15

2 The table shows the weight of the peanuts, walnuts, pecans, and almonds that Mario has to make snack bags of mixed nuts. He wants to put 8 ounces of mixed nuts in each snack bag. Estimate to determine whether or not Mario has enough mixed nuts for 25 snack bags? Explain why your answer is reasonable. [*Hint:* There are 16 ounces in 1 pound.] **6.NS.2, MP 2**

Type of Nut	Weight (lb)
Peanuts	4.2
Walnuts	2.4
Pecans	0.8
Almonds	2.7

3 Noor wants to buy a laptop that costs \$698.99. The tax on the laptop will be \$48.93. Noor has saved \$252. She will use the money she has saved to help pay for the laptop. Noor estimates that if she saves \$70 per month for 6 months, she will have enough money to buy the computer. Is her estimate reasonable? Explain your answer. **6.NS.2, MP 3**

4 **H.O.T. Problem** Show two different ways you can use compatible numbers to estimate the quotient of the problem shown. Which pair of compatible numbers do you think has a quotient closer to the actual quotient? Explain why your answer is reasonable. **6.NS.2, MP 7**

377.5 ÷ 23.15

NAME _______________ DATE _______________ PERIOD _______________

Lesson 7 Multi-Step Problem Solving

Multi-Step Example

The table shows the price for cheese pizzas at the Pizza Parlor. Each added topping is $0.75. Eight friends are sharing 2 small pizzas with mushrooms and 1 large pizza with sausage and peppers. If the friends share the cost equally, how much should each friend pay for the pizzas? **6.NS.3, MP 1**

Pizza Parlor Prices	
Size	**Price**
Small	$8.50
Medium	$10.25
Large	$11.60

Ⓐ $3.45
Ⓑ $3.95
Ⓒ $4.45
Ⓓ $4.95

Use a problem-solving model to solve this problem.

Read to Succeed!
When dividing with decimals, you can use the estimate to help you place the decimal point in the correct location in the quotient.

1 Understand

Read the problem. Circle the information you know.
Underline what the problem is asking you to find.

2 Plan

What will you need to do to solve the problem? Write your plan in steps.

Step 1 Find the total cost of the pizza.

Step 2 Write the division problem. Estimate the quotient.

Step 3 Divide to find how much each friend should pay.

3 Solve

Use your plan to solve the problem. Show your steps.

Multiply and add.

2 small pizzas → 2($8.50 + $0.75) = ______

1 large pizza → $11.60 + 2($0.75) = ______

Total cost → $18.50 + $13.10 = ______

Divide.

$31.60 ÷ 8

Estimate.

$32 ÷ 8 = ______

Divide.

$$\begin{array}{r} \boxed{} \\ 8\overline{)31.60} \\ -24 \\ \hline 76 \\ -72 \\ \hline 40 \\ -40 \\ \hline 0 \end{array}$$

So, ______ is the correct answer. Fill in that answer choice.

4 Check

How do you know your solution is reasonable?

____

____

Lesson 7 *(continued)*

Use a problem-solving model to solve each problem.

1 The table shows the price for some bracelets at Jewelry Gems. Charms can be added to a bracelet for an additional $2.50 each. Maggie buys 2 silver bracelets with 3 charms each and 1 gold bracelet with 4 charms. What is the average price of the bracelets? **6.NS.3, MP 1**

Bracelet Prices	
Type	**Price**
Bronze	$15.95
Silver	$21.75
Gold	$28.25

Ⓐ $12.75
Ⓑ $29.75
Ⓒ $31.75
Ⓓ $32.25

2 The table shows the weight of a Labrador retriever puppy. Was the puppy's average monthly weight gain greater between the ages 2 months and 4 months or between 8 months and 10 months? How much greater? **6.NS.3, MP 1**

Labrador Retriever Puppy	
Age (months)	**Weight (lb)**
2	15.4
4	30.8
6	44.8
8	52.9
10	57.3

3 Waylen painted a house in 15 hours. He was paid $35.75 per hour for painting it. He used some of his earnings to buy four new paintbrushes that each cost the same amount. If Waylen had $468.69 of his earnings left, how much did each paintbrush cost? **6.NS.3, MP 2**

4 **H.O.T. Problem** Find the quotient for 124.66 ÷ 23. Then explain how you can use a pattern to find the quotients for these problems. Write the quotients. **6.NS.3, MP 7**

1246.6 ÷ 23   12.466 ÷ 23   1.2466 ÷ 23

NAME_______________ DATE _______________ PERIOD _______

Chapter 3

Lesson 8 **Multi-Step** Problem Solving

Multi-Step Example

The table shows the cost of produce per pound at a farmer's market. Mr. Gonzalez bought 0.75 pound of pears and 3.5 pounds of plums. What was his change from $10? **6.NS.3,** MP 1

Produce	Cost per Pound
Pears	$0.98
Oranges	$1.29
Carrots	$1.18
Plums	$1.49

Ⓐ $4.05

Ⓑ $5.95

Ⓒ $6.33

Ⓓ $10.50

Use a problem-solving model to solve this problem.

1 Understand

Read the problem. Circle the information you know.
Underline what the problem is asking you to find.

2 Plan

What will you need to do to solve the problem? Write your plan in steps.

Step 1 Determine the total amount spent on pears and the total amount spent on plums.

Step 2 Subtract the amounts spent on pears and plums from $10.00.

3 Solve

Use your plan to solve the problem. Show your steps.

Pears: 0.75 × $______ = $______

Plums: 3.5 × $______ = $______

Total spent: $______ + $______ = $______

$10.00 − $______ = $______

So, Mr. Gonzalez would receive $______ in change. The correct answer is ______.

Read to Succeed!
When rounding, wait to round until the end of the problem.

4 Check

How do you know your solution is accurate?

____

____

____

NAME ______________________ DATE ____________ PERIOD ________

Lesson 8 *(continued)*

Use a problem-solving model to solve each problem.

1 The table shows the cost per yard of different types of fabric. Sierra bought 2.5 yards of nylon and 4.5 yards of cotton. What is the change from $5? Round to the nearest cent. **6.NS.3, MP 1**

Fabric	Cost per Yard
Cotton	$0.50
Linen	$0.25
Rayon	$0.125
Nylon	$0.75

Ⓐ $0.87
Ⓑ $2.00
Ⓒ $2.25
Ⓓ $4.13

2 The rectangle represents Demarco's garden. For each square foot, he needs to use 2 scoops of fertilizer. How many scoops does he use? **6.NS.3, MP 2**

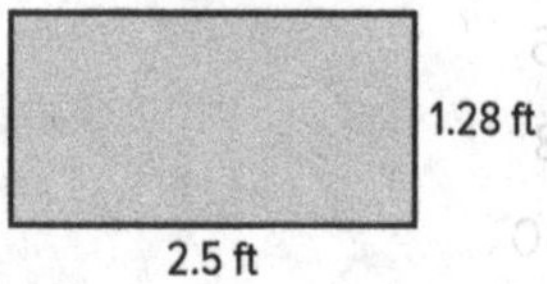

____

3 Abigail runs at a constant rate of 7.057 miles per hour. Jamal runs at a constant rate of 6.4 miles per hour. At these rates, how much farther did Abigail run than Jamal after the first 0.5 hour? Round your answer to the nearest hundredth. (*Hint*: Distance = rate × time) **6.NS.3, MP 1**

____

4 H.O.T. Problem A name brand cereal is on sale and costs $4.50 for an 18.2-oz box. The grocery store version of the cereal costs $4.03 for a 13.1-oz box. Which cereal costs less per ounce? **6.NS.3, MP 4**

____

____

____

NAME ______________________ DATE ______________ PERIOD ________

Lesson 1 **Multi-Step** Problem Solving

Multi-Step Example

Jake made a drawing of the vegetable garden he wants to plant. Estimate the area of the vegetable garden.
Preparation for **6.NS.1, MP 1**

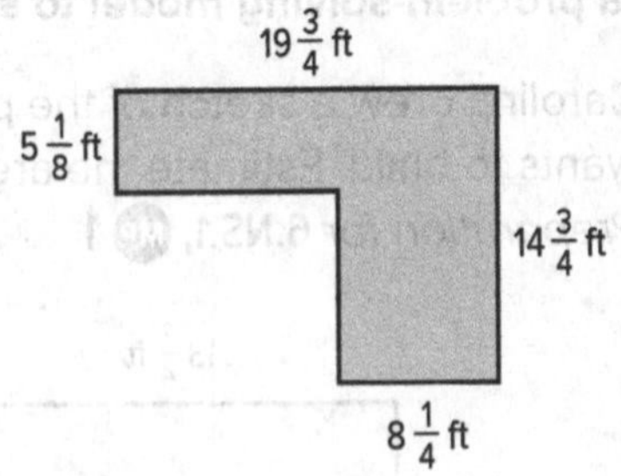

Ⓐ about 120 ft^2
Ⓑ about 180 ft^2
Ⓒ about 200 ft^2
Ⓓ about 300 ft^2

Use a problem-solving model to solve this problem.

1 Understand

Read the problem. Circle the information you know.
Underline what the problem is asking you to find.

Read to Succeed!
To find the area of a rectangle, multiply length by width.

2 Plan

What will you need to do to solve the problem? Write your plan in steps.

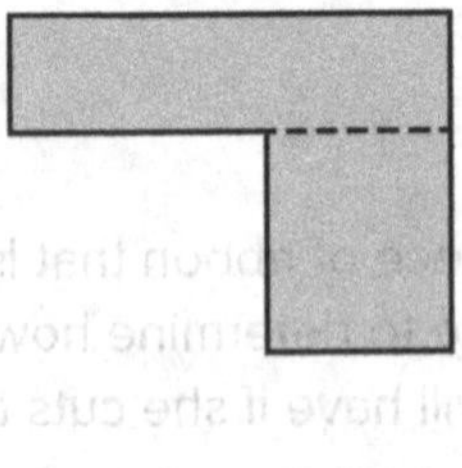

Step 1 Separate the garden into two rectangles. Round each mixed number to the nearest whole number. Subtract to find the estimated width of the smaller rectangle.

Step 2 Multiply to find the area of each rectangle.

Step 3 Add the two areas to find the estimated area of the garden.

3 Solve

Use your plan to solve the problem. Show your steps.

Round: $19\frac{3}{4} \rightarrow$ ______ $5\frac{1}{8} \rightarrow$ ______ $14\frac{3}{4} - 5 \rightarrow$ ______ $8\frac{1}{4} \rightarrow$ ______

Multiply: $20 \times 5 = 100$ and $10 \times 8 = 80$ Add: $100 + 80 = 180$
The garden is about 180 square feet.

So, ______ is the correct answer . Fill in that answer choice.

4 Check

How do you know your solution is reasonable?

____

____

____

Lesson 1 *(continued)*

Use a problem-solving model to solve each problem.

1 Carolina drew a sketch of the patio she wants to build. Estimate the area of the patio. *Preparation for* **6.NS.1, MP 1**

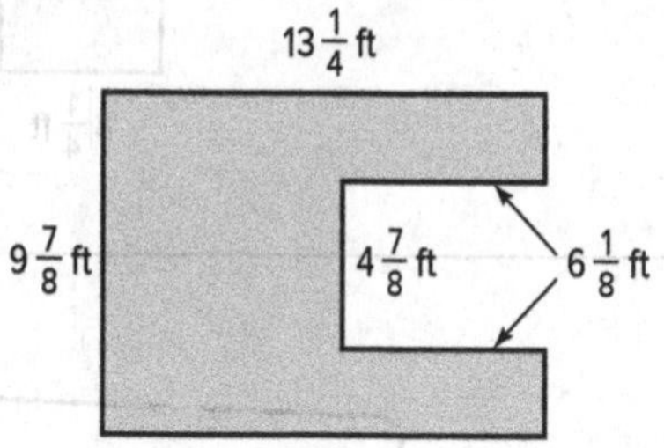

Ⓐ about 100 ft²
Ⓑ about 130 ft²
Ⓒ about 160 ft²
Ⓓ about 190 ft²

2 Layton bought $3\frac{1}{4}$ pounds of grapes. He and his friends ate $\frac{5}{6}$ of the grapes he bought. Jaylinn bought $5\frac{7}{8}$ pounds of grapes. She and her friends ate $\frac{2}{5}$ of the grapes she bought. Estimate to determine who has more grapes left? Explain your answer. *Preparation for* **6.NS.1, MP 4**

____

____

____

____

____

____

3 Kuni has a piece of ribbon that is $45\frac{1}{2}$ yards long. Estimate to determine how many more pieces she will have if she cuts the ribbon into $2\frac{7}{8}$-yard strips than if she cuts the ribbon into $4\frac{1}{4}$-yard strips. *Preparation for* **6.NS.1, MP 2**

____

4 H.O.T. Problem Drawings of a square room and a rectangular room are shown. Estimate to determine how the areas of the floors compare. Which room do you think has the greater actual area? Explain how you know. *Preparation for* **6.NS.1, MP 3**

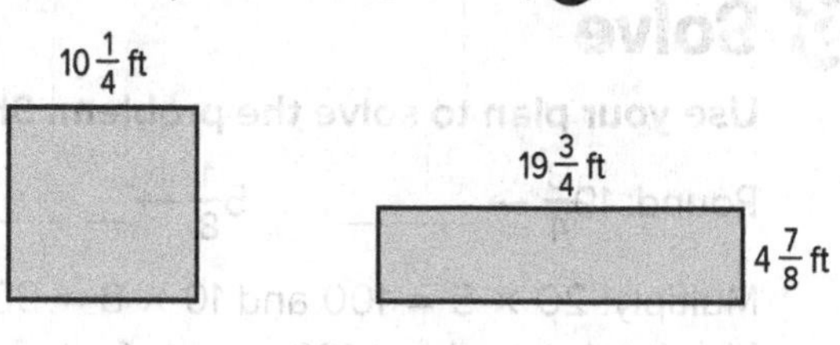

____

____

____

____

____

____

____

Lesson 2 **Multi-Step** Problem Solving

Multi-Step Example

Sophia is $\frac{3}{4}$ as tall as Mandy. Alexis is $\frac{5}{6}$ as tall as Mandy. What is the difference in height between Sophia and Alexis?

Preparation for **6.NS.1, MP 1**

Girl	Height (ft)
Sophia	
Mandy	5
Alexis	

Ⓐ $\frac{1}{12}$ foot

Ⓑ $\frac{5}{12}$ foot

Ⓒ $4\frac{1}{6}$ feet

Ⓓ $3\frac{3}{4}$ feet

Use a problem-solving model to solve this problem.

1 Understand

Read the problem. Circle the information you know.
Underline what the problem is asking you to find.

2 Plan

What will you need to do to solve the problem? Write your plan in steps.

Step 1 Determine the heights of Sophia and Alexis.

Step 2 Subtract to determine the difference between Sophia's height and Alexis's height.

Read to Succeed!
When subtracting mixed numbers, remember to regroup when necessary.
$4\frac{2}{12} = 3\frac{14}{12}$

3 Solve

Use your plan to solve the problem. Show your steps.

Sophia: $\frac{3}{4} \times 5 = \frac{\square}{4} = \square\frac{\square}{4}$ feet tall

Alexis: $\frac{5}{6} \times 5 = \frac{\square}{6} = \square\frac{\square}{6}$ feet tall

$4\frac{1}{6} - 3\frac{3}{4} = 4\frac{2}{12} - 3\frac{9}{12}$

$= 3\frac{14}{12} - 3\frac{9}{12}$ or $\frac{5}{12}$

So, Alexis is $\frac{5}{12}$ feet taller than Sophia. The correct answer is ____.

Fill in that answer choice.

4 Check

How do you know your solution is accurate?

____

____

Lesson 2 *(continued)*

Use a problem-solving model to solve each problem.

1 In June, Arturo spent 20 hours walking dogs in his neighborhood. In July, he spent $\frac{4}{5}$ as much time walking dogs. In August, he spent $\frac{9}{10}$ as much time walking dogs as he did in June. How many more hours did Arturo walk dogs in August than July? *Preparation for* **6.NS.1, MP 1**

Ⓐ $\frac{1}{10}$ hour

Ⓑ $\frac{18}{25}$ hour

Ⓒ 2 hours

Ⓓ 16 hours

2 The table shows the number of students in three classes at Hammond Middle School. Of all these students, $\frac{3}{8}$ plan to play in the school band and $\frac{1}{4}$ plan to play sports. How many more students plan to play in the band than play sports? *Preparation for* **6.NS.1, MP 2**

Class	Total Students
Ms. Chen	33
Mr. Rice	28
Ms. Lang	35

____

3 During a read-a-thon, 40 students read as many books as they could in one month. The circle graph below shows the fraction of students that read less than 10 books, 10 to 20 books, and more than 20 books. How many more students read more than 20 books versus less than 10 books? *Preparation for* **6.NS.1, MP 4**

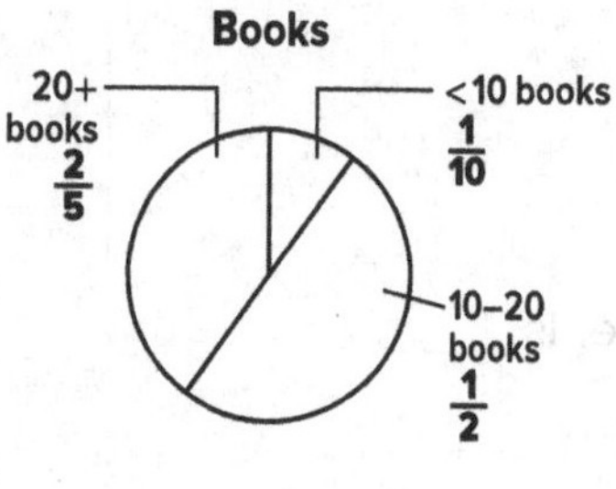

____

4 **H.O.T. Problem** Kisho wants to make ten dozen chocolate chip cookies. He needs $\frac{3}{4}$ cup granulated sugar for the recipe for one dozen cookies. He only has $\frac{1}{4}$ cup of sugar. How many cookies can he make with this amount of sugar? *Preparation for* **6.NS.1, MP 7**

____

NAME________________________ DATE ________________ PERIOD ________

Lesson 3 Multi-Step Problem Solving

Multi-Step Example

Ella had $\frac{1}{3}$ left of a wall to paint in her bedroom. She painted dots on $\frac{1}{4}$ of what was left to paint and stripes on $\frac{1}{2}$ of it. What is the area of the region that Ella has not yet painted? *Preparation for* **6.NS.1, MP 1**

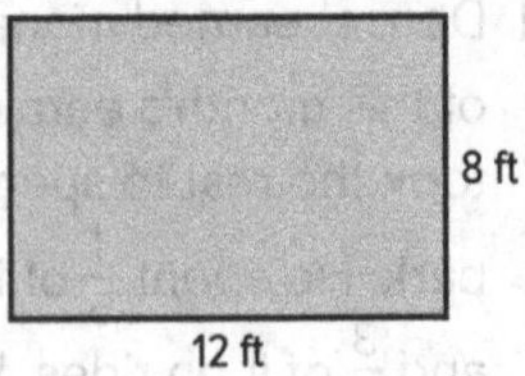

Ⓐ 88 ft² Ⓒ 8 ft²

Ⓑ 16 ft² Ⓓ 1 ft²

Use a problem-solving model to solve this problem.

1 Understand

Read the problem. Circle the information you know.
Underline what the problem is asking you to find.

2 Plan

What will you need to do to solve the problem? Write your plan in steps.

Step 1 Divide the width into thirds. Divide one of the thirds into fourths. Determine the length of the area not painted.

Step 2 Multiply to find the area.

3 Solve

Use your plan to solve the problem. Show your steps.

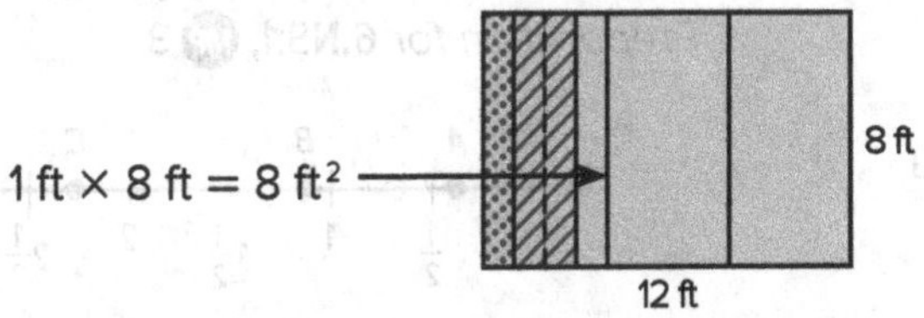

Read to Succeed!
The area of a rectangle is found by using the formula $A = l \cdot w$.

So, Ella has ☐ square feet left to paint. The correct choice is ☐.

Fill in that answer choice.

4 Check

How do you know your solution is accurate?

____

____

____

Lesson 3 *(continued)*

Use a problem-solving model to solve each problem.

1 Denzel earned money after school. He put $\frac{1}{2}$ of this month's earnings into savings. He took the rest to spend at the amusement park. He spent $\frac{1}{5}$ of this amount on popcorn and $\frac{3}{4}$ of it on rides. What fraction of his earnings did he take to the park but not spend on rides or popcorn? *Preparation for* **6.NS.1, MP 1**

Ⓐ $\frac{1}{40}$

Ⓑ $\frac{11}{20}$

Ⓒ $\frac{1}{10}$

Ⓓ $\frac{3}{8}$

2 The table shows how Mura spends her free time on a typical Saturday. If she has 6 hours of free time, how many hours does she spend playing board games or going to the park? *Preparation for* **6.NS.1, MP 6**

Activity	Fraction of Free Time
Board games	$\frac{1}{10}$
Park	$\frac{2}{5}$
Piano	$\frac{3}{7}$
Reading	$\frac{1}{14}$

____

3 Ricardo needs to pave the two rectangular sections shown. Determine the total area that Ricardo needs to pave. *Preparation for* **6.NS.1, MP 4**

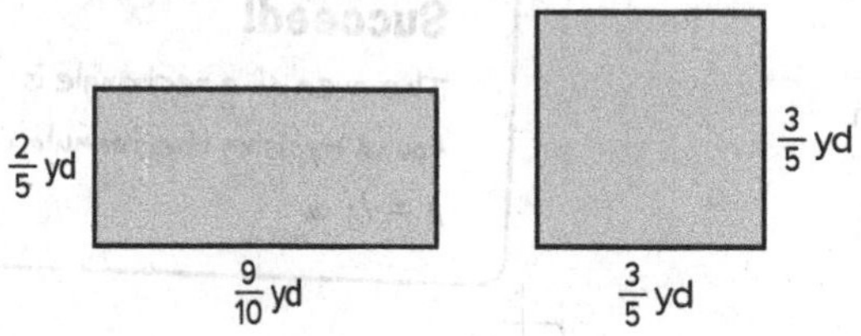

____

4 **H.O.T. Problem** Without multiplying, determine where the product of $2 \times \frac{12}{7} \times \frac{1}{7}$ is located on the number line. Choose *A*, *B*, or *C*. Justify your reasoning. *Preparation for* **6.NS.1, MP 3**

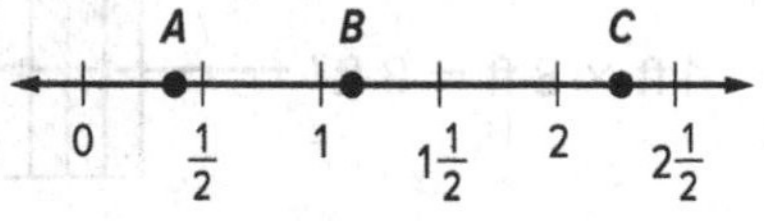

____

____

____

Lesson 4 **Multi-Step** Problem Solving

Multi-Step Example

On Saturday, Ishan rode his bike $5\frac{1}{2}$ miles. Ama rode her bike $1\frac{1}{4}$ times as far as Ishan. Joseph rode his bike $1\frac{2}{5}$ times as far as Ama. How many more miles did Joseph ride than Ama? *Preparation for* **6.NS.1, MP 1**

Ⓐ $1\frac{1}{10}$ mi Ⓒ $6\frac{7}{8}$ mi

Ⓑ $2\frac{3}{4}$ mi Ⓓ $9\frac{5}{8}$ mi

Use a problem-solving model to solve this problem.

1 Understand

Read the problem. Circle the information you know. Underline what the problem is asking you to find.

2 Plan

What will you need to do to solve the problem? Write your plan in steps.

Step 1 Multiply to find the number of miles Ama rode. Then multiply to find the number of miles Joseph rode.

Step 2 Subtract to find how many more miles Joseph rode than Ama.

Read to Succeed!
When you multiply fractions and mixed numbers, remember to simplify your answer.

3 Solve

Use your plan to solve the problem. Show your steps.

Miles Ama rode: $5\frac{1}{2} \times 1\frac{1}{4} = \frac{11}{2} \times \frac{5}{4} = \frac{55}{8} = \square\frac{\square}{\square}$

Miles Joseph rode: $6\frac{7}{8} \times 1\frac{2}{5} = \frac{55}{8} \times \frac{7}{5} = \frac{77}{8} = \square\frac{\square}{\square}$

Find the difference: $9\frac{5}{8} - 6\frac{7}{8} = 8\frac{13}{8} - 6\frac{7}{8} = 2\frac{6}{8} = \square\frac{\square}{\square}$

So, ______ is the correct answer. Fill in that answer choice.

4 Check

How do you know your solution is reasonable?

____

____

Lesson 4 *(continued)*

Use a problem-solving model to solve each problem.

1 It took Everett $4\frac{3}{4}$ hours to write his science report. Hudson took $2\frac{2}{3}$ times as long as Everett to write his report. Brannon took $1\frac{1}{2}$ times as long as Hudson to write his report. How many more hours did it take Brannon to write his report than Hudson? *Preparation for* **6.NS.1, MP 1**

Ⓐ $1\frac{1}{6}$ hr

Ⓑ $6\frac{1}{3}$ hr

Ⓒ $12\frac{2}{3}$ hr

Ⓓ 19 hr

2 Horaclo's garden is shown below. He needs $1\frac{1}{3}$ scoops of fertilizer for each square foot of the garden. How many scoops of fertilizer does Horaclo need for the entire garden? *Preparation for* **6.NS.1, MP 4**

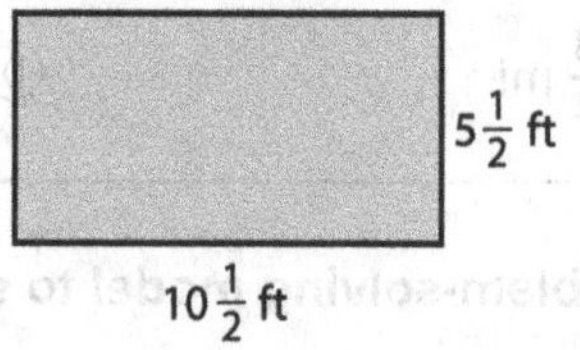

____

3 Jan and Dan work part time. Jan earns $9.50 an hour. Dan earns $8.25 an hour. The table shows how many hours they worked on Monday, Wednesday, and Friday. Who earned more money? How much more? *Preparation for* **6.NS.1, MP 2**

Day	Jan's Time (hr)	Dan's Time (hr)
Monday	$3\frac{1}{3}$	$4\frac{2}{5}$
Wednesday	$4\frac{1}{2}$	$2\frac{3}{4}$
Friday	$2\frac{2}{3}$	$4\frac{1}{4}$

____

4 **H.O.T. Problem** Without multiplying, explain which expression has the greater product. *Preparation for* **6.NS.1, MP 3**

$8\frac{1}{2} \times 7\frac{7}{8}$ or $8\frac{1}{2} \times 7\frac{2}{9}$

____

____

____

____

____

NAME_________________________________ DATE _______________ PERIOD _________

Lesson 5 Multi-Step Problem Solving

Multi-Step Example

The table shows the amount of water each athlete drinks during soccer practice. How many quarts of water are needed for these five athletes during practice? **6.RP.3, 6.RP.3d, MP 1**

Athlete	Amount (c)
Deon	2
Sierra	1.5
Carmen	3.5
Mia	3
Ella	2

Use a problem-solving model to solve this problem.

1 Understand

Read the problem. Circle the information you know.
Underline what the problem is asking you to find.

2 Plan

What will you need to do to solve the problem? Write your plan in steps.

Step 1 Determine the total number of cups drank during practice.

Step 2 Convert cups to quarts.

Read to Succeed!
There are 2 cups in a pint and 2 pints in a quart.

3 Solve

Use your plan to solve the problem. Show your steps.

The total amount drank is 2 + 1.5 + 3.5 + 3 + 2 or ______ cups.

______ cups = ______ pints = ______ quarts

So, the team drank ____ quarts of water.

4 Check

How do you know your solution is accurate?

____

____

____

Lesson 5 *(continued)*

Use a problem-solving model to solve each problem.

1 William collects metal to sell to a recycling plant. The table shows the amount of metal he has collected over several days. He needs to collect 4 tons before he can take the load to the recycling plant. How many more pounds does he need to reach 4 tons? **6.RP.3, 6.RP.3d, MP 1**

Day	Metal (lb)
Monday	2,500
Tuesday	1,375
Wednesday	2,550
Thursday	1,075

2 Joaquin drank 6 glasses of water each containing 10 fluid ounces. His goal was to drink 2 quarts. How many more fluid ounces does he have to drink to reach his goal? **6.RP.3, 6.RP.3d, MP 6**

3 A football team needs to travel 80 yards from their current location to their opponent's end zone to score a touchdown. The team is now 6 feet away from their opponent's end zone, ready to score the touchdown. How many feet have they already traveled down the field? **6.RP.3, 6.RP.3d, MP 2**

4 **H.O.T. Problem** The dimensions of a rectangle are given in the diagram. Jane wanted to know the area in square meters. She used two different methods to determine the area. Which method is correct, and why? **6.RP.3, 6.RP.3d, MP 3**

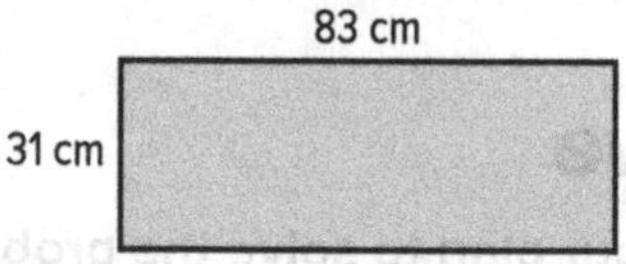

Method 1 83 cm × 31 cm = 2,573 sq cm
2,573 cm = 25.73 m
The area of the rectangle is 25.73 sq m.

Method 2 83 cm = 0.83 m
31 cm = 0.31 m
0.83 m × 0.31 m = 0.2573 sq m
The area of the rectangle is 0.2573 sq m.

Lesson 6 Multi-Step Problem Solving

Multi-Step Example

The table shows the ingredients needed to make one batch of salad dressing. A chef has 3 tablespoons of minced garlic. She made the greatest number of batches possible. How many tablespoons of garlic were left? **6.NS.1, MP 1**

Ingredient	Amount
Oil	1 cup
Vinegar	$\frac{3}{4}$ cup
Minced garlic	$\frac{2}{3}$ tbsp

Ⓐ $\frac{1}{2}$ tablespoon

Ⓑ $\frac{1}{3}$ tablespoon

Ⓒ $\frac{2}{3}$ tablespoon

Ⓓ $\frac{5}{6}$ tablespoon

Use a problem-solving model to solve this problem.

1 Understand

Read the problem. Circle the information you know.
Underline what the problem is asking you to find.

2 Plan

What will you need to do to solve the problem? Write your plan in steps.

Step 1 Divide to determine the number of batches made.

Step 2 Subtract to determine the amount remaining.

Read to Succeed!
When dividing by fractions, multiply by the reciprocal.

3 Solve

Use your plan to solve the problem. Show your steps.

The chef could make $3 \div \frac{2}{3}$ or ☐ $\frac{☐}{☐}$ batches.

Since she made ☐ full batches, there is $\frac{☐}{☐}$ batch left.

One-half of a batch uses $\frac{1}{2} \times \frac{2}{3}$ or ____ tablespoon. The correct choice is ____. Fill in that answer choice.

4 Check

How do you know your solution is accurate?

____

____

____

Lesson 6 *(continued)*

Use a problem-solving model to solve each problem.

1 The table shows the time it takes each person to build a house of cards. If there are 2 hours available to make houses of cards, how many more houses can Fina make than Logan? **6.NS.1, MP 1**

Person	Time (hours)
Jenna	$\frac{1}{4}$
Logan	$\frac{1}{3}$
Fina	$\frac{1}{5}$

Ⓐ 3
Ⓑ 4
Ⓒ 5
Ⓓ 6

2 Aria made 9 pounds of fudge. She separates the fudge into $\frac{3}{4}$-pound portions. She sells each portion for $6.50. If she sells all the fudge, how much money will she make? **6.NS.1, MP 2**

____

3 Robert and Judi are ordering lasagnas for a party. Robert ordered 10 large lasagnas. Each lasagna is cut into tenths. Judi ordered 6 smaller lasagnas. Each lasagna was cut into eighths. How many pieces did they order in all? **6.NS.1, MP 2**

____

4 **H.O.T. Problem** Cadence is making gift bags filled with different colored beads for her jewelry party. She fills the bags using a mixture of $1\frac{1}{2}$ pounds pink beads, $\frac{3}{4}$ pound purple beads, and $1\frac{1}{4}$ pounds green beads. She divides the mixture into 8 packages. How much is in each package? **6.NS.1, MP 4**

____

NAME _______________ DATE _______________ PERIOD _______________

Lesson 7 Multi-Step Problem Solving

Multi-Step Example

Alfonso is making snack bags with different types of nuts as shown in the table. Each snack bag contains $\frac{1}{8}$ pound of one type of nut. How many more whole servings of walnuts can he make than peanuts? 6.NS.1, MP 1

Type of Nut	Weight (lb)
Almonds	$\frac{1}{2}$
Cashews	$\frac{1}{4}$
Peanuts	$\frac{2}{5}$
Walnuts	$\frac{3}{4}$

Ⓐ 1
Ⓑ 2
Ⓒ 3
Ⓓ 6

Use a problem-solving model to solve this problem.

1 Understand

Read the problem. Circle the information you know.
Underline what the problem is asking you to find.

2 Plan

What will you need to do to solve the problem? Write your plan in steps.

Step 1 Divide to determine the number of servings of walnuts and peanuts.

Step 2 Subtract to determine how many more servings of walnuts than peanuts.

3 Solve

Use your plan to solve the problem. Show your steps.

Walnuts: $\frac{3}{4} \div \frac{1}{8} = \frac{3}{4} \cdot \frac{8}{1}$ or ☐ servings

Peanuts: $\frac{2}{5} \div \frac{1}{8} = \frac{2}{5} \cdot \frac{8}{1}$ or ☐ $\frac{☐}{☐}$ servings

So, Alfonso made 6 − 3 or ____ more whole servings of walnuts than peanuts. The correct choice is ____.

Read to Succeed!
The number of whole servings of peanuts is 3 because $\frac{1}{5}$ is not a whole serving.

4 Check

How do you know your solution is accurate?

____

____

____

Lesson 7 *(continued)*

Use a problem-solving model to solve each problem.

1 Anabella is using ribbon to decorate the edge of a picture frame with a length of $\frac{1}{4}$ yard and a width of $\frac{1}{6}$ yard. She will only use one color to decorate the frame. Each color of ribbon is available in lengths as shown in the table. How many more strips of green ribbon than blue ribbon would she need for the frame? **6.NS.1, MP 1**

Color	Strip Length (yd)
Black	$\frac{1}{2}$
Blue	$\frac{2}{3}$
Green	$\frac{1}{6}$

Ⓐ 5
Ⓑ 3
Ⓒ 2
Ⓓ 1

2 Camillo is decorating birthday cards with glitter to send to his friends. The table shows the different colors of glitter that he has. He will mix all these colors together, and then use $\frac{1}{4}$ tube of glitter on each card. How many birthday cards can he decorate? **6.NS.1, MP 4**

Color	Tubes
Red	2
Yellow	$\frac{3}{5}$
Purple	$\frac{3}{8}$
Pink	$\frac{1}{2}$

____

3 Stephanie usually jogs $\frac{3}{4}$ mile every day. She decides that she wants to sprint for a part of this distance. She will jog for $\frac{1}{2}$ of $\frac{3}{4}$ mile and will sprint the rest, but she only sprints $\frac{1}{8}$ mile at a time before resting. How many sprints will Stephanie do each day? **6.NS.1, MP 2**

____

4 **H.O.T. Problem** Without doing any calculations, which expression does not have the same value as $\frac{1}{2} \div \frac{2}{3}$? Explain. **6.NS.1, MP 7**

A	$\frac{1}{2} \div \frac{4}{6}$
B	$\frac{1}{2} \times \frac{3}{2}$
C	$\frac{3}{6} \times \frac{3}{2}$
D	$\frac{3}{6} \div \frac{6}{4}$

____
____
____
____
____

Lesson 8 Multi-Step Problem Solving

Multi-Step Example

The table shows the side lengths of four square mirrors. How many times greater is the area of mirror B than the area of mirror C? **6.NS.1, MP 1**

Mirror	Side Length (ft)
A	$1\frac{1}{4}$
B	$2\frac{1}{2}$
C	$1\frac{3}{4}$
D	$3\frac{1}{6}$

Ⓐ $2\frac{2}{49}$

Ⓑ $3\frac{1}{16}$

Ⓒ $3\frac{3}{16}$

Ⓓ $6\frac{1}{4}$

Use a problem-solving model to solve this problem.

1 Understand

Read the problem. Circle the information you know.
Underline what the problem is asking you to find.

2 Plan

What will you need to do to solve the problem? Write your plan in steps.

Step 1 Use the formula $A = \ell \cdot w$ to determine the area of each mirror.

Step 2 Divide to determine how many times greater mirror B is than mirror C.

3 Solve

Use your plan to solve the problem. Show your steps.

Read to Succeed! To determine the area of a square, multiply the length times the width.

Mirror B: $2\frac{1}{2} \cdot 2\frac{1}{2} = \frac{5}{2} \cdot \frac{5}{2} = \frac{25}{4}$ or ☐$\frac{☐}{☐}$ square feet

Mirror C: $1\frac{3}{4} \cdot 1\frac{3}{4} = \frac{7}{4} \cdot \frac{7}{4} = \frac{49}{16}$ or ☐$\frac{☐}{☐}$ square feet

So, Mirror B is $6\frac{1}{4} \div 3\frac{1}{16}$ or ______ times larger than Mirror C.

Choice ______ is correct. Fill in that answer choice.

4 Check

How do you know your solution is accurate?

____

____

____

Lesson 8 *(continued)*

Use a problem-solving model to solve each problem.

1 The table shows the dimensions of two fenced-in areas at a dog park. How many times greater is the area enclosed by the wood fence than the area enclosed by the metal fence? **6.NS.1, MP 1**

Fence	Length (yd)	Width (yd)
Wood	$6\frac{1}{2}$	$2\frac{1}{4}$
Metal	$2\frac{3}{4}$	$2\frac{1}{2}$

Ⓐ $1\frac{1}{9}$

Ⓑ $2\frac{7}{55}$

Ⓒ $2\frac{4}{11}$

Ⓓ $7\frac{3}{4}$

2 Mylie has $35\frac{3}{4}$ yards of red ribbon and $30\frac{1}{3}$ yards of green ribbon. She cuts the red ribbon into strips that are each $3\frac{1}{4}$ yards long and the green ribbon into strips that are each $2\frac{1}{6}$ yards long. How many more green strips than red strips does she have? **6.NS.1, MP 2**

____

3 On Saturday, Justine studied $1\frac{1}{4}$ times as long as Shantel and $1\frac{3}{4}$ times as long as Nicole. If Justine studied $3\frac{1}{2}$ hours on Saturday, how much longer did Shantel study than Nicole on Saturday? Express your answer as a number of hours in decimal notation. **6.NS.1, MP 2**

____

4 **H.O.T. Problem** Without dividing, explain whether $1\frac{1}{2} \div 3\frac{1}{4} \div 2\frac{5}{6}$ is greater or less than $2\frac{5}{6} \div 3\frac{1}{4} \div 1\frac{1}{2}$. **6.NS.1, MP 7**

____

____

____

NAME______________________________ DATE ______________ PERIOD ________

Lesson 1 **Multi-Step** Problem Solving

Multi-Step Example

Golf scores are measured by the number of strokes over or under par. Scores over par can be represented by a positive integer. Scores under par can be represented by a negative integer. The table shows the golf scores of four players in a golf tournament. Which golfer's score is represented by point *B* on the number line below? **6.NS.6c, MP 1**

Golfer	Score
Chase	2 under par
Augustus	3 over par
Etu	1 under par
Miles	1 over par

Ⓐ Chase
Ⓑ Augustus
Ⓒ Etu
Ⓓ Miles

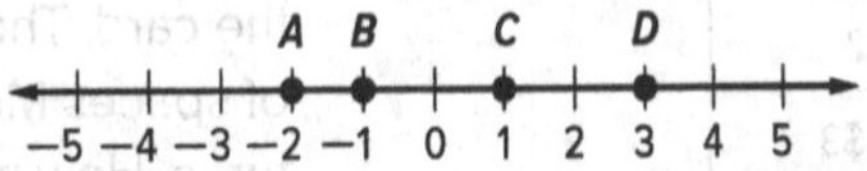

Use a problem-solving model to solve this problem.

1 Understand

Read the problem. Circle the information you know.
Underline what the problem is asking you to find.

2 Plan

What will you need to do to solve the problem? Write your plan in steps.

Step 1 Determine the number located at point ______.

Step 2 Determine the corresponding value in the table.

3 Solve

Use your plan to solve the problem. Show your steps.

Point B is located at ______. −1 is 1 __________ par.

The golfer that scored −1 is ______.

So, the correct answer is ______. Fill in that answer choice.

Read to Succeed!
Numbers to the left of zero on a number line are negative. Numbers to the right are positive.

4 Check

How do you know your solution is accurate?

____

____

____

Lesson 1 *(continued)*

Use a problem-solving model to solve each problem.

1 The table shows the changes in the value of four stocks over one day. Which point on the number line represents the change in value of Stock *R*? **6.NS.6c, MP 1**

Stock	Change in Value
Stock *Q*	Up $2
Stock *R*	Down $3
Stock *S*	Down $1
Stock *T*	Up $3

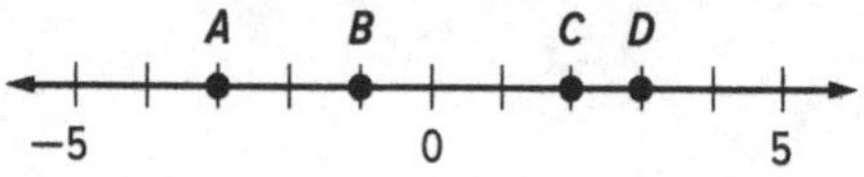

Ⓐ point *A*
Ⓑ point *B*
Ⓒ point *C*
Ⓓ point *D*

2 Monique is playing a board game where players move about the board using numbered cards. If the card is green, you move forward the number of spaces indicated on the card. If the card is red, you move backward the number of spaces indicated on the card. The number line shows the number of spaces Monique moved in her first five turns. How many red cards did Monique get in her first five turns? **6.NS.6c, MP 3**

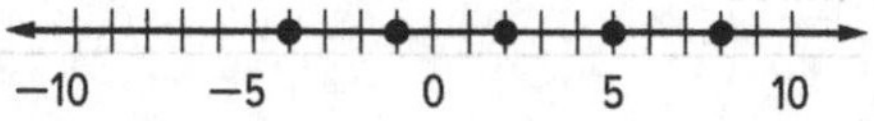

____

3 A football team has four chances, called downs, to gain at least 10 yards. The number line shows the number of yards gained and lost in four downs of a football game.

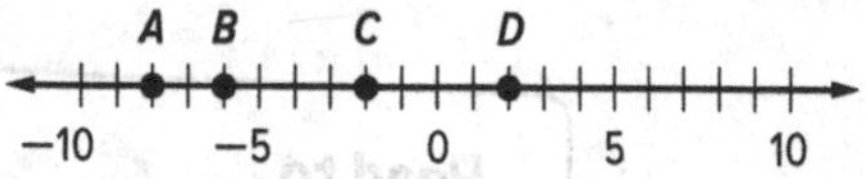

The table shows which downs correspond to the points on the graph. During which down did the team gain 2 yards? **6.NS.6c, MP 3**

Down	Point
1	*B*
2	*C*
3	*A*
4	*D*

____

4 **H.O.T. Problem** The temperature outside is 4°F. The temperature drops 6°F. Between which two points is the location of the temperature after the change? **6.NS.6c, MP 2**

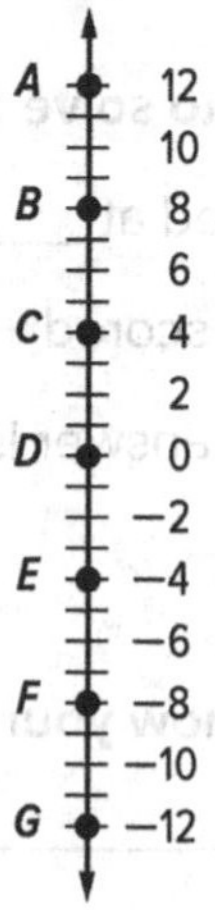

____

NAME ______________________ DATE ______________ PERIOD ________

Lesson 2 **Multi-Step** Problem Solving

Multi-Step Example

The graph shows the freezing and boiling points of water in degrees Fahrenheit. How much greater is the absolute value of the boiling point of water than the absolute value of the freezing point of water? **6.NS.7c, MP 1**

Ⓐ −32

Ⓑ 0

Ⓒ 180

Ⓓ 244

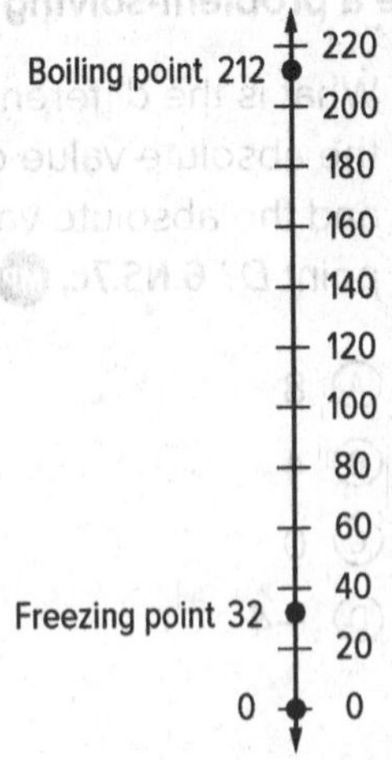

Use a problem-solving model to solve this problem.

1 Understand

Read the problem. Circle the information you know.
Underline what the problem is asking you to find.

2 Plan

What will you need to do to solve the problem? Write your plan in steps.

Step 1 Determine the absolute value of the boiling point of water and the absolute value of the freezing point of water.

Step 2 Subtract to find how much greater.

3 Solve

Use your plan to solve the problem. Show your steps.

Boiling point of water $|212| =$ ______

Freezing point of water $|32| =$ ______

So, the absolute value of the boiling point of water is ______ − ______ or ______ degrees greater than the absolute value of the freezing point of water. Choice C is correct. Fill in that answer choice.

Read to Succeed!
Absolute value is the distance a number is from zero and is always positive.

4 Check

How do you know your solution is accurate?

____

____

____

Lesson 2 *(continued)*

Use a problem-solving model to solve each problem.

1 What is the difference between the absolute value of point *B* and the absolute value of point *D*? **6.NS.7c, MP 1**

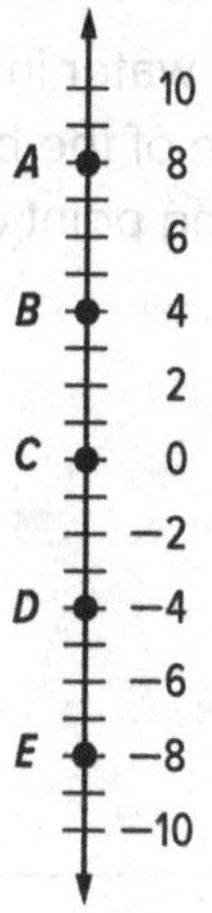

Ⓐ 8

Ⓑ 4

Ⓒ 0

Ⓓ −4

2 The table shows the account balances of five students.

Student	Balance ($)
Yen	−9
Mark	11
Aisha	−3
Wendy	10
Ross	6

What is the difference between the absolute value of Wendy's balance and the absolute value of Yen's balance, in dollars? 6.NS.7c, MP 2

____

3 The graph shows the path Chante walked, beginning at point *A*. The distance between two tick marks represents 1 meter. Chante walked 3 meters east to point *B* and then 5 meters east to point *C*. How many meters west must Chante walk from point *C* to be at the point represented by the opposite of *C*? **6.NS.6a, MP 4**

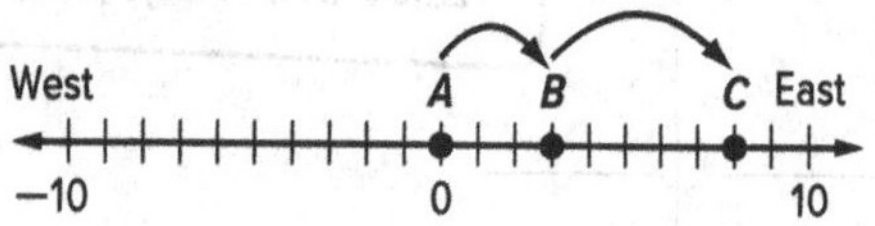

____

4 **H.O.T. Problem** Is the opposite of the absolute value of a number *always*, *sometimes*, or *never* equal to the absolute value of the opposite of a number? Explain your response and give examples. **6.NS.7c, MP 3**

____

____

____

NAME______________________________ DATE ______________ PERIOD ________

Lesson 3 **Multi-Step** Problem Solving

Multi-Step Example

The table shows the freezing points in degrees Celsius of four substances. Which substance(s) have greater freezing points than aniline? **6.NS.7, MP 1**

Substance	Freezing Point (°C)
Aniline	−6
Acetic acid	17
Acetone	−95
Water	0

Ⓐ water only

Ⓑ acetic acid only

Ⓒ acetic acid, acetone, and water

Ⓓ acetic acid, water

Use a problem-solving model to solve this problem.

1 Understand

Read the problem. Circle the information you know.
Underline what the problem is asking you to find.

2 Plan

What will you need to do to solve the problem? Write your plan in steps.

Step 1 Locate the numbers on a ____________.

Step 2 Compare the numbers based on their location on the number line.

3 Solve

Use your plan to solve the problem. Show your steps.

Locate the values on a number line.

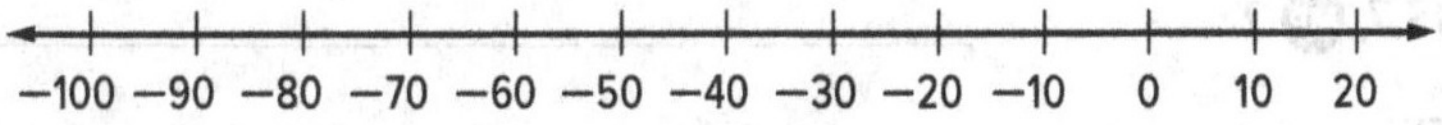

Compare the numbers.
The numbers greater than −6 are 0 and 17, which correspond to water and acetic acid. Choice ______ is correct. Fill in that answer choice.

Read to Succeed!
Make sure to read the directions carefully.

4 Check

How do you know your solution is accurate?

____

____

____

Lesson 3 *(continued)*

Use a problem-solving model to solve each problem.

1 Golf scores are measured as over or under par. The winner has the least score. The table shows the golf scores of five players.

Golfer	Score
Jamal	2 under par
Zaire	1 over par
Dante	even
Alexandra	3 under par
Ajay	4 over par

Which lists the players in order from first place to fifth place? **6.NS.7, MP 1**

Ⓐ Dante, Zaire, Jamal, Alexandra, Ajay

Ⓑ Alexandra, Jamal, Dante, Zaire, Ajay

Ⓒ Ajay, Zaire, Dante, Jamal, Alexandra

Ⓓ Dante, Alexandra, Jamal, Zaire, Ajay

2 The table shows the rise and fall in the value of a certain stock over five days. Which day shows the greatest fall in stock value? **6.NS.7, MP 2**

Day	Change in Stock Value ($)
1	$-1\frac{1}{8}$
2	$4\frac{3}{8}$
3	$6\frac{1}{2}$
4	$-3\frac{1}{4}$
5	$1\frac{3}{4}$

____

3 When a football player causes a penalty during a game, the team can lose 5, 10, or 15 yards on the play. The table shows the players, by jersey number, and the number of penalty yards the team was given based on each player's penalties. How many players caused more penalty yards than the player with jersey number 10? **6.NS.7, MP 2**

Player (jersey number)	Penalty Yards
12	−15
8	−25
28	−30
17	−10
10	−20
48	−5

____

4 H.O.T. Problem Order the numbers from greatest to least. Explain how you know which number is the greatest. **6.NS.7, MP 3**

$\frac{1}{2}$, $-|-3|$, -0.5, $|-2|$, -1

____

____

____

____

Lesson 4 **Multi-Step** Problem Solving

Multi-Step Example

There are 84 chairs that need to be set up in the school's auditorium. Students were asked to write an expression to show how to find the number of chairs needed in each of 12 rows. The table shows samples of expressions given by students. Which sample(s) result in the correct solution? *Preparation for* **6.NS.6c,** MP 1

Sample	Expression
A	$12 \div 84$
B	$84 \div 12$
C	$\frac{12}{84}$
D	$\frac{84}{12}$

Ⓐ Sample A

Ⓑ Sample B

Ⓒ Samples A and C

Ⓓ Samples B and D

Use a problem-solving model to solve this problem.

1 Understand

Read the problem. Circle the information you know.
Underline what the problem is asking you to find.

2 Plan

What will you need to do to solve the problem? Write your plan in steps.

Step 1 Determine the number of chairs needed in each row.

Step 2 Determine whether each sample's expression results in the correct solution.

Read to Succeed!
Make sure to read all choices given when answering multiple choice questions.

3 Solve

Use your plan to solve the problem. Show your steps.

There are 84 ÷ ____, or ____ chairs needed in each row.

Sample A $12 \div 84 \approx$ ____ **Sample C** $\frac{12}{84} \approx$ ____

Sample B $84 \div 12 =$ ____ **Sample D** $\frac{84}{12} =$ ____

Samples ____ and ____ result in the correct solution. So, choice ____ is correct. Fill in that answer choice.

4 Check

How do you know your solution is accurate?

____

____

____

NAME ____________________ DATE ____________ PERIOD ________

Lesson 4 *(continued)*

Use a problem-solving model to solve each problem.

1 Selina has 8 cups of sugar to use for baking pies. She would like to bake 12 pies. She knows she can divide to find how much sugar to use for each pie. Which expression is equivalent to $8 \div 12$?
Preparation for **6.NS.6c, MP 1**

Ⓐ 2×3

Ⓑ 3×2

Ⓒ $\frac{2}{3}$

Ⓓ $\frac{3}{2}$

2 Alivia needs to classify the numbers in the table using the Venn diagram shown.

Numbers
$\lvert -6 \rvert, \frac{3}{4}, 1.55, 11 \div 6, -2, 0.\overline{7}, \frac{2}{3}, 6 \div 2$

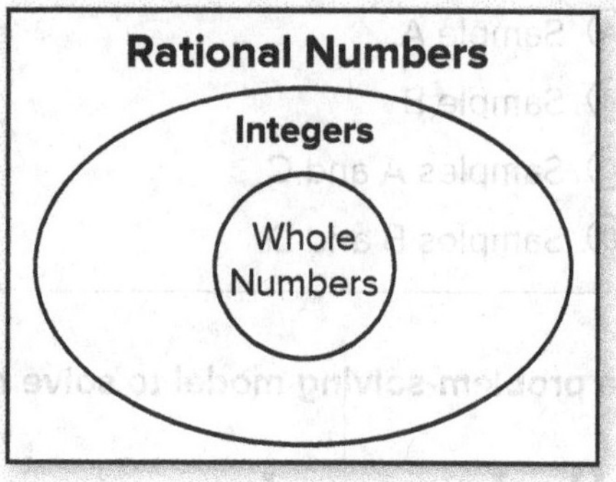

She decides to first color-code the numbers by highlighting the numbers that represent *terminating* decimals in yellow and *repeating* decimals in red. How many numbers should Alivia highlight in red?
Preparation for **6.NS.6c, MP 8**

____

3 The table shows the perimeters of different equilateral triangles. How many of the triangles have side lengths that are terminating decimals?
Preparation for **6.NS.6c, MP 8**

Perimeter (in.)
11
$15.\overline{6}$
18
21
$21.\overline{33}$
23.3

____

4 **H.O.T. Problem** Which number is greater, $-\frac{1}{3}$ or -0.3? Use the number line to help explain your answer. **6.NS.7a, MP 3**

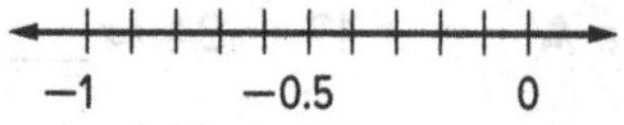

____

____

____

____

NAME ______________________________ DATE ______________ PERIOD ________

Lesson 5 **Multi-Step** Problem Solving

Multi-Step Example

The table shows the difference between the actual amount of rainfall, in inches, that a city received over four weeks and the average amount that it usually receives during those weeks. Which shows the weeks in order of the differences from least to greatest? **6.NS.7b,** **MP 1**

Week	Difference (in.)
1	$\frac{1}{3}$
2	–1.6
3	0.3
4	$-1\frac{1}{2}$

Ⓐ 3, 1, 4, 2

Ⓑ 2, 4, 3, 1

Ⓒ 1, 3, 4, 2

Ⓓ 2, 4, 1, 3

Use a problem-solving model to solve this problem.

1 Understand

Read the problem. Circle the information you know.
Underline what the problem is asking you to find.

2 Plan

What will you need to do to solve the problem? Write your plan in steps.

Step 1 Compare the negative rational numbers.

Step 2 Compare the positive rational numbers.

Step 3 Order the rational numbers.

Read to Succeed!
When ordering from least to greatest, remember negative numbers are less than positive numbers.

3 Solve

Use your plan to solve the problem. Show your steps.

–1.6 ◯ $-1\frac{1}{2}$ and $\frac{1}{3}$ ◯ 0.3

So, the correct order of weeks is 2, 4, 3, 1. Choice ______ is correct.
Fill in that answer choice.

4 Check

How do you know your solution is accurate?

____

____

____

Lesson 5 *(continued)*

Use a problem-solving model to solve each problem.

1 In last year's diving competition, Stefani's average score per dive was 9.55 points. The table shows the difference between her average score and her actual scores for her first four dives from this year's competition. Which of the following lists the dives in order of the differences from greatest to least? **6.NS.7b, MP 1**

Dive	Difference (points)
1	$\frac{1}{4}$
2	−0.35
3	$-\frac{3}{10}$
4	0.4

Ⓐ 4, 1, 3, 2
Ⓑ 1, 3, 2, 4
Ⓒ 4, 1, 2, 3
Ⓓ 2, 3, 1, 4

2 The table shows the heights, in feet, of five classmates. How many of these classmates are taller than $5\frac{1}{2}$ feet? **6.NS.7b, MP 2**

Student	Height (ft)
Mario	5.6
Phong	$5\frac{1}{3}$
Travis	$5\frac{5}{6}$
Zack	5.45
Tavon	$5\frac{3}{5}$

____

3 The table shows the dimensions, in centimeters, of four rectangles. How many centimeters wider is the rectangle with the greatest perimeter than the rectangle with the least perimeter? **6.NS.7b, MP 2**

Rectangle	Width (cm)	Length (cm)
A	6.25	8.9
B	6.3	8.73
C	6.5	8.7
D	6.6	8.5

____

4 H.O.T. Problem Jaquan has lengths of colored string as shown in the table. He finds a piece of yellow string that is 0.2 yard shorter than the blue string. He lays all four pieces of string end to end from left to right in order of length, beginning with the shortest piece. Between which two colors is the yellow string? **6.NS.7b, MP 2**

Color	Length (yd)
Red	$3\frac{2}{5}$
Blue	$3\frac{5}{8}$
Green	3.5

____

Lesson 6 Multi-Step Problem Solving

Multi-Step Example

Samantha drew a map of the park in her neighborhood. She put a point on the grid for the playground. The fountain is located at the reflection of the location of the playground across the *y*-axis. The picnic tables are located at the reflection of the location of the fountain across the *x*-axis. What ordered pair describes the location of the picnic tables? **6.NS.6b, MP 1**

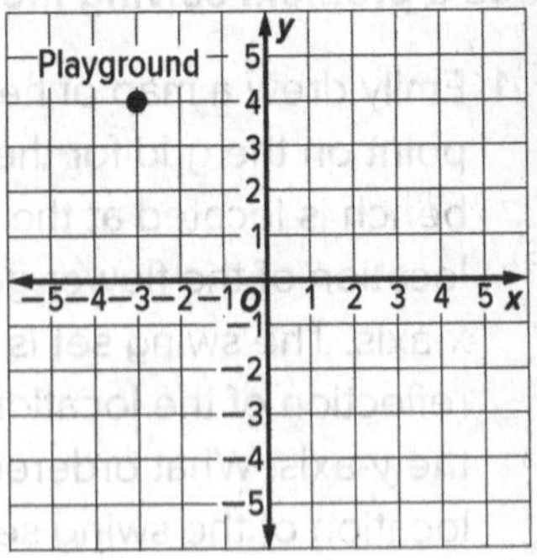

Use a problem-solving model to solve this problem.

1 Understand

Read the problem. Circle the information you know.
Underline what the problem is asking you to find.

2 Plan

What will you need to do to solve the problem? Write your plan in steps.

Step 1 Identify the ordered pair for the playground.

Step 2 Identify the reflection across the *y*-axis for the location of the fountain.

Step 3 Identify the reflection across the *x*-axis for the location of the picnic tables.

Read to Succeed!
When locating points on a grid, begin at the origin and move horizontally along the x-axis and then move vertically along or parallel to the y-axis.

3 Solve

Use your plan to solve the problem. Show your steps.

The ordered pair for the playground is ________.

The reflection across the *y*-axis for the location of the fountain is ________.

The reflection across the *x*-axis for the location of the picnic tables is ________.

So, the location of the picnic tables is at ________.

4 Check

How do you know your solution is reasonable?

____

____

____

NAME ______________________________ DATE ______________ PERIOD ________

Lesson 6 *(continued)*

Use a problem-solving model to solve each problem.

1 Emily drew a map of her backyard. She put a point on the grid for the flower garden. The bench is located at the reflection of the location of the flower garden across the *x*-axis. The swing set is located at the reflection of the location of the bench across the *y*-axis. What ordered pair describes the location of the swing set? **6.NS.6b, MP 1**

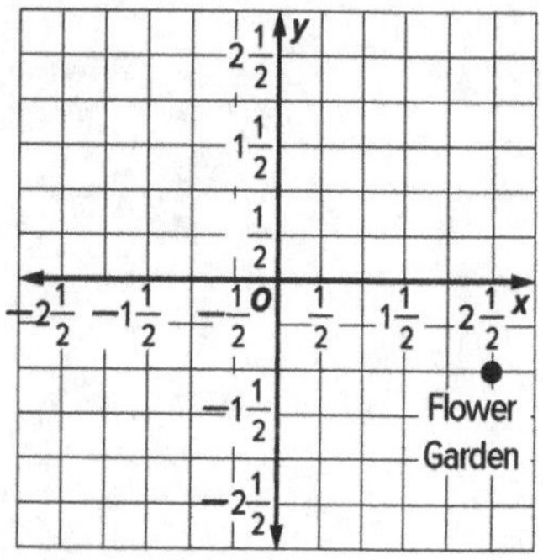

____

2 Jorge identified the ordered pair that is a reflection of (3, −2) across the *y*-axis. Juliana identified the ordered pair that is a reflection of (−3, 2) across the *x*-axis. Camila identified the ordered pair that is a reflection of (−3, −2) across the *y*-axis. Who identified a point inside the square? **6.NS.6b, MP 1**

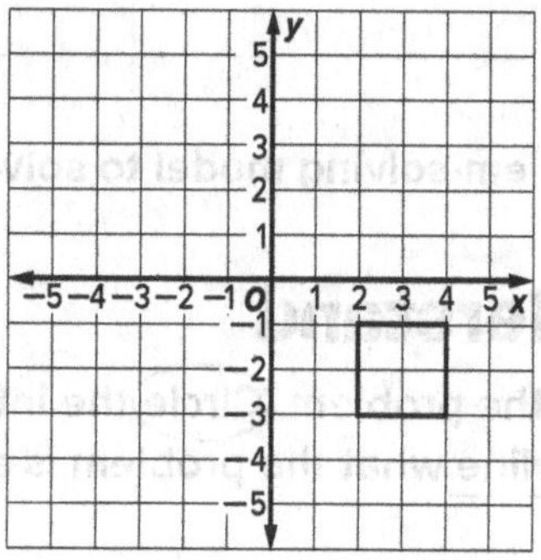

____

3 Carlos says all ordered pairs that have a 0 as either the *x*-coordinate or the *y*-coordinate are on the *x*-axis. Is he correct? Explain. **6.NS.6c, MP 3**

____

____

____

____

4 H.O.T. Problem Write each ordered pair from the box in the correct column of the table shown. Explain how you knew where to write the ordered pairs. **6.NS.6b, MP 7**

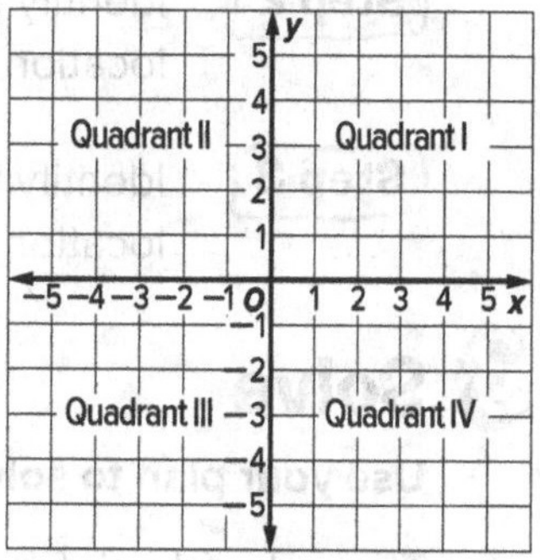

(2, 6)	(−4, −3)	(1, −7)	(−8, 9)
$(5\frac{1}{3}, -3\frac{5}{6})$	(−1.5, 2.2)	$(-6\frac{2}{5}, -3)$	$(1\frac{1}{2}, 2)$

Quadrant I	Quadrant II	Quadrant III	Quadrant IV

____

____

____

NAME______________________________________ DATE __________________ PERIOD __________

Lesson 7 **Multi-Step** Problem Solving

Multi-Step Example

The table shows the locations for several different places around town. The grid shows a map of the town, and each square on the grid represents one city block. Ben needs to go to the dry cleaner, which is 5 blocks north of the library. Where on the grid must he go? **6.NS.8,** MP 1

Place	Location
Bank	(5, −4)
Grocery	(−3, 0)
Library	(0, −3)
Post Office	(−4, 5)

Ⓐ (0, −8)

Ⓑ (5, −3)

Ⓒ (0, 2)

Ⓓ (0, 5)

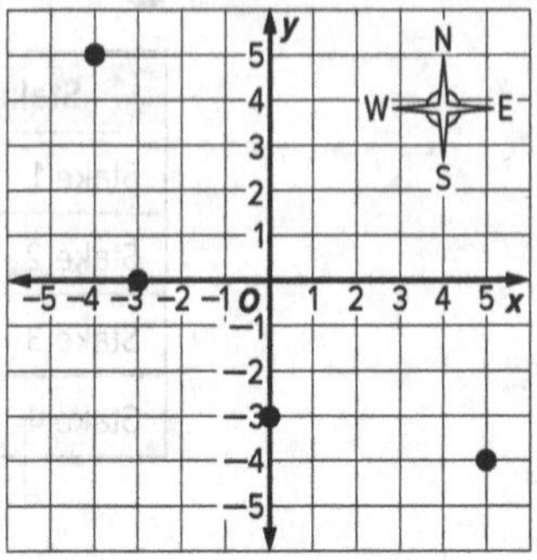

Use a problem-solving model to solve this problem.

1 Understand

Read the problem. Circle the information you know.
Underline what the problem is asking you to find.

2 Plan

What will you need to do to solve the problem? Write your plan in steps.

Step 1 Determine the dot on the grid that corresponds to the ________.

Step 2 Determine the ordered pair of the location ___ blocks _____ of the library.

3 Solve

Use your plan to solve the problem. Show your steps.

The library is located at (____, ____). Five blocks north would be the ordered pair (____, ____). Choice ___ is correct. Fill in that answer choice.

Read to Succeed!
The x-coordinate of an ordered pair tells you left or right and the y-coordinate tells you up or down.

4 Check

How do you know your solution is accurate?

____

____

____

Lesson 7 *(continued)*

Use a problem-solving model to solve each problem.

1 Rosa is currently at (4, −2) on the map. To which place in the table is she the closest? **6.NS.8,** MP 1

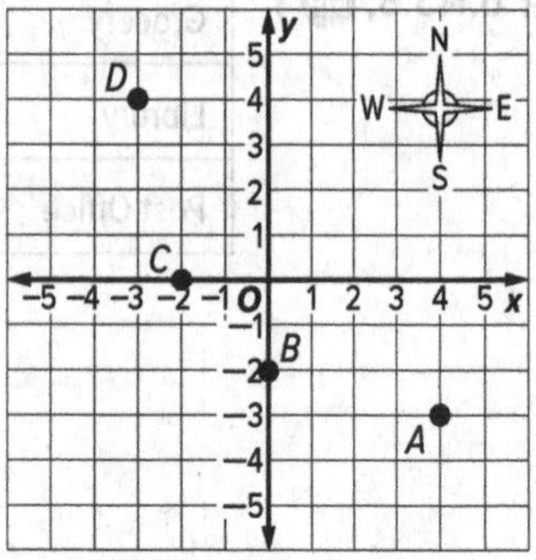

Place	Location
Stadium	*A*
Playground	*B*
Mall	*C*
Hospital	*D*

Ⓐ Stadium

Ⓑ Playground

Ⓒ Mall

Ⓓ Hospital

2 Josie posts stakes at the following locations. She ties rope to the stakes to section off a rectangle. Each unit represents 1 foot. What is the perimeter of the rectangle in feet? **6.NS.8,** MP 2

Stake	Location
Stake 1	(2, 0)
Stake 2	(2, −4)
Stake 3	(−1, −4)
Stake 4	(−1, 0)

____

3 Eduardo drives from *A* to *B*. Each unit on the map represents 10 miles. How many miles does he drive? **6.NS.8,** MP 4

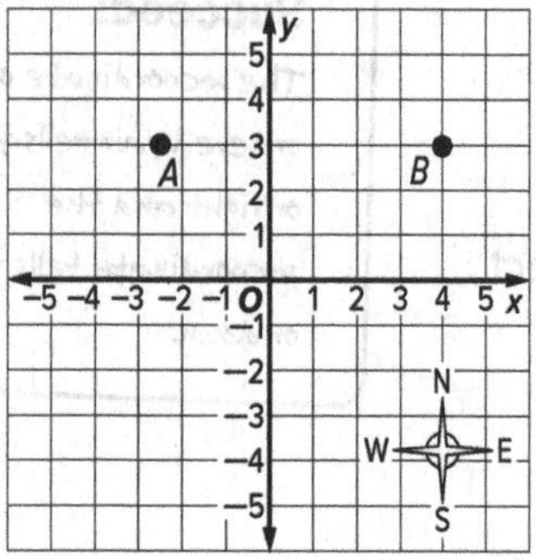

____

4 **H.O.T. Problem** Catalina plots point *C* at $\left(4, -1\frac{1}{2}\right)$. She also plots the point with the opposite *y*-coordinate and labels the point as *D*. What is the distance from *C* to *D*? **6.NS.8,** MP 2

____

NAME ________________ DATE ________________ PERIOD ________

Chapter 6

Lesson 1 Multi-Step Problem Solving

Multi-Step Example

Delmar is studying the reproduction rate of a specific type of bacteria. He places 3 cells in a Petri dish and records the number of bacteria over time. He notices a pattern, which is shown in the table. Predict the number of bacteria in the Petri dish after 25 hours. Express the answer using exponents. Then evaluate to determine the number of bacteria. **6.EE.1, MP 1**

Number of Hours	Number of Bacteria
5	3 × 3
10	3 × 3 × 3
15	3 × 3 × 3 × 3
20	3 × 3 × 3 × 3 × 3

Ⓐ 3^6; 729　　Ⓑ 6^3; 216　　Ⓒ 3^5; 243　　Ⓓ 3^6; 18

Use a problem-solving model to solve this problem.

1 Understand

Read the problem. Circle the information you know.
Underline what the problem is asking you to find.

2 Plan

What will you need to do to solve the problem? Write your plan in steps.

Step 1 Determine the pattern in the table.

Step 2 Express the answer using an exponent based on the pattern and evaluate.

3 Solve

Use your plan to solve the problem. Show your steps.

Every 5 hours, the number of bacteria triples. So, the number of bacteria after 25 hours is 3 × 3 × 3 × 3 × 3 × 3 or 3^6.

Since 3^6 = ______, choice ____ is correct. Fill in that answer choice.

Read to Succeed!
The exponent tells you how many times a base is used as a factor.

4 Check

How do you know your solution is accurate?

____

____

____

NAME________________________ DATE ______________ PERIOD ________

Lesson 1 *(continued)*

Use a problem-solving model to solve each problem.

1 Sonia is studying the reproduction rate of fleas. She places 2 fleas in an enclosed habitat and records the number of eggs each day. She notices a pattern, which is shown in the table. Predict the number of eggs in the habitat on the 5th day. Express the answer using exponents. Then evaluate to determine the number of eggs. **6.EE.1, MP 1**

Number of Days	Number of Eggs
1	5×5
2	$5 \times 5 \times 5$
3	$5 \times 5 \times 5 \times 5$
4	$5 \times 5 \times 5 \times 5 \times 5$

Ⓐ 6^5; 7,776

Ⓑ 5^5; 3,125

Ⓒ 5^6; 15,625

Ⓓ 6^6; 46,656

2 Elena has a fish tank that holds 2^5 gallons of water. How many fluid ounces of water does the fish tank hold? (*Hint:* 1 gal = 128 fl oz) **6.EE.1, MP 2**

____

3 Faith is turning 12 this year. She asks her parents to give her $1 on her birthday and to double that amount for her next birthday. If she continues with this pattern, how much money will Faith get on her 20th birthday? **6.EE.1, MP 7**

Birthday	Amount ($)
12th	2^0
13th	2^1
14th	2^2
15th	2^3

____

4 **H.O.T. Problem** In any cube, the length, width, and height each have the same measure. The volume of the cube below can be found by calculating a^3, where a is the length of a side. Suppose a is 8 inches. How many gallons of water would the cube hold if 231 cubic inches is equal to 1 gallon? Round the answer to the nearest tenth of a gallon. List the steps you used to find your answer. **6.EE.1, MP 1**

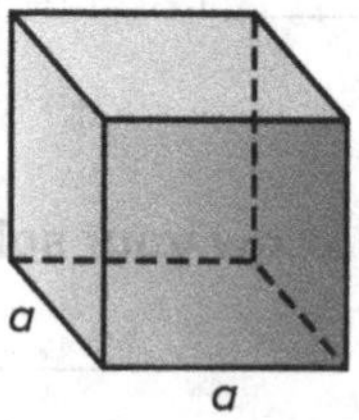

____

____

____

____

Lesson 2 Multi-Step Problem Solving

Multi-Step Example

An art store sells art kits that include crayons and a sketch pad. The table shows the number of crayons and sketch pad pages in each art kit size. A school buys 30 small, 10 large, and 24 medium art kits. Then they return 18 medium art kits. How many crayons do they have in all? **6.EE.1,** MP 1

Art Kit Size	Number of Crayons	Sketch Pad Pages
Small	16	20
Medium	24	40
Large	68	100

Ⓐ 3,240
Ⓑ 2,560
Ⓒ 1,736
Ⓓ 1,304

Use a problem-solving model to solve this problem.

1 Understand

Read the problem. Circle the information you know.
Underline what the problem is asking you to find.

2 Plan

What will you need to do to solve the problem? Write your plan in steps.

Step 1 Write an expression to represent the situation.

Step 2 Evaluate the expression.

Read to Succeed!
When evaluating the expression, remember to follow the order of operations. Perform the operations in the parentheses first then add.

3 Solve

Use your plan to solve the problem. Show your steps.

The school bought a total of 30 small, 6 medium, and 10 large kits.

(30 × ____) + (6 × ____) + (10 × ____) = ____

So, the school has ____ crayons. Choice ____ is correct.

Fill in that answer choice.

4 Check

How do you know your solution is accurate?

____

____

____

NAME________________________ DATE ______________ PERIOD ________

Lesson 2 *(continued)*

Use a problem-solving model to solve each problem.

1 The table shows the chocolate chip cookies in each container type. Yesterday, the bakery sold 4 tubs, 10 baskets, and 12 boxes. However, 3 of the boxes were returned. How many total cookies were sold by the end of the day? **6.EE.1, MP 1**

Container Type	Chocolate Chip Cookies
Basket	16
Box	36
Tub	48

Ⓐ 160
Ⓑ 192
Ⓒ 432
Ⓓ 676

2 The table shows the number of salt and yogurt pretzels that come in different sized boxes. A store orders 3 small boxes and 5 large boxes and then sells them at $2.00 per salt pretzel and $2.50 per yogurt pretzel. How much will the store make if they sell all the pretzels? **6.EE.1, MP 2**

Box Size	Salt Pretzel	Yogurt Pretzel
Small	36	12
Large	48	24

____

3 A luxury line of furniture at a store sells couches for $4,000, reclining chairs for $2,040, and loveseats for $2,800. This luxury line went on sale where the price of each piece of furniture was divided by 4. During the sale, how much would 2 couches, 1 reclining chair, and 3 loveseats cost? **6.EE.1, MP 2**

____

4 **H.O.T. Problem** Determine the value of the expression below. Explain your answer. **6.EE.1, MP 3**

$$\frac{120 - (3^3 - 36 \div 2)^2}{10 + 3(10 - 6) - 27 \div 3}$$

____

____

____

____

____

Lesson 3 **Multi-Step** Problem Solving

Multi-Step Example

The table shows the dimensions of three picture frame sizes available at a framing shop. What is the total perimeter, in inches, of two small frames and three large frames? The perimeter of a rectangle is $2\ell + 2w$, where ℓ is the length and w is the width. **6.EE.2c, MP 1**

Picture Frame Size	Length (in.)	Width (in.)
Small	3	5
Medium	5	7
Large	8	10

Use a problem-solving model to solve this problem.

1 Understand

Read the problem. Circle the information you know.
Underline what the problem is asking you to find.

Read to Succeed!
Remember that 2ℓ is the same as $2 \times \ell$ and $2w$ is $2 \times w$.

2 Plan

What will you need to do to solve the problem? Write your plan in steps.

Step 1 Determine the perimeter of a small frame and a large frame.

Step 2 Determine the total perimeter for the five frames.

3 Solve

Use your plan to solve the problem. Show your steps.

The perimeter of a small frame is 2(☐) + 2(☐) or ☐ inches.

The perimeter of a large frame is 2(☐) + 2(☐) or ☐ inches.

The total perimeter of the five frames is 2(16) + 3(36) or ☐ inches.

4 Check

How do you know your solution is accurate?

____

____

____

Lesson 3 *(continued)*

Use a problem-solving model to solve each problem.

1 The table shows different dog carrier sizes available at a pet store. What is the total perimeter, in feet, of one large, two extra-small, and three medium dog carriers? Represent the situation with an expression. **6.EE.2c, MP 1**

Carrier Size	Length (in.)	Width (in.)
Extra-Small	19	13
Small	24	18
Medium	30	19
Large	36	23

2 Gabby is going to cover her ruler shown below with construction paper for an art project. The area of the ruler can be found using the expression ℓw, where ℓ is the length and w is the width of the rectangle. What is the area, in square inches, of the construction paper needed if 4 rulers are used in her art project? Represent the situation with an expression. **6.EE.2c, MP 4**

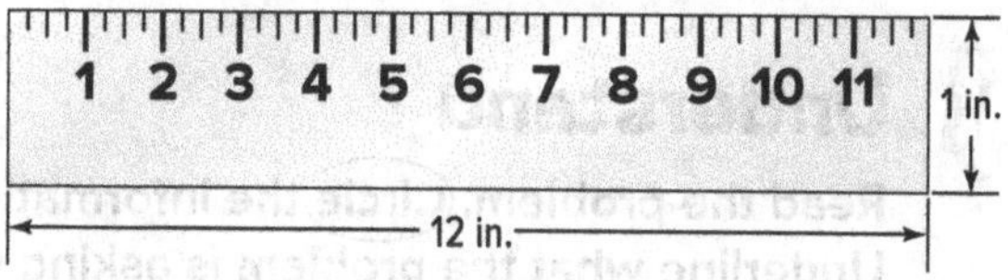

3 Calvin is filling a sandbox with sand. The volume of the sandbox can be found using the expression ℓwh where ℓ is the length, w is the width, and h is the height. What is the volume, in cubic inches, of two of these sandboxes? Represent the situation with an expression. (*Hint: There are 12 × 12 × 12 cubic inches in a cubic foot.*) **6.EE.2c, MP 4**

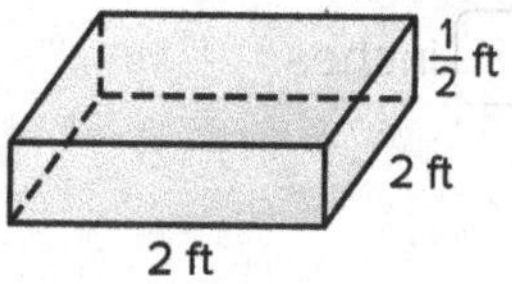

4 **H.O.T. Problem** Write an expression for the perimeter of the irregular-shaped figure shown below. Explain. **6.EE.6, MP 4**

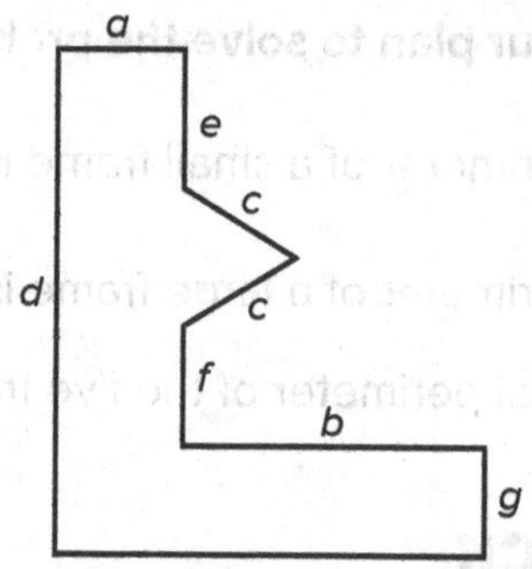

Lesson 4 Multi-Step Problem Solving

Multi-Step Example

The cost of tickets at a movie theater is shown in the table. Write an expression to represent the total cost of tickets using the variables in the table. Then use the expression to find the total ticket cost in dollars for 2 adults, 3 children, and 1 senior. **6.EE.6, MP 1**

Type of Ticket	Number of Tickets	Cost ($)
Adult	a	8
Child	c	5
Senior	s	6

Ⓐ $a + c + s$; $6

Ⓑ $(8 + 5 + 6)(a + c + s)$; $114

Ⓒ $8a + 5c + 6s$; $37

Ⓓ $8a + 5c + 6s$; $19

Use a problem-solving model to solve this problem.

1 Understand

Read the problem. Circle the information you know. Underline what the problem is asking you to find.

2 Plan

What will you need to do to solve the problem? Write your plan in steps.

Step 1 Represent the situation with an expression.

Step 2 Substitute the values given and evaluate.

Read to Succeed!
Follow the order of operations when evaluating expressions. Multiply first then add.

3 Solve

Use your plan to solve the problem. Show your steps.

The cost of a adult tickets is ☐, c child tickets is ☐, and s senior tickets is ☐.

The total for any number of tickets is ☐ + ☐ + ☐.

So, the total for 2 adults, 3 children, and 1 senior is $8(2) + 5(3) + 6(1)$ or ☐.

Choice ☐ is correct. Fill in that answer choice.

4 Check

How do you know your solution is accurate?

____

____

____

NAME______________________________ DATE ______________ PERIOD ________

Lesson 4 *(continued)*

Use a problem-solving model to solve each problem.

1 Shelby is attending classes at a university that charges $175 per credit, plus an application fee of $45. Write an expression that represents the total cost of tuition based on the number of credits *c*. Then use the expression to determine the tuition amount if Shelby plans on taking 14 credits. **6.EE.6, MP 1**

Ⓐ $45c + 175$; $805

Ⓑ $175c + 45$; $2,495

Ⓒ $(175 + 45)c$; $3,080

Ⓓ $\frac{175}{c} + 45$; $57.50

2 Michael is working on his budget and decides to allocate a percentage of each paycheck as described in the circle graph. Write an expression to determine the total amount he would deposit in the bank *b*. How much does he deposit if his check is $800? **6.EE.3, MP 2**

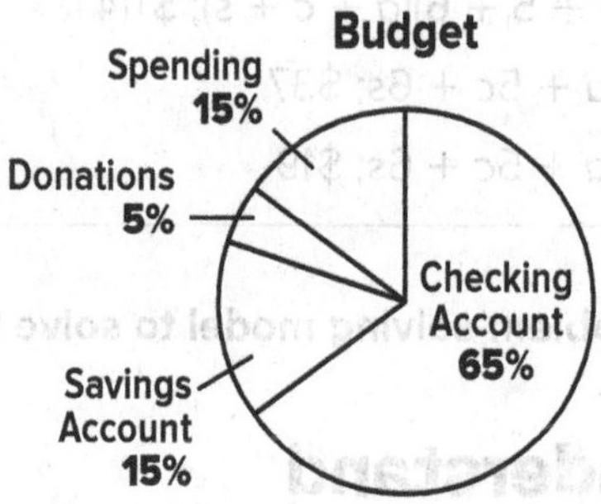

____

3 Brad is designing an A-frame house and needs to include angle measures on the diagram below. The sum of angles 1, 2, and 3 is 180°. The sum of angles 3 and 4 is 180°. The measure of angle 1 is 50° and the measure of angle 2 is 20°. Write an equation that can be used to find the measure of angle 4. **6.EE.6, MP 4**

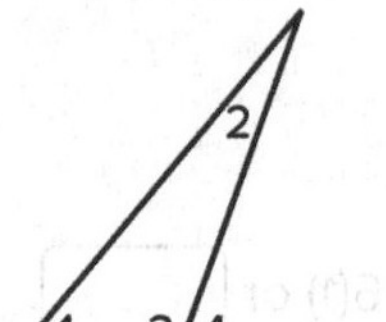

____

____

4 **H.O.T. Problem** Write an expression to find the area of the shaded region. **6.EE.6, MP 4**

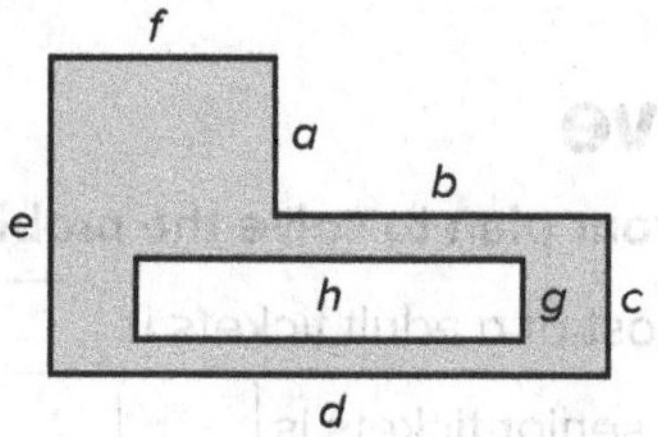

____

NAME______________________________ DATE ____________________ PERIOD ________

Lesson 5 Multi-Step Problem Solving

Multi-Step Example

Marla calculated the sum 1 + 2 + 3 + 4 + 5 + … + 12 + 13 + 14 by performing the additions shown in the table and then multiplying 15 × 7. What property did Marla use to make the addition problem easier to compute? **6.EE.3, MP 1**

Addends	Sum
1 + 14	15
2 + 13	15
3 + 12	15
4 + 11	15
5 + 10	15
6 + 9	15
7 + 8	15

Ⓐ Commutative Property of Addition

Ⓑ Commutative Property of Multiplication

Ⓒ Associative Property of Addition

Ⓓ Associative Property of Multiplication

Use a problem-solving model to solve this problem.

1 Understand

Read the problem. Circle the information you know.
Underline what the problem is asking you to find.

2 Plan

What will you need to do to solve the problem? Write your plan in steps.

Step 1 Determine the pattern shown in the table.

Step 2 Determine the property used.

3 Solve

Use your plan to solve the problem. Show your steps.

When Marla reordered the addends, she used the first and last, second and second-to-last addend, and so on. When reordering the addends, she used the Commutative Property of Addition.

So, choice ______ is correct. Fill in that answer choice.

Read to Succeed!
The Commutative Property says the order in which two numbers are added or multiplied does not matter.
The Associative Property says the grouping of the numbers when added or multiplied does not matter.

4 Check

How do you know your solution is accurate?

____

____

____

Lesson 5 *(continued)*

Use a problem-solving model to solve each problem.

1 The table shows the number of students that chose different sports based on gender. Kevin wants to determine the total number of boys surveyed. Which of the following expressions uses the Associative Property to help him mentally determine the sum? **6.EE.3, MP 1**

Sport	Number of Girls	Number of Boys
Basketball	14	11
Football	2	19
Lacrosse	7	5
Soccer	13	12
Swimming	12	13

Ⓐ $14 + 2 + 7 + 13 + 12$

Ⓑ $(11 + 19) + [5 + (12 + 13)]$

Ⓒ $(11 + 5) + (19 + 12 + 13)$

Ⓓ $2(13) + 2(12)$

2 The table shows pairs of equivalent expressions. Which expression does not model one of the different properties of number operations? **6.EE.3, MP 7**

$9 + 7 = 7 + 9$	$9 + 0 = 9$
$10 + 10 = 5 + 15$	$(8 + 6) + 0 = 8 + (6 + 0)$
$0 \times 8 = 8 \times 0$	$8 \times 1 = 8$

____

3 Four students were each given 48 blocks to arrange to make a rectangular prism. For two of these students, the length and width model the Commutative Property. What is the product of the length and width that they used? **6.EE.3, MP 7**

Student	Length	Width	Height
Lucy	12	4	1
Jose	4	4	3
Maria	3	4	4
Tony	4	12	1

____

4 **H.O.T. Problem** How could you use the Commutative and Associative Properties to quickly find the value of the expression $28 + (9^2 + 72)$? **6.EE.3, MP 3**

____

____

____

____

____

____

Lesson 6 Multi-Step Problem Solving

Multi-Step Example

Wen is buying bottles of apple juice and wants to mentally calculate how much they will cost. He buys 5 bottles of juice at $2.15 each. Which of the following shows equivalent expressions using the Distributive Property? How much change will he receive from $20? **6.EE.3, MP 1**

Ⓐ 5(2.15) = 5(2) + 0.15; $10.75

Ⓑ 5(2.15) = 5(2) − 0.15; $9.25

Ⓒ 5(2.15) = 5(2) + 5(0.15); $9.25

Ⓓ 5(2.15) = 5(2) − 5(0.15); $10.75

Use a problem-solving model to solve this problem.

1 Understand

Read the problem. Circle the information you know.
Underline what the problem is asking you to find.

2 Plan

What will you need to do to solve the problem? Write your plan in steps.

Step 1 Generate equivalent expressions using the Distributive Property.

Step 2 Determine the total spent and subtract from $20 to find the change he will receive.

Read to Succeed!
The number 2.15 can be expressed as 2 + 0.15.

3 Solve

Use your plan to solve the problem. Show your steps.

Five bottles of apple juice will cost 5(______). An equivalent expression, using the Distributive Property, is 5(2) + 5(0.15).

The total spent would be $______. Wen will receive $20.00 − $______ or $______ in change.

So, choice ______ is correct. Fill in that answer choice.

4 Check

How do you know your solution is accurate?

____

____

____

Lesson 6 *(continued)*

Use a problem-solving model to solve each problem.

1 Cole is connecting 3 benches to make one long bench. Each bench has a length of $5\frac{1}{2}$ feet. Which of the following show equivalent expressions using the Distributive Property? **6.EE.3, MP 1**

Ⓐ $3\left(5\frac{1}{2}\right) = 3(5) + 3\left(\frac{1}{2}\right)$

Ⓑ $3\left(5\frac{1}{2}\right) = 3(5) - 3\left(\frac{1}{2}\right)$

Ⓒ $3\left(5\frac{1}{2}\right) = 3(5) + \left(\frac{1}{2}\right)$

Ⓓ $3\left(5\frac{1}{2}\right) = 5(3) + 5\left(\frac{1}{2}\right)$

2 The total area of two rectangles can be calculated using different methods. What number can be substituted for x so that each expression will show the total area of the two rectangles? **6.EE.3, MP 4**

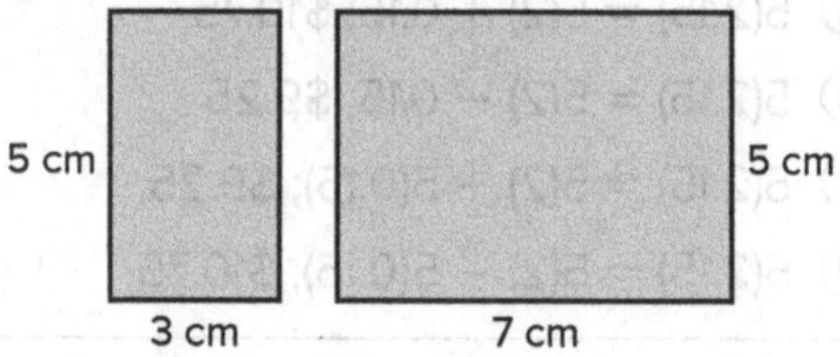

$$x(3 + 7) = x(3) + x(7)$$

____

3 Elijah bought 5 more candles, c, than Bianca. The candles cost $2 each. What number should be replaced for x to give an equivalent expression for the total amount Elijah and Bianca spent? **6.EE.3, MP 2**

$$2c + 2(c + 5) = xc + 10$$

____

4 **H.O.T. Problem** Marta needs to buy 8 notebooks that cost $3.75 each. Two friends show a shortcut for mentally calculating the total price. Evaluate each student's method. **6.EE.3, MP 3**

Juanita	8(3 + 0.75)
Joe	8(4 − 0.25)

____

____

____

____

Lesson 7 Multi-Step Problem Solving

Multi-Step Example

When buying frozen yogurt, there are many choices for the toppings. Which simplified expression represents the price of 2 cones with fruit and 3 cones with candy and syrup? **6.EE.2, MP 1**

Frozen Yogurt	Price ($)
Cone	x
Candy Topping	add $0.75
Syrup Topping	add $0.50
Fruit Topping	add $1.00

Ⓐ $5x + 2.25$

Ⓑ $5x + 5.75$

Ⓒ $2x + 2$

Ⓓ $2x + 4.50$

Use a problem-solving model to solve this problem.

1 Understand

Read the problem. Circle the information you know.
Underline what the problem is asking you to find.

2 Plan

What will you need to do to solve the problem? Write your plan in steps.

Step 1 Determine the expressions for each type of cone.

Step 2 Combine like terms to simplify the expressions.

Read to Succeed!
Remember to combine like terms when simplifying expressions. Like terms have the same variable.

3 Solve

Use your plan to solve the problem. Show your steps.

Two cones with fruit: $2(x + 1.00) =$ ☐ + ☐

Three cones with candy and syrup: $3(x + 0.75 + 0.50) =$ ☐ + ☐

Adding the two expressions and simplifying results in ☐ + ☐.

Choice ☐ is correct. Fill in that answer choice.

4 Check

How do you know your solution is accurate?

____

____

____

Lesson 7 *(continued)*

Use a problem-solving model to solve each problem.

1 At a taco stand, chicken tacos are available with a choice of 3 different toppings. Which simplified expression represents the price of 2 tacos with cheese and lettuce, and 1 taco with cheese, lettuce, and sour cream? **6.EE.2, MP 1**

Item	Price ($)
Chicken taco	x
Cheese	$0.25
Lettuce	$0.15
Sour cream	$0.45

Ⓐ $2x + 0.40$
Ⓑ $3x + 1.25$
Ⓒ $3x + 1.65$
Ⓓ $3x + 2.55$

2 Treven and Dylan each bought fruit at the market. Treven bought 3 pears and 7 apples. Dylan bought 4 pears and 9 apples and spent $3.00 on oranges. If p represents the cost of each pear and a represents the cost of each apple, the expression below represents the total cost. What is the value of x? **6.EE.2b, MP 2**

$$xp + 16a + 3$$

____

3 The length of a rectangle is x feet more than its width. The area of the rectangle is $5x + 25$. What is the width of the rectangle? **6.EE.3, MP 2**

____

4 **H.O.T. Problem** Julian asked x students about their favorite color, and displayed the results in the circle graph. Claire thinks that the same number of people voted for orange and red as did green and purple. Write three equal expressions that show that she is correct. **6.EE.3, MP 4**

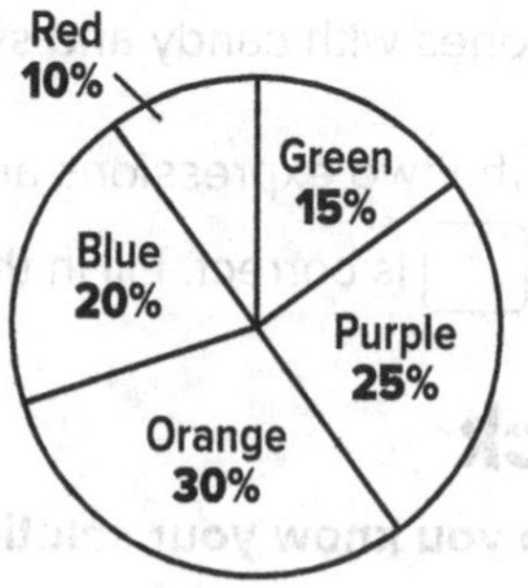

____

Lesson 1 Multi-Step Problem Solving

Multi-Step Example

Sandi bought a sandwich and a milkshake. She spent $12 in all. The equation $s + 4.25 = 12$ can be used to determine the cost of the sandwich. Which of the following is the solution to the equation? **6.EE.5, MP 1**

Item	Cost ($)
Sandwich	s
Milkshake	4.25

Ⓐ $7.75
Ⓑ $8.75
Ⓒ $16
Ⓒ $16.25

Use a problem-solving model to solve this problem.

1 Understand

Read the problem. Circle the information you know.
Underline what the problem is asking you to find.

2 Plan

What will you need to do to solve the problem? Write your plan in steps.

Step 1 Test each answer choice to determine if a true number sentence is created.

3 Solve

Use your plan to solve the problem. Show your steps.

$7.75 + 4.25 = 12$ ✓

$8.75 + 4.25 \neq 12$ ✗

$16 + 4.25 \neq 12$ ✗

$16.25 + 4.25 \neq 12$ ✗

Since ______ $+ 4.25 = 12$, choice ______ is correct. Fill in that answer choice.

Read to Succeed!
Substitute each answer choice for s and add to determine if the number sentence is true.

4 Check

How do you know your solution is accurate?

____

____

____

NAME________________________ DATE______________ PERIOD__________

Lesson 1 *(continued)*

Use a problem-solving model to solve each problem.

1 Kaleena paid \$26.25 to rent a kayak for 3 hours. The equation $3x = 26.25$ can be used to determine the amount she paid per hour. Which of the following is the solution to the equation? **6.EE.5, MP 1**

Ⓐ \$8.66

Ⓑ \$8.75

Ⓒ \$23.25

Ⓓ \$29.25

2 Shalah went to the grocery store with \$50 in cash and bought the items shown in the table. In the equation $s + x = 50$, *s* represents the amount she spent, and *x* represents the amount she had left. Does she have \$14.25, \$28.50, or \$35.75 left? **6.EE.5, MP 2**

Item	Milk	Eggs	Ham
Cost	\$3.25	\$2.50	\$8.50

3 Kirsten had 65 inches of wire. After she used some for a project, she had 14 inches left. The equation $x + 14 = 65$ can be used to determine the amount of wire that she used in inches. Did she use 79, 51, or 37 inches of wire? Plot the solution on the number line. **6.EE.5, MP 2**

4 **H.O.T. Problem** George made 12 ounces of rice. He had 4 ounces left after he finished eating. Use *guess, check, and revise* strategy to solve the equation $12 - r = 4$ to find *r*, the number of ounces of rice that he ate. Then convert your answer to find how many cups of rice George ate. **6.EE.5, MP 6**

r	$12 - r = 4$	Are Both Sides Equal?
5		
6		
7		
8		

Lesson 2 Multi-Step Problem Solving

Multi-Step Example

A bookstore has a sale on mysteries. The table shows the cost of each book format. Abigail has $70 to spend. She bought two paperbacks, one hardcover, and one audio book. Which addition equation can be used to determine how much more money Abigail still has to spend? **6.EE.7, MP 1**

Book	Cost ($)
Hardcover	19
Paperback	8
E-book	10
Audio Book	25

Ⓐ $70 + 60 = x$

Ⓑ $60 + x = 70$

Ⓒ $70 + x = 60$

Ⓓ $52 + 70 = x$

Use a problem-solving model to solve this problem.

1 Understand

Read the problem. Circle the information you know.
Underline what the problem is asking you to find.

2 Plan

What will you need to do to solve the problem? Write your plan in steps.

Step 1 Determine how much Abigail spent.

Step 2 Write an equation to represent the situation.

3 Solve

Use your plan to solve the problem. Show your steps.

Abigail spent 2($____) + $____ + $____, or $____.

Let x = the amount she has left.

So, ____ + x = ____. Choice ____ is correct. Fill in that answer choice.

Read to Succeed!
The total she has to spend is the sum so it will be by itself in the equation.

4 Check

How do you know your solution is accurate?

____

____

____

Lesson 2 *(continued)*

Use a problem-solving model to solve each problem.

1 Kisho has $45 to spend at a pizza shop for a pizza party. The table shows the cost of each size of pizza. He bought four small pizzas and one large pizza. Which addition equation can be used to determine how much more money he still has to spend? **6.EE.7, MP 1**

Pizza Size	Cost ($)
Small	5
Medium	8
Large	10
Extra Large	15

Ⓐ $x + 45 = 30$

Ⓑ $x - 30 = 45$

Ⓒ $x + 30 = 45$

Ⓓ $x - 45 = 30$

2 Miguel has $2\frac{1}{2}$ hours to work on his homework. The table shows how much time he spent working on his English homework and his math homework. Write and solve an addition equation that can be used to find how much time, in minutes, he has left to work on his science project if he wants to take a 15-minute snack break. **6.EE.7, MP 7**

Homework	Time Spent (min)
English	45
Math	28
Science project	?

____

____

3 Two rectangular rugs in Winona's bedroom are shown below. The larger rug is 32 square feet bigger than the smaller rug. The equation $x + 32 = 35$ can be used to find the area of the smaller rug. What is the width of the smaller rug, in feet? **6.EE.7, MP 6**

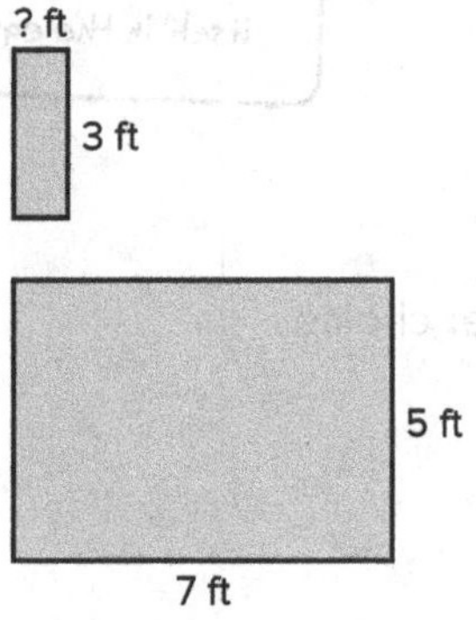

____

4 **H.O.T. Problem** The perimeter of the triangle below is 58 centimeters. Write an addition equation to determine the unknown length, x. Explain how you can use the Subtraction Property of Equality to solve the equation. Write the unknown length. **6.EE.7, MP 3**

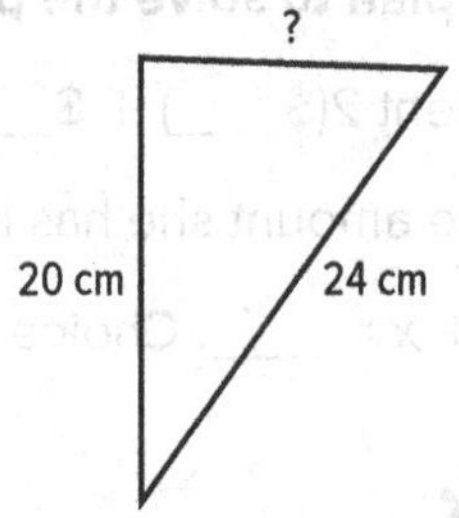

____

____

____

____

____

____

Lesson 3 **Multi-Step** Problem Solving

Multi-Step Example

Tyson withdrew money from his savings account to go shopping. The circle graph shows how he spent the money. He has \$326 left in his savings account. Which subtraction equation can be used to determine how much Tyson had in his account before he withdrew the money? **6.EE.7,** MP 1

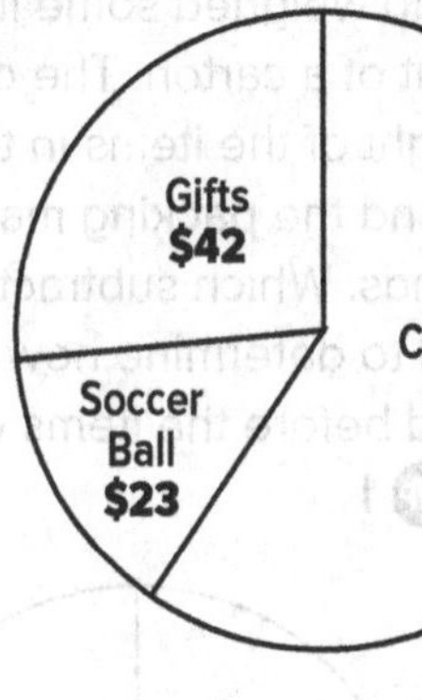

Ⓐ $326 - 160 = m$

Ⓑ $160 - 95 = m$

Ⓒ $326 - m = 65$

Ⓓ $m - 160 = 326$

Use a problem-solving model to solve this problem.

1 Understand

Read the problem. Circle the information you know.
Underline what the problem is asking you to find.

2 Plan

What will you need to do to solve the problem? Write your plan in steps.

Step 1 Determine how much Tyson spent.

Step 2 Write an equation to represent the situation.

Read to Succeed!

The amount he has left in his account after withdrawing money to shop is the difference. So it will be by itself in the equation.

3 Solve

Use your plan to solve the problem. Show your steps.

Tyson spent \$___ + \$___ + \$___, or \$____ in all.

Let m = the amount of money he had in his account before he withdrew the money.

So, m − ____ = ____. Choice __ is correct. Fill in that answer choice.

4 Check

How do you know your solution is accurate?

____

____

____

NAME______________________________ DATE ______________ PERIOD ________

Lesson 3 *(continued)*

Use a problem-solving model to solve each problem.

1 Fernando weighed some items as he took them out of a carton. The circle graph shows the weight of the items in the carton. The carton and the packing material weigh 1.5 pounds. Which subtraction equation can be used to determine how much the carton weighed before the items were taken out? **6.EE.7, MP 1**

Ⓐ $9.5 - 1.5 = w$

Ⓑ $9.5 - w = 1.5$

Ⓒ $1.5 - w = 9.5$

Ⓓ $w - 9.5 = 1.5$

2 A grocery store is having a "two for the price of one" sale for several items. The cost of these items is shown in the table. Davion bought two boxes of cereal and two loaves of bread. He also paid \$0.39 in tax. The equation $x - (4.50 + 1.99 + 0.39) = 13.12$ can be used to determine how much money he gave the cashier. Determine how much money, in dollars, Davion gave the cashier if he received \$13.12 in change. **6.EE.7, MP 2**

Item	Cost (\$)
Bread	1.99
Cereal	4.50
Orange Juice	3.00

____

3 Nina used $12\frac{1}{3}$ yards of ribbon to make hair bows and 5 yards of ribbon to wrap gifts. She has $22\frac{2}{3}$ yards of ribbon left. Write and solve a subtraction equation that can be used to find how many yards of ribbon she had to start. **6.EE.7, MP 2**

____

____

____

4 **H.O.T. Problem** What value of x makes the following equation true? Write a real-world problem that could be modeled with this equation. **6.EE.7, MP 7**

$$x - \frac{1}{8} = \frac{1}{12}$$

____

____

____

____

____

NAME________________________________ DATE ____________ PERIOD ________

Lesson 4 Multi-Step Problem Solving

Multi-Step Example

The table shows the number of miles Diego and some friends traveled each day and the amount of time it took. Write an equation that can be used to determine the average speed, r, at which they traveled. What property would you use to solve the equation? 6.EE.7, MP 1

Day	Distance (miles)	Time (hours)
1	110	1.9
2	90	1.5
3	105	1.8
4	120	2.1

Use a problem-solving model to solve this problem.

1 Understand

Read the problem. Circle the information you know.
Underline what the problem is asking you to find.

2 Plan

What will you need to do to solve the problem? Write your plan in steps.

Step 1 Write an equation to represent the situation.

Step 2 Determine the property used to solve the equation.

Read to Succeed!
When solving problems dealing with distance, rate, and time, the equation is $d = rt$, where d is the distance, r is the rate, and t is the time.

3 Solve

Use your plan to solve the problem. Show your steps.

The group drove a total of 425 miles in 7.3 hours.

$d = r \cdot t$ Distance, rate, time equation

____ $= r \cdot$ ____ Substitute known values.

To solve the equation, you would use the ________ Property of Equality.

4 Check

How do you know your solution is accurate?

____

____

____

Lesson 4 *(continued)*

Use a problem-solving model to solve each problem.

1 The table shows the distance Catrell biked each day and his rate. Write an equation that can be used to determine the average time, t, he spent riding his bike each day. What property would you use to solve the equation? **6.EE.7, MP 1**

Day	Distance (miles)	Rate (miles/hour)
1	15	10
2	18	14
3	21	16
4	24	12

2 The model below shows the relationship between a gallon and pints. Use the model to write an equation to determine the number of pints given the gallons. Use the equation to convert $\frac{2}{3}$ gallon into pints. Round the answer to the nearest tenth. **6.EE.7, MP 2**

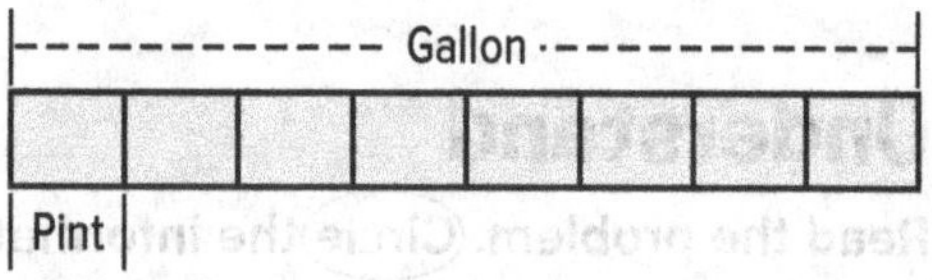

3 Linh decides to put $\frac{1}{6}$ of her paycheck into her savings account. Use the model below to write an equation that represents the amount of her paycheck in terms of the amount put into savings. Then use the equation to determine the amount of her paycheck. **6.EE.7, MP 2**

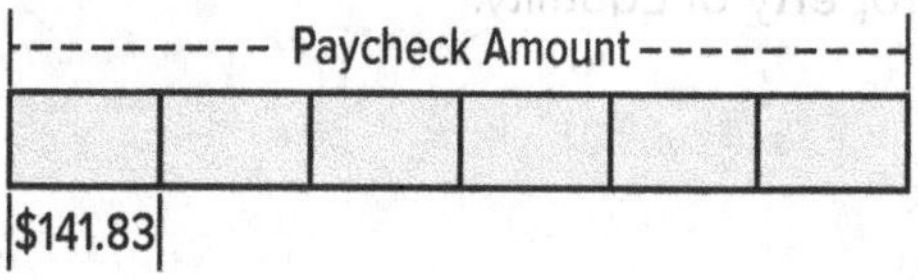

4 **H.O.T. Problem** In the diagram below, $\angle ABC$ is divided into 4 angles of equal measure and $m\angle ABC = 140°$. Write and solve two equations that can be used to find the degree measure of $\angle EBF$. Identify any properties of equalities that you used to solve either equation. **6.EE.7, MP 2**

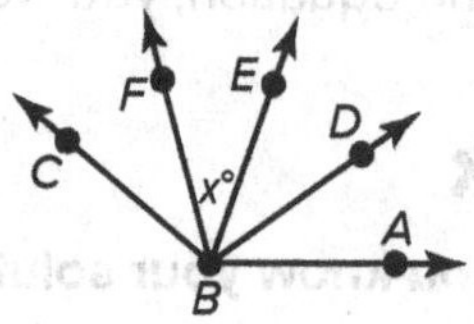

Lesson 5 Multi-Step Problem Solving

Multi-Step Example

Julian read a book in 4 days. The table shows the time he read each day. He read an average of 12 pages per hour. Which equation can be used to find the total number of pages p he read? What property would you use to solve the equation? 6.EE.7, MP 1

Day	Time (h)
1	1.2
2	1.7
3	2.6
4	2.5

Ⓐ $8 = p + 12$; Subtraction Property of Equality

Ⓑ $8 = p - 12$; Addition Property of Equality

Ⓒ $8 = 12p$; Division Property of Equality

Ⓓ $8 = \frac{p}{12}$; Multiplication Property of Equality

Use a problem-solving model to solve this problem.

1 Understand

Read the problem. Circle the information you know.
Underline what the problem is asking you to find.

2 Plan

What will you need to do to solve the problem? Write your plan in steps.

Step 1 Write an equation to represent the problem.

Step 2 Determine the property used to solve the equation.

Read to Succeed!
When you are writing an equation to solve a problem, it is helpful to draw a bar diagram first to model the problem.

3 Solve

Use your plan to solve the problem. Show your steps.

Julian spent a total of 8 hours reading. He read an average of 12 pages per hour.

The total time is equal to the total number of pages divided by the average number of pages.

______ $= \frac{p}{12}$ To solve the equation, you could use the ______________ Property of Equality and multiply each side of the equation by ______.

Choice ______ is the correct answer. Fill in that answer choice.

4 Check

How do you know your solution is accurate?

____

____

Lesson 5 *(continued)*

Use a problem-solving model to solve each problem.

1 It took Roberto 3 weeks to read a book for his English class. The table shows the time he spent reading the book each week. He read an average of 15 pages per hour. Which equation can be used to determine the total number of pages p he read? What property would you use to solve the equation? **6.EE.7, MP 1**

Week	Time (h)
1	8.5
2	6
3	3.5

Ⓐ $18 = \frac{p}{15}$; Division Property of Equality

Ⓑ $18 = \frac{p}{15}$; Multiplication Property of Equality

Ⓒ $18 = 15p$; Division Property of Equality

Ⓓ $18 = 15p$; Multiplication Property of Equality

2 Olivia's little sister divided 40 stickers evenly among some friends. Each friend received 4 animal stickers, 2 flower stickers, and 2 plant stickers. Write and solve a multiplication equation to find how many friends received stickers. **6.EE.7, MP 2**

____

3 Latisha swam on Monday, Wednesday, and Friday for 2 weeks and on Tuesday and Thurday for 1 week. She swam an average of 25 laps each day she swam. Write and solve a division equation to find how many laps she swam in all. **6.EE.7, MP 2**

____

4 **H.O.T. Problem** Timothy and Heather each solved the equation $\frac{12}{x} = 3$. Who solved the problem correctly? Explain how you know. **6.EE.7, MP 3**

Timothy	Heather
$\frac{12}{x} = 3$	$\frac{12}{x} = 3$
$36 = x$	$12 = 3x$
	$4 = x$

____

____

____

____

____

NAME______________________________ DATE ______________ PERIOD ________

Lesson 1 Multi-Step Problem Solving

Multi-Step Example

Sandra places wooden sculptures on top of a 3-foot-tall stand. The table can be used to compare the height of a sculpture to the height including the stand. What is the total height, in inches, of a 5-foot-tall sculpture on a stand? **6.EE.9, MP 1**

Height of Sculpture (ft) (x)	$x + 3$	Height on Stand (ft) (y)
2		
3		
5		

Ⓐ 8 in.
Ⓑ 10 in.
Ⓒ 60 in.
Ⓓ 96 in.

Use a problem-solving model to solve this problem.

1 Understand

Read the problem. Circle the information you know.
Underline what the problem is asking you to find.

2 Plan

What will you need to do to solve the problem? Write your plan in steps.

Step 1 Use the table to determine the height including the stand.

Step 2 Convert the height to inches by multiplying by 12.

3 Solve

Use your plan to solve the problem. Show your steps.

Height of Sculpture (ft) (x)	$x + 3$	Height on Stand (ft) (y)
2		
3		
5		

Read to Succeed!
There are 12 inches in 1 foot. To convert feet to inches, multiply by 12.

So, the height on the stand is ☐ feet. In inches, this is 8 × 12 or ☐ inches. Choice ☐ is correct. Fill in that answer choice.

4 Check

How do you know your solution is accurate?

____

____

____

Lesson 1 *(continued)*

Use a problem-solving model to solve each problem.

1 Eduardo has a $5 coupon for groceries. The amount he owes can be found by subtracting 5 from the total cost of his groceries. He buys $42 worth of groceries and pays with a $50 bill. How much change does he receive? **6.EE.9, MP 1**

Cost of Groceries (x)	$x - 5$	Amount Owed (y)
$15		
$25		
$42		

Ⓐ $8
Ⓑ $13
Ⓒ $37
Ⓓ $82

2 Sam plays a trivia game. He earns 6 points for each question that he answers correctly. He creates a table to show this relationship. What is x when $y = 30$? **6.EE.9, MP 2**

Number of Correct Answers (x)	$6(x)$	Points Earned (y)
		6
		30

3 Catarina invites friends to come over for breakfast. She made 50 muffins. If there are x guests, then each guest can have $50 \div x$ muffins. She creates a table to show the relationship. If $y = 2$, what is x? **6.EE.9, MP 2**

Number of Guests (x)	$50 \div (x)$	Muffins per Guest (y)
		5
		2

______

4 **H.O.T. Problem** If the side length of a square is x, then the perimeter of the square is $4x$. Complete the table to find the perimeters for the squares with sides lengths as shown. Describe the relationship shown in the table. **6.EE.9, MP 7**

Side Length (x)	$4(x)$	Perimeter (y)
$\frac{1}{3}$		
$\frac{1}{2}$		
$1\frac{1}{2}$		

______

______

______

______

Lesson 2 Multi-Step Problem Solving

Multi-Step Example

Autumn and Bennett painted signs for a school campaign. The table shows the total number of signs painted, based on the number of hours spent painting. How many more signs did Autumn paint than Bennett after 6 hours? **6.EE.2,** MP 1

Hours	Autumn	Bennett
1	6	4
2	9	6
3	12	8
4	15	10

Use a problem-solving model to solve this problem.

1 Understand

Read the problem. Circle the information you know.
Underline what the problem is asking you to find.

2 Plan

What will you need to do to solve the problem? Write your plan in steps.

Step 1 Determine the rule for each person.

Step 2 Use the rule to determine the number of signs made at 6 hours.

Step 3 Subtract to determine how many more signs Autumn painted.

Read to Succeed!
After determining the algebraic rule for each person, test your rule by using the independent and dependent quantities from the table.

3 Solve

Use your plan to solve the problem. Show your steps.

Autumn: $3(x) + 3$ Bennett: $2(x) + 2$

$3(\square) + 3 = \square$ $2(\square) + 2 = \square$

So, Autumn painted $\square - \square$ or $\square$ more signs.

4 Check

How do you know your solution is accurate?

____

____

NAME ______________________ DATE ______________ PERIOD ________

Lesson 2 *(continued)*

Use a problem-solving model to solve each problem.

1 The table shows the number of boxes Sandra and Conisha can fill with canned food during a food drive, based on the number of hours worked. How many more boxes can Conisha fill than Sandra after 7 hours? **6.EE.2, MP 1**

Hours	Sandra	Conisha
1	4	8
2	7	16
3	10	24
4	13	32

____

2 The table shows the number of miles Lan and Bailey ran each of the last six days. How many more miles did Lan run than Bailey on the sixth day? **6.EE.2, MP 7**

Days	Lan	Bailey
1	1.10	5.0
2	2.20	4.1
3	3.30	3.2
4	4.40	2.3
5	?	?
6	?	?

____

3 The table shows the amount it costs to ride a go-kart, based on the number of hours. The rule to find the total cost is $a(x) + b$. What is the sum of a and b? **6.EE.2, MP 7**

Time (x)	Amount ($)
$1\frac{1}{2}$	13
2	15
3	19
5	27

____

4 **H.O.T. Problem** The table shows two rules. Describe each rule. What is the relationship between Rule 1 and Rule 2? **6.EE.2, MP 7**

Position	Rule 1: Value of Term	Rule 2: Value of Term
1	1	1
2	4	8
3	9	27
4	16	64

____

____

____

____

____

____

NAME_______________________________ DATE ______________ PERIOD ________

Lesson 3 Multi-Step Problem Solving

Multi-Step Example

Kya's take-home pay, y, on a given day equals her earnings, x, minus $3 for parking. The relationship can be represented by the equation $y = x - 3$. Make a table of values and plot the ordered pairs. Which ordered pair will NOT be on the graph?
6.EE.9, MP 1

Ⓐ (70, 67) Ⓒ (82, 79)
Ⓑ (51, 54) Ⓓ (93, 90)

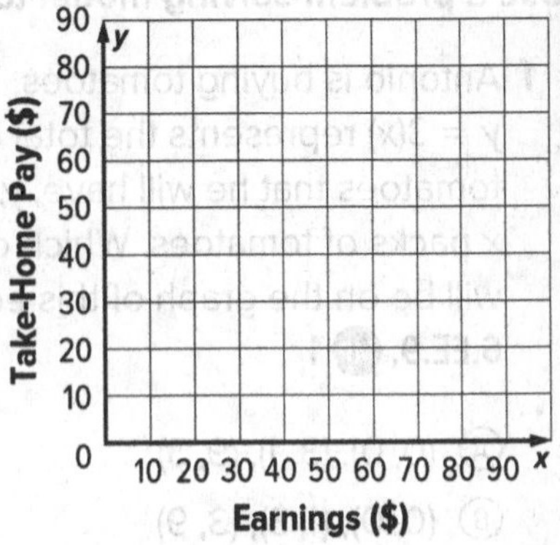

Use a problem-solving model to solve this problem.

1 Understand

Read the problem. Circle the information you know.
Underline what the problem is asking you to find.

2 Plan

What will you need to do to solve the problem? Write your plan in steps.

Step 1 Make a table to show the independent and dependent quantities.

Step 2 Write and graph the ordered pairs.

3 Solve

Use your plan to solve the problem. Show your steps.

x	$x - 3$	y

Read to Succeed!
Use the x-coordinates in the table to check the y-coordinates of the answer choices.

The ordered pair (51, 54) is not on the graph. So, choice ____ is correct.
Fill in that answer choice.

4 Check

How do you know your solution is accurate?

____

____

Lesson 3 *(continued)*

Use a problem-solving model to solve each problem.

1 Antonio is buying tomatoes. The equation $y = 3(x)$ represents the total number of tomatoes that he will have, y, if he buys x packs of tomatoes. Which ordered pairs will be on the graph of this equation? **6.EE.9, MP 1**

Ⓐ (0, 0), (3, 1), (9, 3)

Ⓑ (0, 0), (1, 3), (3, 9)

Ⓒ (0, 3), (1, 4), (3, 10)

Ⓓ (0, 0), (2, 6), (3, 6)

2 The table shows the amount of commission, y, that Fernando earns compared to his weekly sales, x. If the equation for this situation is $y = ax$, what is the value of a? **6.EE.9, MP 7**

Sales (x)	Commission (y)
$250	$75
$175	$52.50
$80	$24

____

3 The graph compares the number of snow cones purchased to the total cost of the snow cones. If the equation for the situation is $y = ax$, what is the value of a? As the number of snow cones increases, does the total cost increase or decrease? **6.EE.9, MP 7**

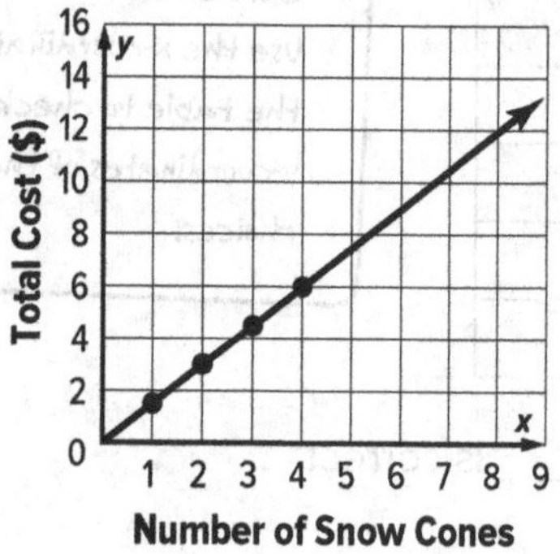

____

4 **H.O.T. Problem** The perimeter of a rectangle is 30. The dimensions are unknown. Write an equation that gives the length, y, in terms of the width, x. Complete the table of values to help you. **6.EE.9, MP 7**

Width (x)	Length (y)

____

____

NAME________________ DATE ____________ PERIOD ________

Lesson 4 Multi-Step Problem Solving

Multi-Step Example

Brian's baseball team is hosting a tournament to help raise money to buy new uniforms. Each team pays $100 to compete, but some of the money will be used to pay for uniforms. Use the table to help you find the amount of money raised for uniforms if 6 teams compete. **6.EE.9, MP 1**

Number of Teams Competing	Money for Uniforms ($)
1	50
2	100
3	

Ⓐ $50 Ⓒ $300

Ⓑ $250 Ⓓ $600

Use a problem-solving model to solve this problem.

1 Understand

Read the problem. Circle the information you know.
Underline what the problem is asking you to find.

2 Plan

What will you need to do to solve the problem? Write your plan in steps.

Step 1 Determine the rule that represents the situation.

Step 2 Use the rule to determine the amount of money raised if 6 teams compete.

Read to Succeed!
Looking for a pattern in the table is another strategy.

3 Solve

Use your plan to solve the problem. Show your steps.

Each dependent quantity is ☐ times the independent quantity.

The rule is $y =$ ☐.

So, the amount raised when 6 teams compete is 50(☐) or $☐.

Choice ☐ is correct. Fill in that answer choice.

4 Check

How do you know your solution is accurate?

____

____

____

Lesson 4 *(continued)*

Use a problem-solving model to solve each problem.

1 Felicia has an alarm system for her home. She paid $80 to have the alarm installed, and pays $45.50 each month. How much does she spend on the alarm system in the first year? **6.EE.9, MP 1**

Number of Months	Total Cost ($)
1	125.50
2	171

Ⓐ $125.50
Ⓑ $216.50
Ⓒ $626.00
Ⓓ $1,005.50

2 At a craft store, Georgina buys several packs of beads. The equation $y = 1.50x$ represents her total cost y if she buys x packs of beads. If the data are graphed, what is the value of x in $(x, 16.50)$? **6.EE.9, MP 2**

____

3 Montel is playing a math game. He earns points for every correct answer. The points on the graph represent the number of questions he answered correctly x, to his total score y. If the point $(7, y)$ is on the graph, what is y? **6.EE.9, MP 2**

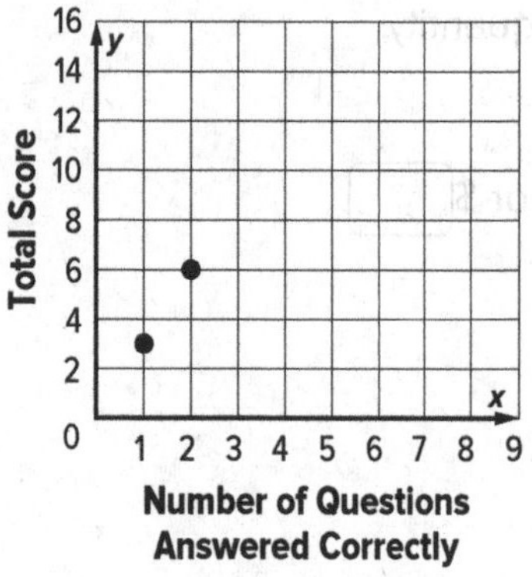

____

4 **H.O.T. Problem** Three friends are throwing a party at a park. They decide to rent a climbing wall for the party. The equation $y = 150x$ represents the total cost y for renting the wall for x hours. If they rented the wall for 5 hours and they split the cost equally, how much does each friend owe? **6.EE.9, MP 4**

____

Lesson 5 **Multi-Step** Problem Solving

Multi-Step Example

Some friends each hope to attend a festival that costs \$65. To earn money, they mowed lawns. Use the inequality $f + s \geq 65$, where f is the Friday earnings and s is the Saturday earnings, to determine who earned enough money to go to the festival. **6.EE.5,** MP 1

	Friday	Saturday
Cody	\$30	\$10
Dominic	\$60	\$0
Emir	\$45	\$10
Fernando	\$20	\$55

Ⓐ Cody
Ⓑ Dominic
Ⓒ Emir
Ⓓ Fernando

Use a problem-solving model to solve this problem.

1 Understand

Read the problem. Circle the information you know.
Underline what the problem is asking you to find.

2 Plan

What will you need to do to solve the problem? Write your plan in steps.

Step 1 Substitute the values for f and s in the inequality.

Step 2 Determine whether the inequality is true.

Read to Succeed!
The inequality $\geq$ means greater than or equal to, so sums must be greater than or equal to 65 for the inequality to be true.

3 Solve

Use your plan to solve the problem. Show your steps.

Cody	Dominic	Emir	Fernando
$30 + 10 \geq 65$	$60 + 0 \geq 65$	$45 + 10 \geq 65$	$20 + 55 \geq 65$
☐ ≥ 65	☐ ≥ 65	☐ ≥ 65	☐ ≥ 65

Since $75 \geq 65$, __________ earned enough money. Choice ☐ is correct.
Fill in that answer choice.

4 Check

How do you know your solution is accurate?

____

____

____

Lesson 5 *(continued)*

Use a problem-solving model to solve each problem.

1 The classmates below each have a cell phone plan that allows 200 text messages per month. Use the inequality $a + b > 200$, where a is the number of texts during Week 1 and b is the number of texts during Week 2, to determine who is already over their monthly budget of 200 texts. **6.EE.5, MP 1**

	Week 1	Week 2
Olivia	25	50
Anna	150	10
Sierra	125	110
Vanesa	100	100

Ⓐ Olivia
Ⓑ Anna
Ⓒ Sierra
Ⓓ Vanesa

2 Rafael has at least 5 feet of wire. He uses 1.5 feet for a project. In the inequality $x + 1.5 \geq 5$, x represents the amount of wire that Rafael has left, in feet. What is the minimum number of inches he has left? **6.EE.5, MP 4**

____

3 Hannah's backpack can hold no more than 30 pounds. She has a laptop that weighs 7 pounds and books that weigh 5 pounds each. What is the maximum number of books that Hannah can carry in her backpack? **6.EE.5, MP 2**

____

4 **H.O.T. Problem** If $x < 11$ and $x \geq 4$, what are the possible whole number values of x? **6.EE.5, MP 2**

____

Lesson 6 Multi-Step Problem Solving

Multi-Step Example

The table shows the heights of 5 people and whether they were allowed to ride a certain roller coaster. Let a represent a person's height in inches. Which inequality represents the heights, in inches, of people allowed to ride this roller coaster? **6.EE.6, MP 1**

Height	Allowed
5 ft 7 in.	Yes
5 ft 3 in.	Yes
4 ft 6 in.	Yes
4 ft 5 in.	No
4 ft 3 in.	No

Ⓐ $a > 46$

Ⓑ $a > 54$

Ⓒ $a \geq 54$

Ⓓ $a < 54$

Use a problem-solving model to solve this problem.

1 Understand

Read the problem. Circle the information you know.
Underline what the problem is asking you to find.

2 Plan

What will you need to do to solve the problem? Write your plan in steps.

Step 1 Determine the minimum height allowed on the roller coaster. Convert the height to inches.

Step 2 Write an inequality that represents the situation.

Read to Succeed!
When the term "minimum" is used, the inequality is greater than or equal to.

3 Solve

Use your plan to solve the problem. Show your steps.

The minimum height that was allowed on the roller coaster was ☐ ft ☐ in. Heights shorter than that were not allowed on the roller coaster.

☐ feet ☐ inches = ☐ inches

So, $a \geq$ ☐. Choice ☐ is correct. Fill in that answer choice.

4 Check

How do you know your solution is accurate?

____

____

____

Lesson 6 *(continued)*

Use a problem-solving model to solve each problem.

1 The table shows the resting heart rate of different people and whether or not it was considered elevated. Let *h* represent the heart rate in beats per minute. Which inequality represents the heart rates that are considered elevated? **6.EE.6, MP 1**

Heart Rate (bpm)	Result
62	Normal
79	Normal
100	Normal
101	Elevated
105	Elevated

Ⓐ $h \leq 100$

Ⓑ $h > 101$

Ⓒ $h < 100$

Ⓓ $h \geq 101$

2 Lorenzo solved the following inequality $x + 3 < 10$ and graphed the solution on a number line. The graph contained the values −3, 2.4, 0, and 7. Which of these values is incorrect? Explain. **6.EE.6, MP 2**

____

3 Marcus has $50 in his wallet. He buys 2 CDs that costs $12 each and wants to buy posters that cost $6 each. The inequality $p \leq 4$ represents the number of posters he can buy. Represent the solutions of the inequality on the number line. **6.EE.6, MP 3**

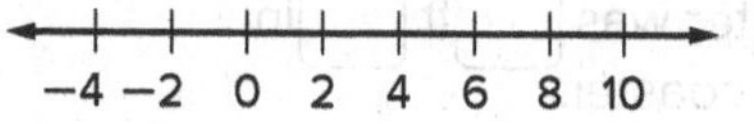

4 **H.O.T. Problem** Marianna wants to run at least 6.2 miles to train for a quarter-marathon. The inequality $m \geq 6.2$, where *m* is the number of miles she has run, represents the runs where she met her goal. Represent the solutions of the inequality on the number line. **6.EE.6, MP 3**

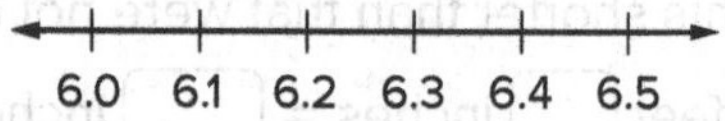

Lesson 7 Multi-Step Problem Solving

Multi-Step Example

The table shows the costs of different size sandwiches at a sandwich shop. Ava has $30 to spend. She spends $5.25 on drinks and buys two Club sandwiches. Write and solve an inequality to find the maximum number of foot long sandwiches she can buy. **6.EE.5, MP 1**

Sandwich Size	Cost ($)
Club	2.75
6-inch	4.00
Foot long	6.25

Use a problem-solving model to solve this problem.

1 Understand

Read the problem. Circle the information you know.
Underline what the problem is asking you to find.

2 Plan

What will you need to do to solve the problem? Write your plan in steps.

Step 1 Write an inequality to represent the situation.

Step 2 Solve the inequality.

> **Read to Succeed!**
> The term "maximum" means that all values that make the inequality true will be less than the given value.

3 Solve

Use your plan to solve the problem. Show your steps.

Let *x* represent the number of foot long sandwiches she can buy.

$[\ ] + 2($[\ ]) + [\ ]x ≤ $[\ ]

drinks; club sandwiches; foot long sandwiches; total she can spend

$5.25 + 2(2.75) + 6.25x \leq 30$ Write the inequality.

[ ] $+ 6.25x \leq 30$ Simplify the constants.

$6.25x \leq$ [ ] Subtract.

$x \leq$ [ ] Divide.

The greatest whole number that is a solution to the inequality is [ ].

4 Check

How do you know your solution is accurate?

____

____

NAME________________________ DATE ________________ PERIOD ________

Lesson 7 *(continued)*

Use a problem-solving model to solve each problem.

1 The table shows the costs of different bed sheet sizes at a home interior store. Phong has $100.25 to spend. He spends $45 on a blanket, and he buys one twin sheet for his brother. Write and solve an inequality to find the maximum number of full sheets he can buy. **6.EE.5, MP 1**

Sheet Size	Cost ($)
Twin	9.99
Full	10.50
Queen	20
King	28.50

____

2 The rectangle below represents a table top that Lakita wants to cover with tiles. Each tile has an area of 7 square inches. To find the maximum number of tiles that will fit the table top, Lakita wants to use the inequality $7x \leq y$. What is the value of y? What is the maximum number of tiles she can use? **6.EE.5, MP 2**

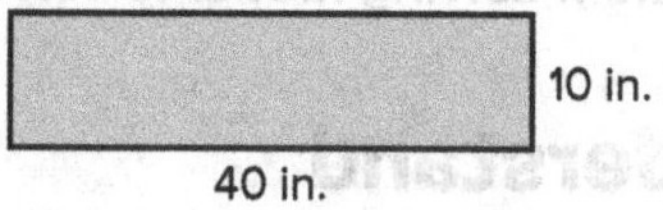

____

3 Jamal wants to sell at least 50 tickets for a school raffle. He sells 10 tickets to his family. His dad is going to take at least $\frac{1}{2}$ of the remaining tickets to sell at work. Write and solve an inequality to find the minimum number of tickets his dad will take to work. Represent the solution on the number line. **6.EE.6, MP 3**

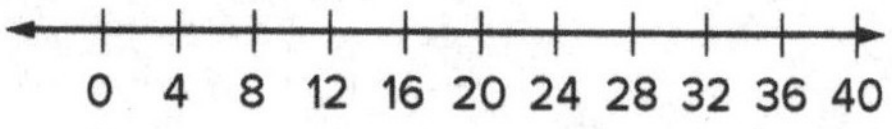

____

4 **H.O.T. Problem** Write a word problem for the one-step inequality $3x \leq 15$. **6.EE.5, MP 2**

____

____

____

____

____

Lesson 1 Multi-Step Problem Solving

Multi-Step Example

Li is designing a flower bed for a school project. His design consists of a square inside a parallelogram. The shaded area will be planted with small shrubs. In order to get the correct number of shrubs, Li needs to determine the area of the shaded region once he has the dimensions. Which of the following formulas can be used to find the area of the shaded region? **6.G.1, MP 1**

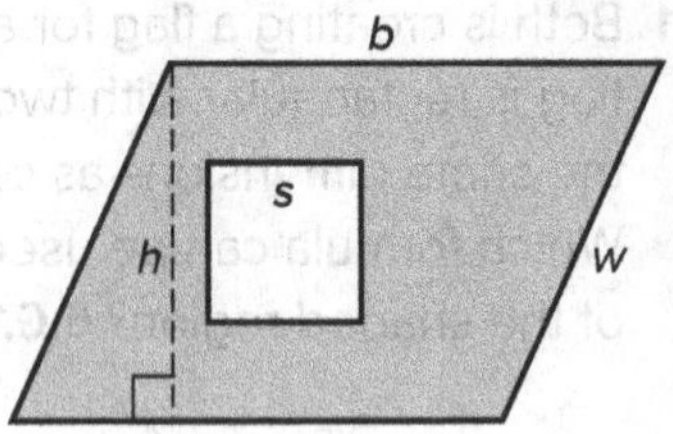

Ⓐ $A = b \cdot h - s^2$ Ⓒ $A = b \cdot w + s^2$

Ⓑ $A = b \cdot w - s^2$ Ⓓ $A = b \cdot h + s^2$

Use a problem-solving model to solve this problem.

1 Understand

Read the problem. Circle the information you know.
Underline what the problem is asking you to find.

Read to Succeed!
Be careful when finding the area of a parallelogram that the slant height is not used. The height of the parallelogram is perpendicular to the base.

2 Plan

What will you need to do to solve the problem? Write your plan in steps.

Step 1 Determine the formula for the area of the parallelogram and square.

Step 2 Subtract the area of the square from the area of the parallelogram.

3 Solve

Use your plan to solve the problem. Show your steps.

Area of Parallelogram − Area of Square = Area of Shaded Region

$b \cdot h \quad - \quad s^2 \quad = \quad b \cdot h - s^2$

The formula is $A = b \cdot h - s^2$, so choice ____ is correct. Fill in that answer choice.

4 Check

How do you know your solution is accurate?

____

____

____

Lesson 1 *(continued)*

Use a problem-solving model to solve each problem.

1 Beth is creating a flag for an art project. The flag is rectangular with two parallelograms of the same dimensions as seen in the diagram. Which formula can be used to find the area of the shaded region? **6.G.1, MP 1**

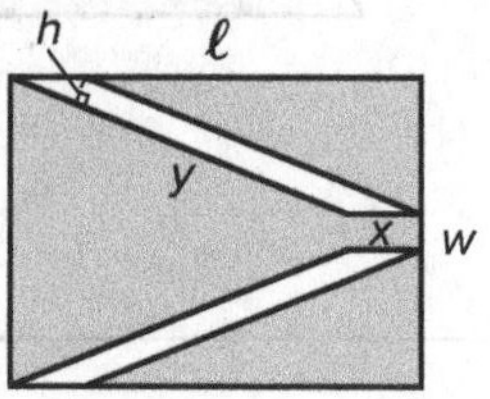

Ⓐ $A = \ell \cdot w - y \cdot h$

Ⓑ $A = \ell \cdot w - x \cdot h$

Ⓒ $A = \ell \cdot w - 2 \cdot x \cdot h$

Ⓓ $A = \ell \cdot w - 2 \cdot y \cdot h$

2 The side of an office building is made of mirrored glass panels in the shape of parallelograms. If one parallelogram-shaped piece of glass has a base of 8.5 feet and a height of 4 feet, determine how many windows there are in an 850 square foot area. **6.G.1, MP 2**

____

3 An area of 861 square feet is used for 4 identical parking spaces as seen in the diagram below. Use the information in the diagram to find the height of one parallelogram-shaped parking space, *h*. **6.G.1, MP 2**

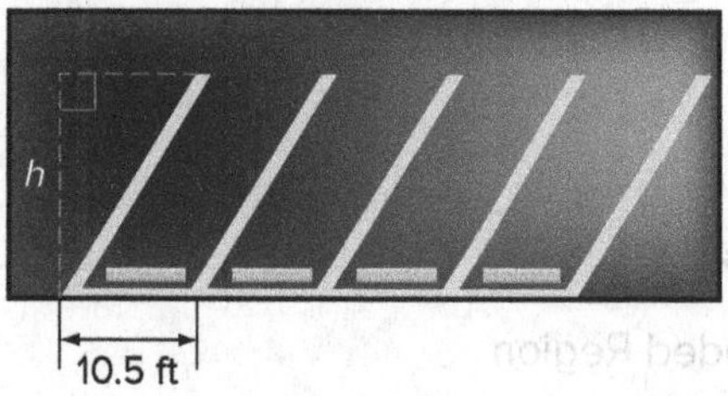

____

4 **H.O.T. Problem** A rectangle is drawn with dimensions 6 units long and 4 units wide. Then a triangle is cut from the rectangle with dimensions shown, and placed on the left side of the rectangle to form a parallelogram. **6.G.1, MP 7**

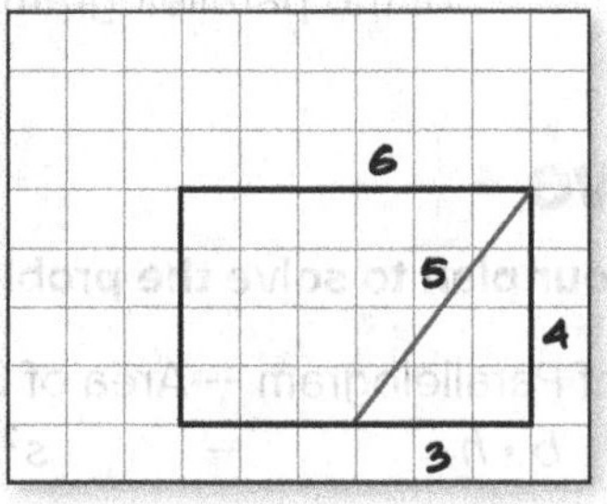

What are the area and perimeter of the original rectangle? What are the area and perimeter of the parallelogram formed by cutting and moving the triangle?

____

____

____

Lesson 2 **Multi-Step** Problem Solving

Multi-Step Example

The triangle on the grid represents a triangular-shaped pillow. What is the area of the triangle? **6.G.1, MP 1**

Ⓐ 270 in^2

Ⓑ 135 in^2

Ⓒ 45 in^2

Ⓓ 15 in^2

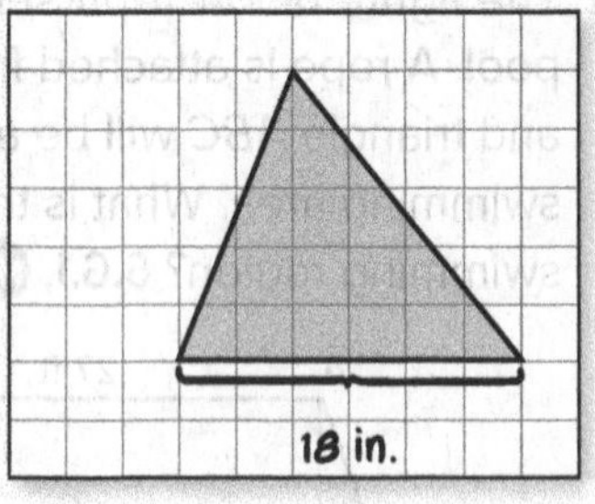

Use a problem-solving model to solve this problem.

1 Understand

Read the problem. Circle the information you know.
Underline what the problem is asking you to find.

2 Plan

What will you need to do to solve the problem? Write your plan in steps.

Step 1 Determine the height of the triangle.

Step 2 Use the formula $A = \frac{1}{2}bh$ to find the area of the triangle.

3 Solve

Use your plan to solve the problem. Show your steps.

Each square on the grid represents 18 ÷ ☐ or ☐ inches.

The height of the triangle is ☐ × ☐ or ☐ inches.

The area of the triangle is $\frac{1}{2}bh$ or $\frac{1}{2}$ · ☐ · ☐.

The area is ☐ square inches.

Choice ☐ is correct. Fill in that answer choice.

Read to Succeed!
Remember to determine the scale of the grid.

4 Check

How do you know your solution is accurate?

____

____

____

Lesson 2 *(continued)*

Use a problem-solving model to solve each problem.

1 The figure below represents a swimming pool. A rope is attached from point *A* to *C*, and triangle *ABC* will be a designated adult swimming area. What is the area of the adult swimming region? **6.G.1, MP 1**

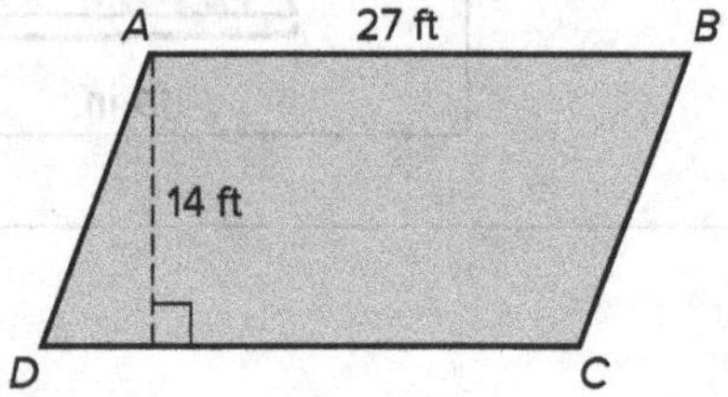

Ⓐ 41 ft^2

Ⓑ 82 ft^2

Ⓒ 189 ft^2

Ⓓ 378 ft^2

2 The triangle on the grid outlines the border of a town. Each square on the grid represents a side length of 1.5 miles. What is the area of the town in square miles? **6.G.1, MP 4**

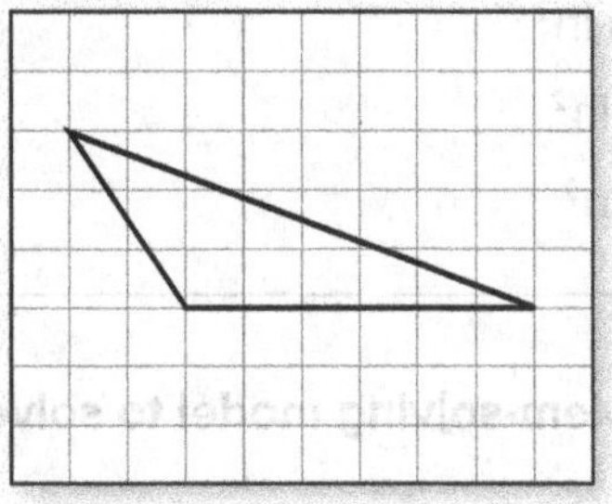

____

3 The table shows the dimensions of three triangles. How much greater is the area of triangle C than triangle A, in square centimeters? **6.G.1, MP 7**

Triangle	Base	Height
A	8.5 cm	6 cm
B	7 cm	7 cm
C	9 cm	6.5 cm

____

4 H.O.T. Problem Marco is going to paint 30% of the triangle. How many square inches will he paint? **6.G.1, MP 2**

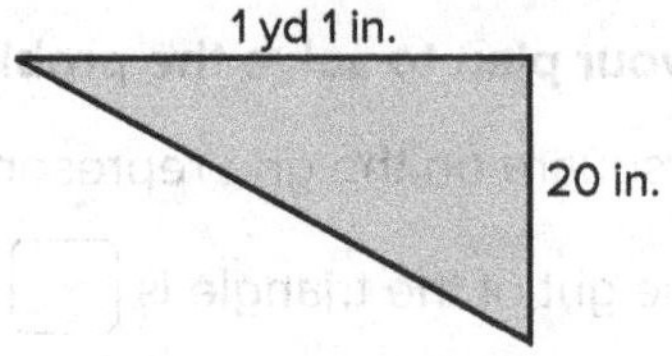

____

Lesson 3 **Multi-Step** Problem Solving

Multi-Step Example

The figure on the grid represents a parking lot. Asphalt for the parking lot costs $8.95 per square foot. How much will it cost to asphalt the parking lot, to the nearest cent? **6.G.1, MP 1**

Ⓐ $290.88　　Ⓒ $18,616.00

Ⓑ $805.50　　Ⓓ $37,232.00

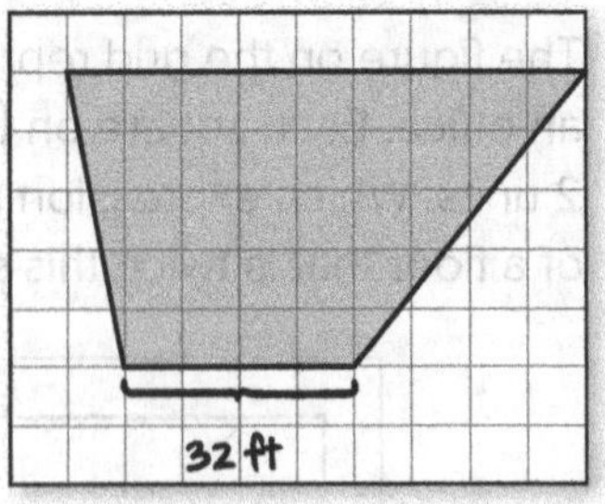
32 ft

Use a problem-solving model to solve this problem.

1 Understand

Read the problem. Circle the information you know.
Underline what the problem is asking you to find.

2 Plan

What will you need to do to solve the problem? Write your plan in steps.

Step 1 Determine the area of the trapezoid.

Step 2 Multiply to determine the cost of the asphalt.

Read to Succeed!
The height of a trapezoid is perpendicular to the two bases.

3 Solve

Use your plan to solve the problem. Show your steps.

Each square on the grid represents a length of 32 ÷ 4 or 8 feet.
The two bases are 32 feet and 72 feet. The height is 40 feet.

$A = \frac{1}{2}h(b_1 + b_2)$　　Area of a trapezoid

$A = \frac{1}{2}(40)(72 + 32)$　　Replace h with 40, b_1 with 72, and b_2 with 32.

$A = 2{,}080$　　Multiply.

So, the cost of the asphalt is $8.95 × 2,080 or $18,616. Choice ____ is correct.
Fill in that answer choice.

4 Check

How do you know your solution is reasonable?

____

____

____

____

Lesson 3 *(continued)*

Use a problem-solving model to solve each problem.

1 The figure on the grid represents the floor of an office. Each square on the grid represents 2 units. Which expression represents an area of a floor that is twice this size? **6.G.1, MP 1**

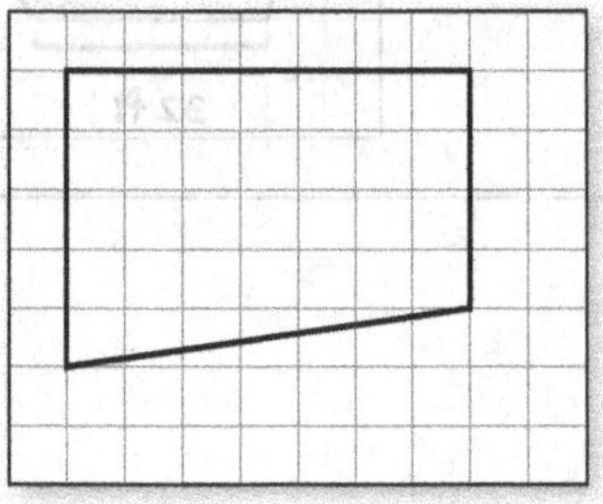

Ⓐ $\frac{1}{2}(7)(4 + 5)$

Ⓑ $7(4 + 5)$

Ⓒ $\frac{1}{2}(14)(8 + 10)$

Ⓓ $(14)(8 + 10)$

2 A farmer spread fertilizer onto the plot of land shown below. He uses 4 scoops of fertilizer per square meter. If he used 312 scoops, what is the height of the trapezoid in meters? **6.G.1, MP 2**

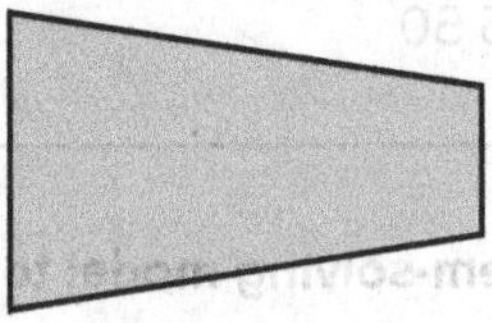

______

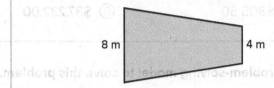

3 In the figure below, the area inside *ABEF* will be colored red. What will be the red area, in square feet? **6.G.1, MP 2**

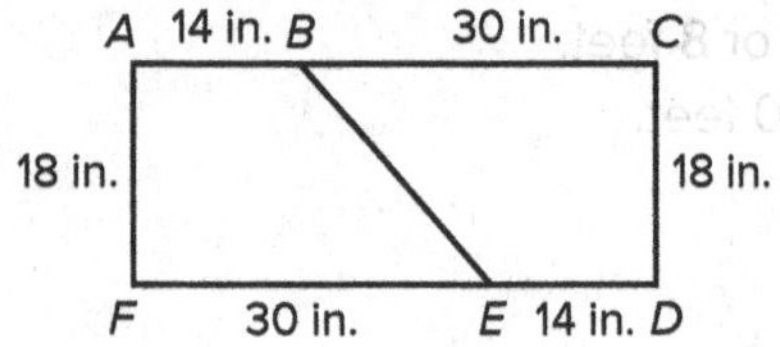

______

4 **H.O.T. Problem** Suppose the bases and height of a trapezoid are all multiplied by 3. How will the area change? **6.G.1, MP 7**

______

Lesson 4 Multi-Step Problem Solving

Multi-Step Example

The side lengths of the smaller triangle are multiplied by the same number to create a larger triangle with a base of 1 foot and height of $\frac{1}{2}$ foot. How many times greater is the area of the larger triangle? 6.G.1, MP 1

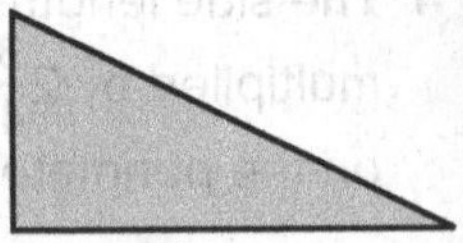

Ⓐ 3
Ⓑ 4
Ⓒ 9
Ⓓ 36

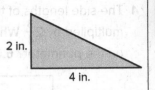

Use a problem-solving model to solve this problem.

1 Understand

Read the problem. Circle the information you know.
Underline what the problem is asking you to find.

2 Plan

What will you need to do to solve the problem? Write your plan in steps.

Step 1 Determine the ______ of the smaller triangle.

Step 2 Convert the dimensions of the larger triangle to ______, determine the area, then compare the ______.

3 Solve

Use your plan to solve the problem. Show your steps.

Determine the area of the smaller triangle.

$\frac{1}{2} \times 2 \times 4 =$ ______

Determine the area of the larger triangle in inches.

1 ft = ______ $\frac{1}{2}$ ft = ______ $\frac{1}{2} \times 12 \times 6 =$ ______

____ ÷ ____ = ____

The area of the larger triangle is ____ times larger.

So, the correct answer is ____. Fill in that answer choice.

Read to Succeed!
Remember to convert the units of the larger triangle to inches before comparing the area of each triangle.

4 Check

How do you know your solution is accurate?

____

____

____

Lesson 4 *(continued)*

Use a problem-solving model to solve each problem.

1 The side lengths of the rectangle below are multiplied by $2\frac{3}{4}$. What effect would this have on the perimeter? **6.G.1, MP 1**

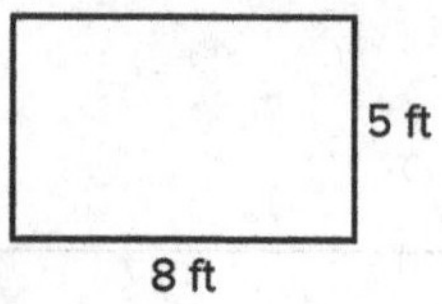

Ⓐ The perimeter will be $1\frac{3}{4}$ times greater.

Ⓑ The perimeter will be $2\frac{3}{4}$ times greater.

Ⓒ The perimeter will be $2\frac{5}{8}$ times greater.

Ⓓ The perimeter will be 3 times greater.

2 The dimensions of the rectangles listed in the table will all be multiplied by 3. What is the combined area of the enlarged rectangles in square meters? **6.G.1, MP 8**

Rectangle	Length (cm)	Width (cm)
Rectangle A	4	5
Rectangle B	6	6
Rectangle C	7	9

____

3 The Pentagon in Washington, D.C. is a regular pentagon. Juan made two scale models of the Pentagon. The perimeter of the larger model is how many times greater than the perimeter of the smaller model? (*Hint:* 1 inch ≈ 2.54 cm) **6.G.1, MP 8**

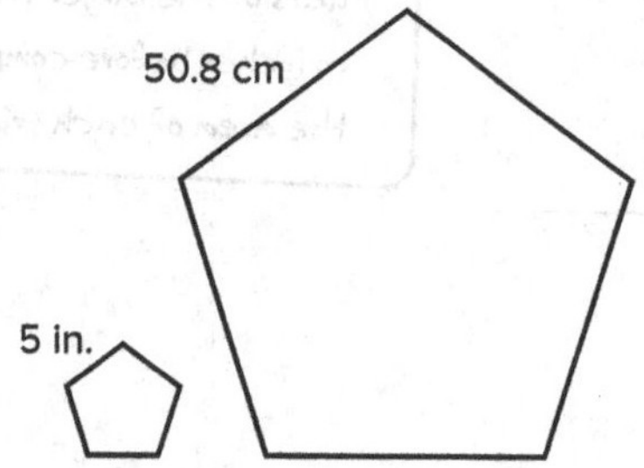

____

4 **H.O.T. Problem** The area of Rectangle B is 5 times greater than the area of Rectangle A. Give possible dimensions for each rectangle. Justify your answer. **6.G.1, MP 3**

____

____

____

____

____

____

____

Lesson 5 Multi-Step Problem Solving

Multi-Step Example

Madeline drew this diagram of her rectangular vegetable garden. What are the coordinates of the vertices of the rectangle she drew? If the length of each grid square on the diagram is 2 yards, what is the area of Madeline's garden? **6.G.1, MP 1**

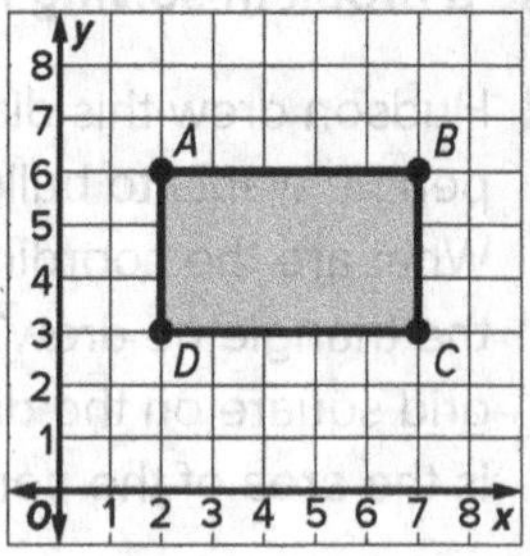

Use a problem-solving model to solve this problem.

Read to Succeed!
Remember, when x-coordinates are the same, you can subtract y-coordinates to find distance. And, when y-coordinates are the same, you can subtract x-coordinates to find distance.

1 Understand

Read the problem. Circle the information you know.
Underline what the problem is asking you to find.

2 Plan

What will you need to do to solve the problem? Write your plan in steps.

Step 1 Write the coordinates of the vertices of the rectangle and then determine the length of each side.

Step 2 Multiply the length and the width of the rectangle by 2 yards to find the length and the width of the garden.

Step 3 Multiply the length by the width to find the area of the garden.

3 Solve

Use your plan to solve the problem. Show your steps.

The coordinates of the vertices are *A*(___, ___), *B*(___, ___), *C*(___, ___), *D*(___, ___).

Length: Subtract *x*-coordinates → $\overline{AB}$: 7 − 2 = ______ and $\overline{CD}$: __________

Width: Subtract *y*-coordinates → $\overline{AD}$: 6 − 3 = ______ and $\overline{BC}$: __________

Multiply the length and the width by 2 yards: 5 × 2 = ______. 3 × 2 = ______

Find the area: ____________. The area of the garden is ______ square yards.

4 Check

How do you know your solution is accurate?

____

____

____

Lesson 5 *(continued)*

Use a problem-solving model to solve each problem.

1 Hudson drew this diagram of the triangular pen he wants to build for his pet rabbits. What are the coordinates of the vertices of the triangle he drew? If the length of each grid square on the diagram is 1.5 feet, what is the area of the pen? **6.G.1, MP 1**

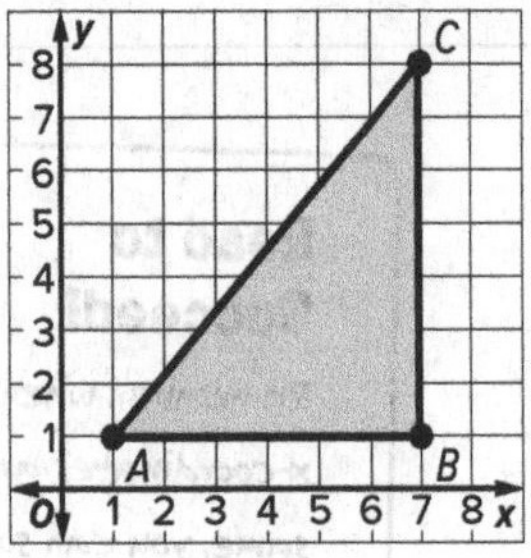

2 Montrel drew a diagram of his rectangular pool. What are the coordinates of the vertices of the pool he drew? The length of each grid square on the diagram is 1 foot. Montrel wants to put a deck around the outside of his pool that is 5 feet wide on all sides. What will be the perimeter of the outside of the deck? **6.G.3, MP 2**

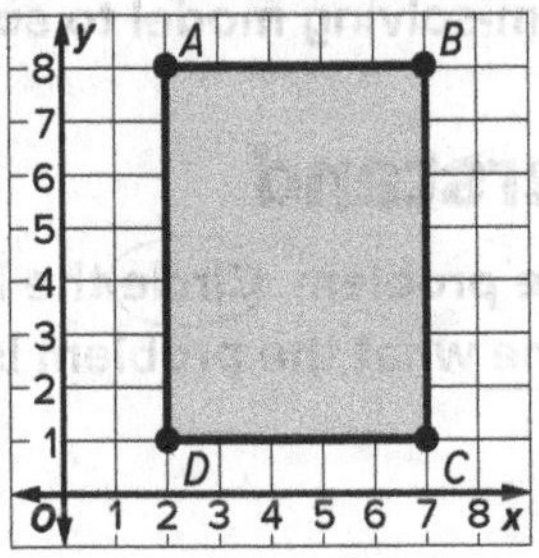

3 The diagram shows the shape and size of two different tiles that Charl wants to use to cover a wall. He wants to cover 72 square feet with each kind of tile. If the length of each grid square on the diagram is 1 foot, how many of each tile does Charl need? **6.G.1, MP 2**

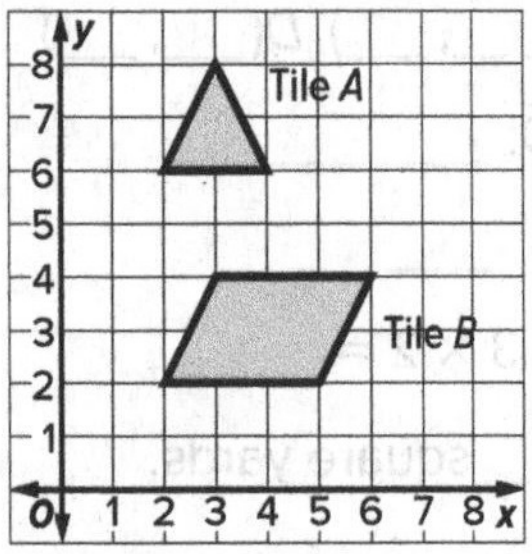

4 **H.O.T. Problem** Vivian drew two diagrams for flower gardens. The length of each grid square on the diagrams is 2 feet. Vivian wants a flower garden that is 64 square feet. Which diagram should she use for her garden? Explain. **6.G.1, MP 3**

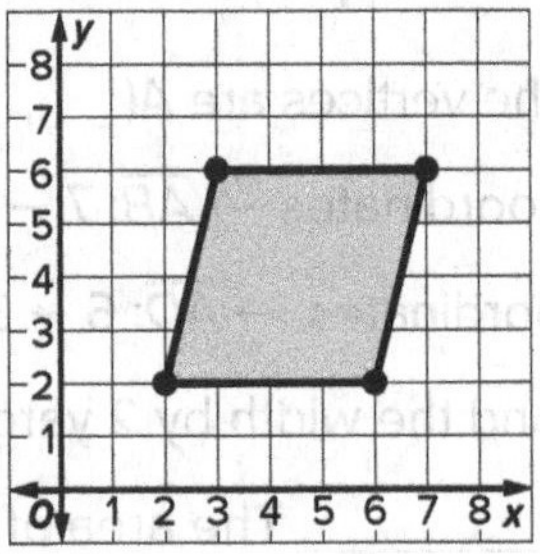

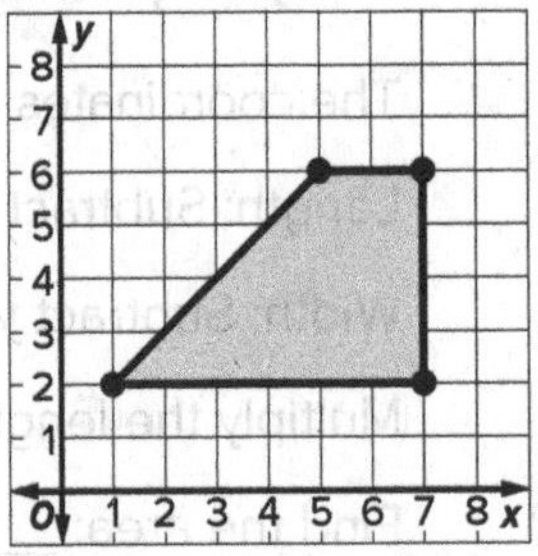

Lesson 6 Multi-Step Problem Solving

Multi-Step Example

The diagram shows a wall in Javier's living room that he wants to paint. Find the total area to be painted. 6.G.1, MP 1

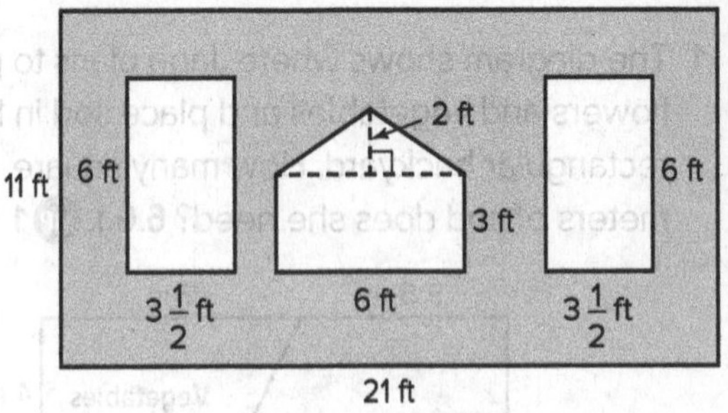

Ⓐ 165 square feet Ⓒ 207 square feet

Ⓑ 189 square feet Ⓓ 231 square feet

Use a problem-solving model to solve this problem.

Read to Succeed!
The window in the center of the wall is a composite figure. You can find the area of a composite figure by separating it into figures for which you know how to find the area.

1 Understand

Read the problem. Circle the information you know. Underline what the problem is asking you to find.

2 Plan

What will you need to do to solve the problem? Write your plan in steps.

Step 1 Find the area of the wall including the windows.

Step 2 Find the areas of the three windows.

Step 3 Subtract the areas of the windows from the total area of the wall.

3 Solve

Use your plan to solve the problem. Show your steps.

$21 \times 11 =$ ____ The area of the wall including the windows is ____ square feet.

Window on the left	Middle window	Window on the right
$6 \times 3\frac{1}{2} =$ ____	$3 \times 6 = 18$; $\frac{1}{2}(6)(2) = 6$ $18 + 6 =$ ____	$6 \times 3\frac{1}{2} =$ ____

$21 + 24 + 21 =$ ____ The total area of the windows is ____ square feet.

$231 - 66 =$ ____ The total area to be painted is ____ square feet.

So, ____ is the correct answer. Fill in that answer choice.

4 Check

How do you know your solution is accurate?

____

____

Lesson 6 *(continued)*

Use a problem-solving model to solve each problem.

1 The diagram shows where Jada plans to plant flowers and vegetables and place sod in her rectangular backyard. How many square meters of sod does she need? **6.G.1, MP 1**

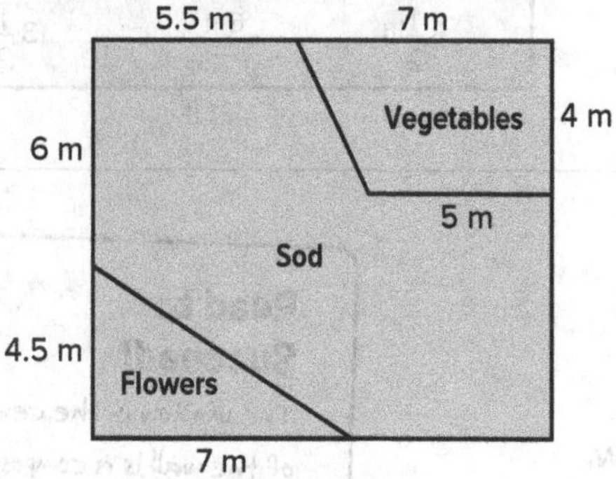

Ⓐ 131.25 square meters

Ⓑ 107.25 square meters

Ⓒ 91.5 square meters

Ⓓ 75.5 square meters

2 There are doors into the closet from both Ella's bedroom and Bethany's bedroom. How many square yards of carpet will it take to cover the floors in the two bedrooms and the closet? **6.G.1, MP 2**

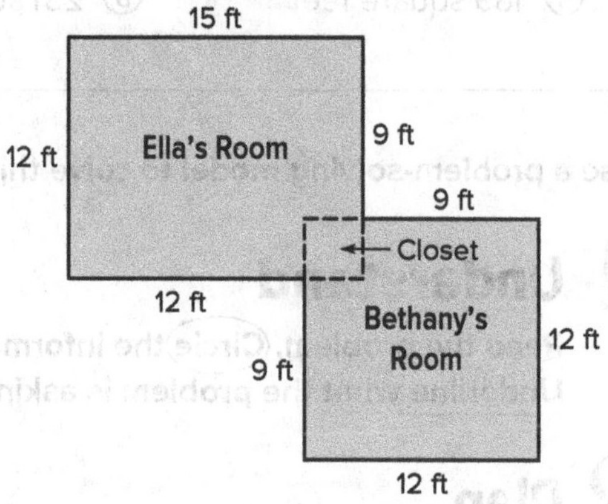

____

3 Ellery drew a diagram of a pen she fenced in for her rabbits. What is the area of the pen? **6.G.1, MP 2**

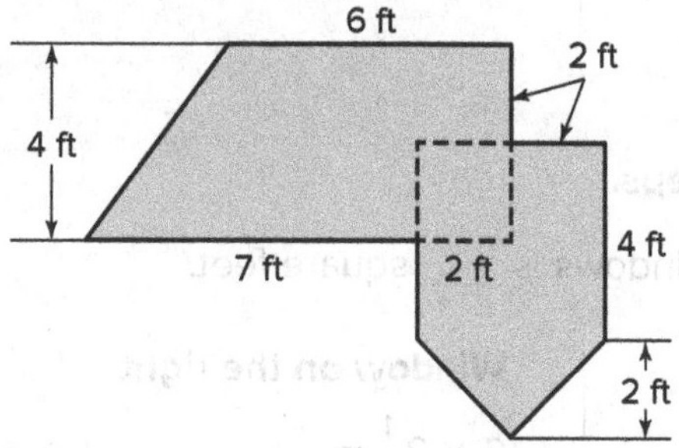

____

4 **H.O.T. Problem** To find the area of the shaded composite figure below, you can find the sum of the areas of the two shaded triangles and the shaded square. Describe another way you can find the area of the shaded composite figure. Then use one of the ways to find the area. **6.G.1, MP 3**

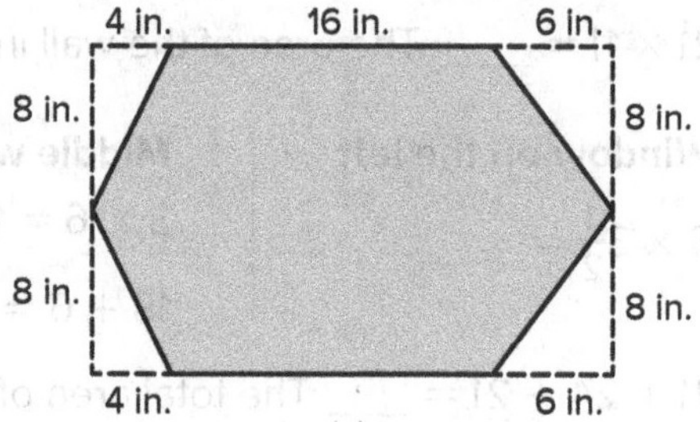

____

____

____

____

____

Lesson 1 Multi-Step Problem Solving

Multi-Step Example

If the fish tank shown is 80% filled with water, how much water is in the tank? 6.G.2, MP 1

Ⓐ 5,772 cubic inches

Ⓑ 4,617.6 cubic inches

Ⓒ 1,154.4 cubic inches

Ⓓ 384.8 cubic inches

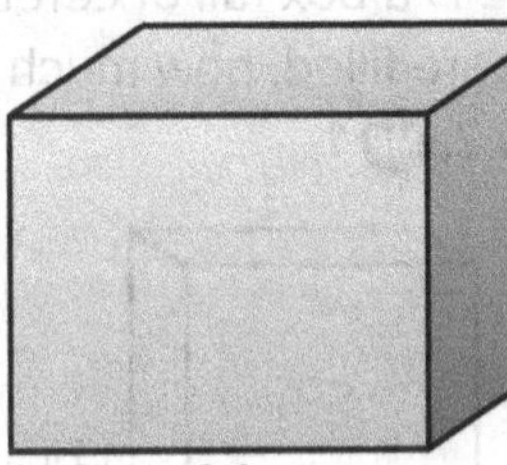

Use a problem-solving model to solve this problem.

1 Understand

Read the problem. Circle the information you know.
Underline what the problem is asking you to find.

2 Plan

What will you need to do to solve the problem? Write your plan in steps.

Step 1 Determine the volume of the tank.

Step 2 Multiply to determine the volume of water in the tank.

Read to Succeed!
The dimensions of the fish tank are given in feet and inches. Convert 2 feet to inches before finding the volume.

3 Solve

Use your plan to solve the problem. Show your steps.

$V = \ell \cdot w \cdot h$ Volume of a rectangular prism

$V = 24 \cdot 13 \cdot 18.5$ $\ell = 24$ in., $w = 13$ in., $h = 18.5$ in.

$V =$ ______ Multiply.

To find 80% of the volume, multiply by 0.80. The volume of water is ______ × 0.80 or ________ cubic inches. Choice ____ is correct. Fill in that answer choice.

4 Check

How do you know your solution is reasonable?

____

____

____

Lesson 1 *(continued)*

Use a problem-solving model to solve each problem.

1 The figure is a box full of cereal. If a case of 24 boxes are filled, how much cereal is there in all? **6.G.2, MP 1**

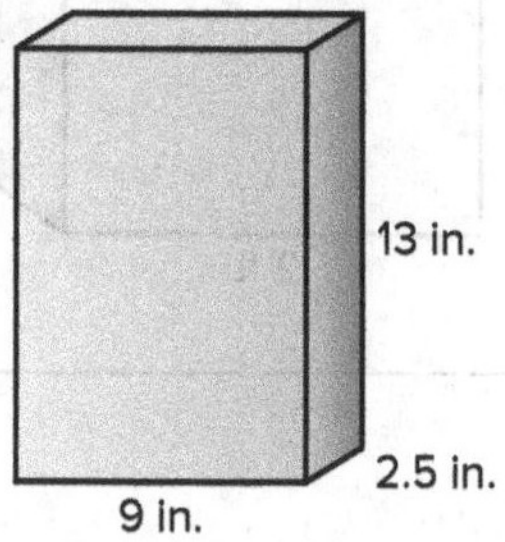

Ⓐ 292.5 cubic inches

Ⓑ 588 cubic inches

Ⓒ 1,176 cubic inches

Ⓓ 7,020 cubic inches

2 A storage cube that has an edge length of 16 centimeters is being packed in a cardboard box with a length of 28 centimeters, a width of 18 centimeters, and a height of 22 centimeters. The extra space is being filled with packing peanuts. How many cubic centimeters of peanuts are needed to fill the space? **6.G.2, MP 2**

____

3 What is the volume of the statue in cubic feet? Round to the nearest hundredth. **6.G.2, MP 7**

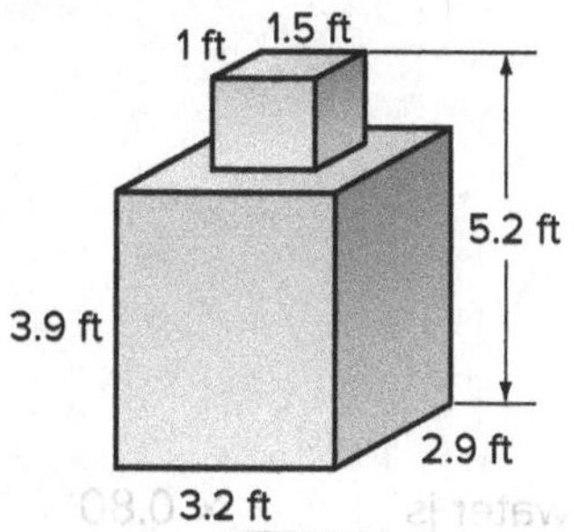

____

4 **H.O.T. Problem** One cube has a side length of 1 millimeter, and another cube has a side length of 1 centimeter. What is the ratio of the smaller volume to the greater volume? Express the numerator and denominator using the same units. Explain how you found your answer. **6.G.2, MP 3**

____

____

____

____

____

Lesson 2 Multi-Step Problem Solving

Multi-Step Example

Angela has a candle in the shape of a triangular prism with the dimensions shown in the drawing. If she burns the candle and reduces the volume by 25%, what will be the volume of the candle that is left?
Extension of **6.G.2, MP 1**

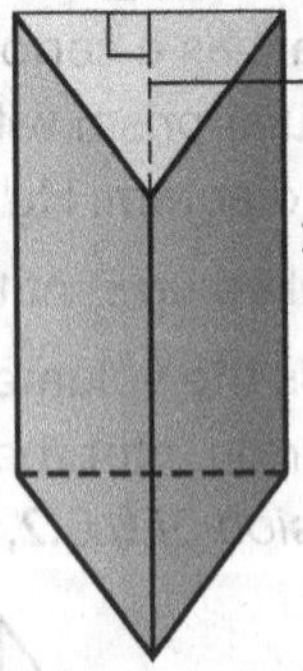

Use a problem-solving model to solve this problem.

Read to Succeed!

Remember, the formula for the volume of a triangular prism is $V = Bh$, where B is the area of the base and h is the height.

1 Understand

Read the problem. Circle the information you know.
Underline what the problem is asking you to find.

2 Plan

What will you need to do to solve the problem? Write your plan in steps.

Step 1 Determine the volume of the candle.

Step 2 Subtract the percent that has burned from 100% to find the percent that will be left.

Step 3 Multiply to determine the volume of the candle that will be left.

3 Solve

Use your plan to solve the problem. Show your steps.

$V = Bh$
$V = \left(\frac{1}{2} \cdot 12 \cdot 8\right)20$
$V =$ ______ cm^3

$100 - 25 =$ ______
______% of the volume will be left.

To find 75% of the volume, multiply by 0.75.
$960 \cdot 0.75 = 720$ cm^3

So, ______ cubic centimeters will be left.

4 Check

How do you know your solution is reasonable?

____

____

Lesson 2 *(continued)*

Use a problem-solving model to solve each problem.

1 Jamaal has a carton in the shape of a triangular prism with the dimensions shown in the diagram. He packs a gift in the carton that takes up $\frac{2}{3}$ of the volume of the carton. What is the volume of the space that is left in the carton after the gift is packed inside? *Extension of* **6.G.2, MP 1**

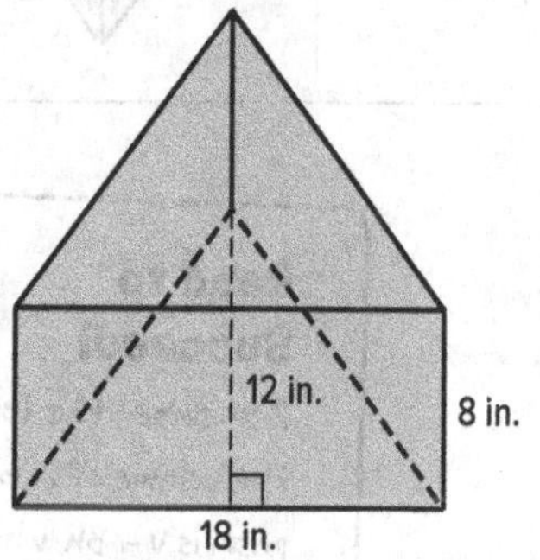

Ⓐ 216 cubic inches

Ⓑ 288 cubic inches

Ⓒ 648 cubic inches

Ⓓ 864 cubic inches

2 The diagram shows the dimensions of a fish pond that is in the shape of a triangular prism. Tom wants to build a fish pond with a depth that is $1\frac{1}{2}$ times greater than the depth of the fish pond shown in the diagram. How many cubic meters of water will be needed to fill Tom's pond to the top? *Extension of* **6.G.2, MP 2**

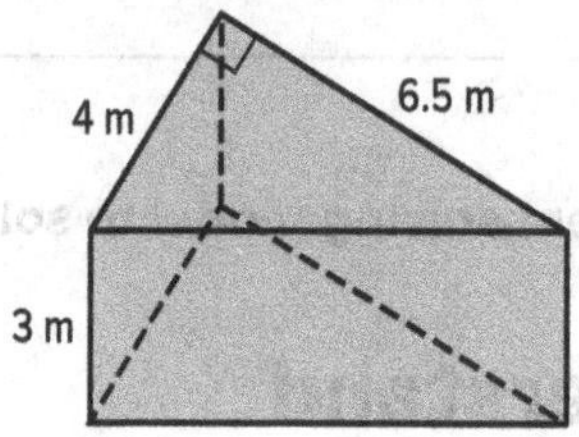

3 The diagram shows the dimensions of two vases. Which vase holds the greater volume of water? How much greater? *Extension of* **6.G.2, MP 2**

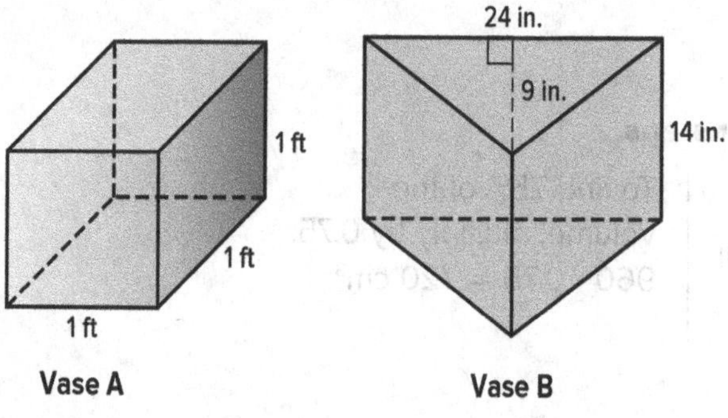

____

4 H.O.T. Problem The diagrams show the dimensions of two sheds. If both sheds have the same volume, what is the missing dimension on the shed on the right? Explain. *Extension of* **6.G.2, MP 3**

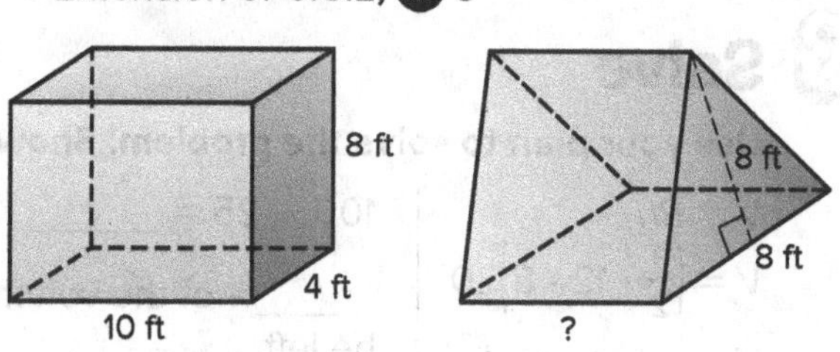

____

____

____

____

____

____

Lesson 3 Multi-Step Problem Solving

Multi-Step Example

Determine the surface area for each package. How much greater is the surface area of package B? **6.G.4, MP 1**

Ⓐ 10 square inches
Ⓑ 12 square inches
Ⓒ 20 square inches
Ⓓ 24 square inches

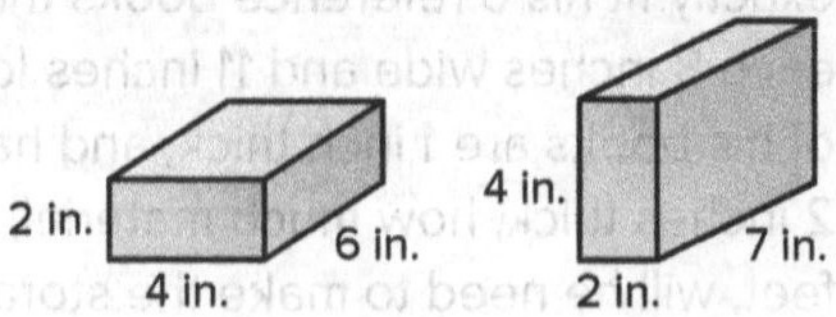

Use a problem-solving model to solve this problem.

1 Understand

Read the problem. Circle the information you know.
Underline what the problem is asking you to find.

2 Plan

What will you need to do to solve the problem? Write your plan in steps.

Step 1 Determine the ______________ for each prism.

Step 2 __________ the surface areas.

> **Read to Succeed!**
> You may need to draw a net of each prism to help you visualize the area of each face.

3 Solve

Use your plan to solve the problem. Show your steps.

Package A surface area:
front and back: 2(2 • 4) = ____
top and bottom: 2(6 • 4) = ____
sides: 2(2 • 6) = ____
____ + ____ + ____ = ____ Add.

Package B surface area:
front and back: 2(2 • 4) = ____
top and bottom: 2(2 • 7) = ____
sides: 2(7 • 4) = ____
____ + ____ + ____ = ____ Add.

____ − ____ = ____ square inches greater Subtract.

So, the correct answer is ____. Fill in that answer choice.

4 Check

How do you know your solution is accurate?

____

____

____

Lesson 3 *(continued)*

Use a problem-solving model to solve each problem.

1 Taro wants to build a storage box that will exactly fit his 6 reference books that are each 8 inches wide and 11 inches long. If half of his books are 1 inch thick, and half are 2 inches thick, how much material, in square feet, will he need to make the storage box? Round your answer to the nearest thousandth. **6.G.4, MP 1**

Ⓐ 3.597 square feet

Ⓑ 3.65 square feet

Ⓒ 4.735 square feet

Ⓓ 5.375 square feet

2 The coordinate grid shows the base of a rectangular prism. If the prism has a surface area of 170 units, what is its height, in units? **6.G.4, MP 7**

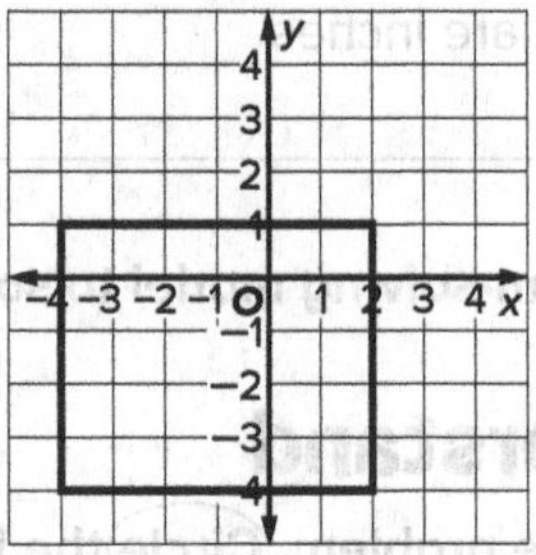

____

3 Each side length of a unit cube measures 2 units and increases by 50% every minute. What is the ratio of the surface area after 3 minutes to the original surface area? Write your answer as a decimal rounded to the nearest tenth. **6.G.4, MP 2**

____

4 **H.O.T. Problem** A chemical company wants to reduce the cost of their shipping containers. The measurements of the containers are shown. They pay for the containers by the amount of material required to make them. If they want to ship the greatest volume of chemicals at the lowest cost, which container should they use? Justify your answer. **6.G.4, MP 3**

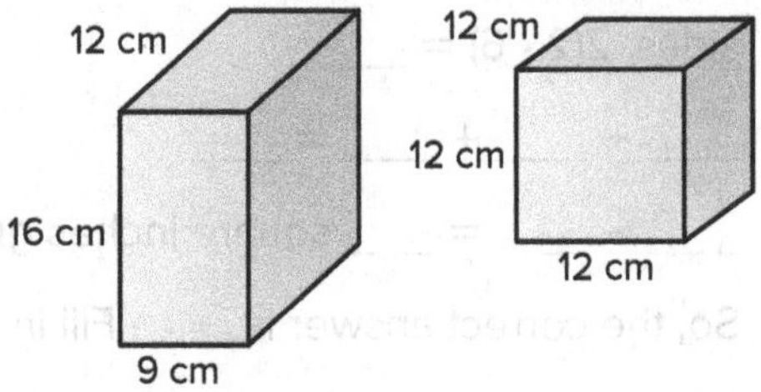

____

____

____

____

____

Lesson 4 Multi-Step Problem Solving

Multi-Step Example

Two play houses at a children's gym are shown at the right. How much greater, in square feet, is the surface area of the larger house than the smaller house? **6.G.4, MP 1**

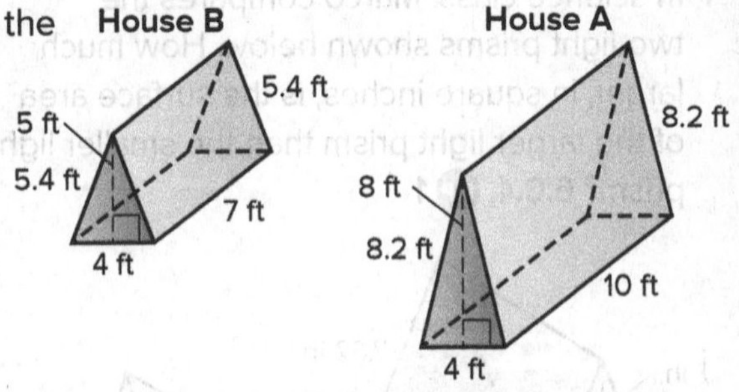

Ⓐ 112.4 square feet
Ⓑ 123.6 square feet
Ⓒ 224 square feet
Ⓓ 236 square feet

Use a problem-solving model to solve this problem.

1 Understand

Read the problem. Circle the information you know.
Underline what the problem is asking you to find.

2 Plan

What will you need to do to solve the problem? Write your plan in steps.

Step 1 Determine the ____________ for each triangular prism.

Step 2 ________ the surface areas.

Read to Succeed!
You may need to draw a net of each prism to help you visualize the area of each face.

3 Solve

Use your plan to solve the problem. Show your steps.

House A surface area:

triangular bases: $\frac{1}{2}(4 \cdot 8)(2) =$ ___

faces: $2(8.2 \cdot 10) =$ ____

$4 \cdot 10 =$ ____

____ + ____ + ____ = ____ Add.

House B surface area:

triangular bases: $\frac{1}{2}(4 \cdot 5)(2) =$ ___

faces: $2(5.4 \cdot 7) =$ ____

$4 \cdot 7 =$ ___

____ + ____ + ___ = ____ Add.

_____ − _____ = _____ square feet Subtract.

So, the correct answer is __. Fill in that answer choice.

4 Check

____

____

NAME________________________________ DATE ______________ PERIOD ________

Lesson 4 *(continued)*

Use a problem-solving model to solve each problem.

1 In science class, Marco compares the two light prisms shown below. How much larger, in square inches, is the surface area of the larger light prism than the smaller light prism? **6.G.4, MP 1**

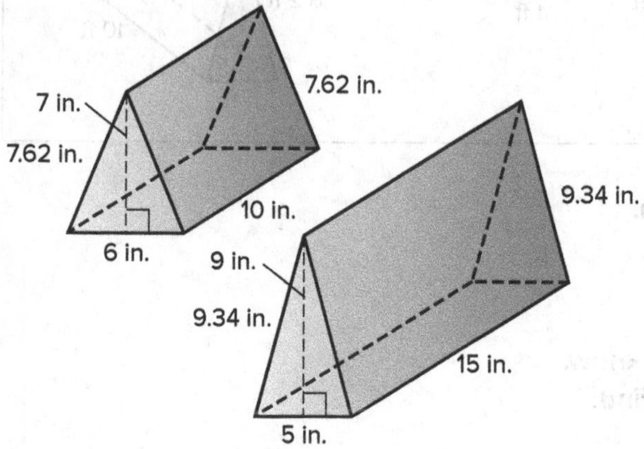

Ⓐ 142.8 square feet

Ⓑ 145.8 square feet

Ⓒ 212.4 square feet

Ⓓ 567.6 square feet

2 The net below represents a portion of a mural on a park sidewalk. If the dimensions are doubled, how many times greater is the surface area of the similar net, in square yards? **6.G.4, MP 7**

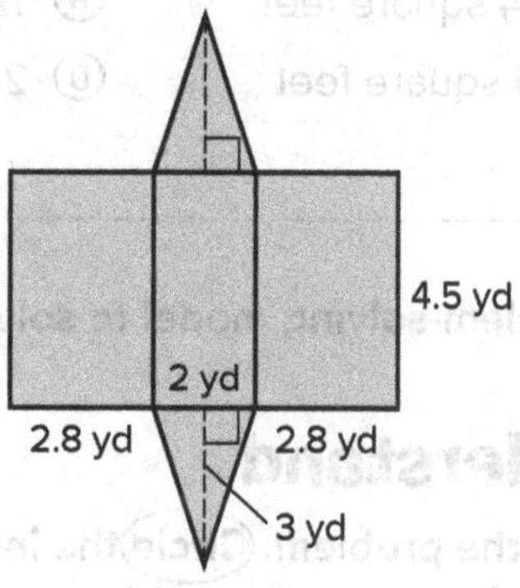

____

3 Gloria purchased a wedge pillow as shown below. She wants to make a pillow case for it. She has 500 square inches of fabric. How many more square inches of fabric does she need for the pillow case? **6.G.4, MP 7**

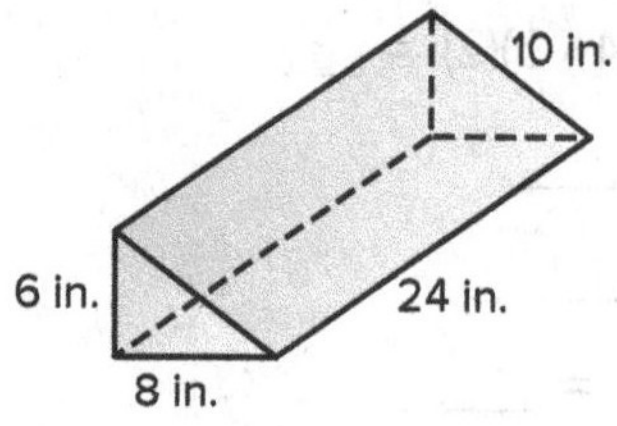

____

4 **H.O.T. Problem** The rectangular prism shown is cut in half diagonally to create the triangular prism. Is the surface area of the right triangular prism equal to one-half the surface area of the rectangular prism? Explain. **6.G.4, MP 3**

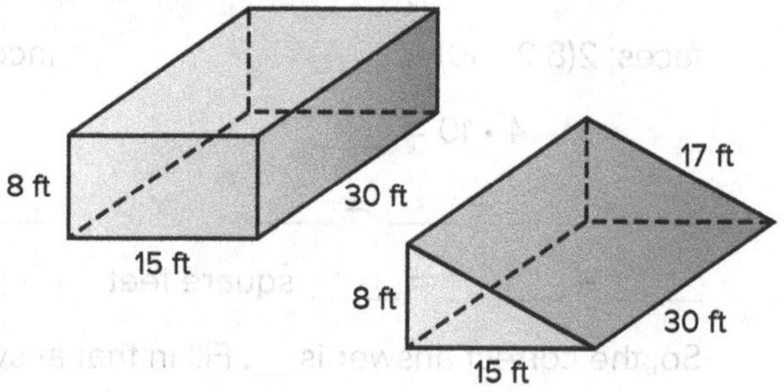

____

____

____

____

NAME_______________________________ DATE _______________ PERIOD _______

Lesson 5 Multi-Step Problem Solving

Multi-Step Example

The table shows the dimensions of three different square pyramids. What is difference between the greatest and least surface area, in square inches? 6.G.4, MP 1

Pyramid	(Base Edge (in.)	Slant Height (in.)
A	2	5
B	5	12
C	3.5	9

Ⓐ 51.25 square inches

Ⓑ 69.75 square inches

Ⓒ 96 square inches

Ⓓ 121 square inches

Use a problem-solving model to solve this problem.

1 Understand

Read the problem. Circle the information you know.
Underline what the problem is asking you to find.

2 Plan

What will you need to do to solve the problem? Write your plan in steps.

Step 1 Determine the ___________ for each square pyramid.

Step 2 ________ the least from the greatest surface areas.

3 Solve

Use your plan to solve the problem. Show your steps.

Determine the area of each base and lateral face:

A: (2)(2) = ___ B: (5)(5) = ___ C: (3.5)(3.5) = ___

A: $\frac{1}{2}$(2)(5) = ___ B: $\frac{1}{2}$(5)(12) = ___ C: $\frac{1}{2}$(3.5)(9) = _____

A: ____ + ____ + ____ + ____ + ____ = ____ Add.

B: ____ + ____ + ____ + ____ + ____ = ____ Add.

C: ______ + ______ + ______ + ______ = ______ Add.

____ − ____ = ____ square inches Subtract.

So, the correct answer is __. Fill in that answer choice.

Read to Succeed!
A square pyramid has four triangular sides. Determine the area of one side, then add the area four times to determine the lateral surface area.

4 Check

How do you know your solution is accurate?

____

____

____

NAME _______________ DATE _______________ PERIOD _______________

Lesson 5 *(continued)*

Use a problem-solving model to solve each problem.

1 The table shows the dimensions of three different square pyramids. What is difference between the greatest and least surface area, in square centimeters? **6.G.4, MP 1**

Pyramid	Base Edge (cm)	Slant Height (cm)
1	5	8.5
2	8	5
3	6	10

Ⓐ 12 square centimeters
Ⓑ 34 square centimeters
Ⓒ 46 square centimeters
Ⓓ 55 square centimeters

2 The net of Alana's crystal square pyramid is shown below. She wants to wrap the pyramid in three layers of tissue paper so she can put it in storage. What is the area, in square centimeters, of tissue paper will she need? **6.G.4, MP 7**

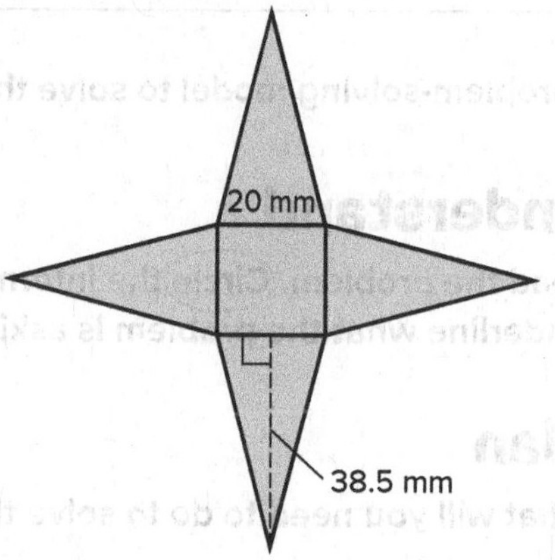

3 The pyramid below represents a sign at the entryway to a state park. The sign is going to be covered using advertisements on a large canvas. The bottom of the sign does not need to be covered since it is on the ground. There will only be advertisements on two lateral faces of the pyramid. Determine the lateral surface area, in square feet, to cover the two sides of the sign. **6.G.4, MP 7**

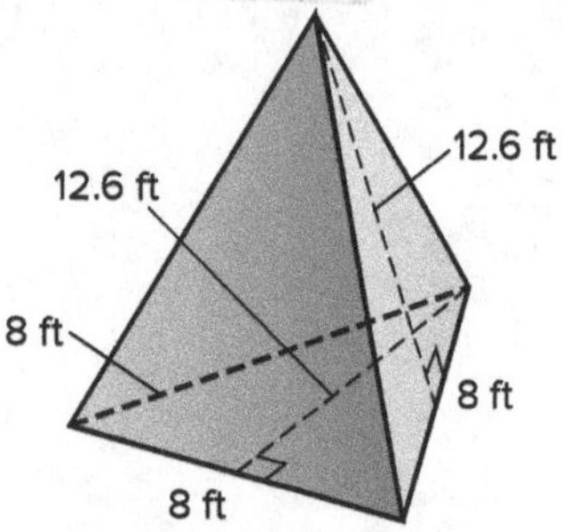

4 **H.O.T. Problem** The square pyramids below are congruent. What is the surface area of the composite figure? Explain. **6.G.4, MP 3**

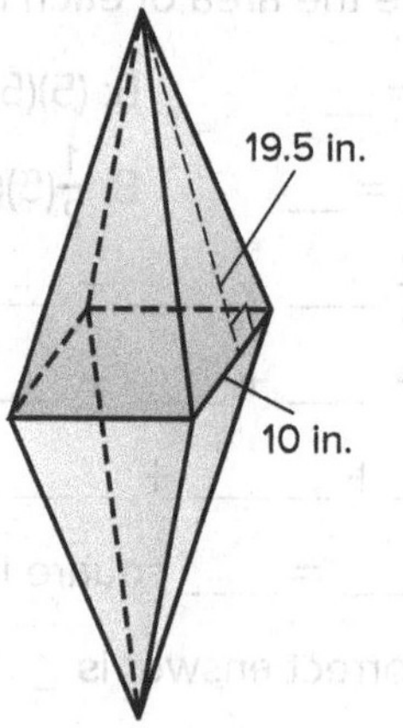

NAME______________________________ DATE ______________ PERIOD ________

Lesson 1 Multi-Step Problem Solving

Multi-Step Example

Edward's family owns a tree farm, which is open every day of the week except Monday. Edward kept track of how many trees were sold each day for two weeks. How much greater was the mean number of trees for Week 2 than for Week 1? 6.SP.3, MP 1

Week 1	
Day	**Trees**
Tuesday	7
Wednesday	12
Thursday	6
Friday	14
Saturday	22
Sunday	17

Week 2	
Day	**Trees**
Tuesday	10
Wednesday	8
Thursday	12
Friday	17
Saturday	31
Sunday	18

Use a problem-solving model to solve this problem.

1 Understand

Read the problem. Circle the information you know.
Underline what the problem is asking you to find.

2 Plan

What will you need to do to solve the problem? Write your plan in steps.

Step 1 Determine the mean for each week.

Step 2 Subtract to find how much greater the mean for Week 2 is than Week 1.

Read to Succeed!
To determine the mean, add each value in the set and divide by the number of values in the set.

3 Solve

Use your plan to solve the problem. Show your steps.

Week 1: $\frac{7 + 12 + 6 + 14 + 22 + 17}{6} =$ ____

Week 2: $\frac{10 + 8 + 12 + 17 + 31 + 18}{6} =$ ____

So, Week 2's mean is 16 − 13 or ___ trees greater.

4 Check

How do you know your solution is reasonable?

____

____

____

Lesson 1 *(continued)*

Use a problem-solving model to solve each problem.

1 The dot plot shows how many minutes Mr. Elliot's piano students said they practiced on the day before their lessons. Imala practiced 31 minutes but forgot to tell Mr. Elliot. If Imala's time were included, by how much time (in minutes) would the mean increase? **6.SP.3, MP 1**

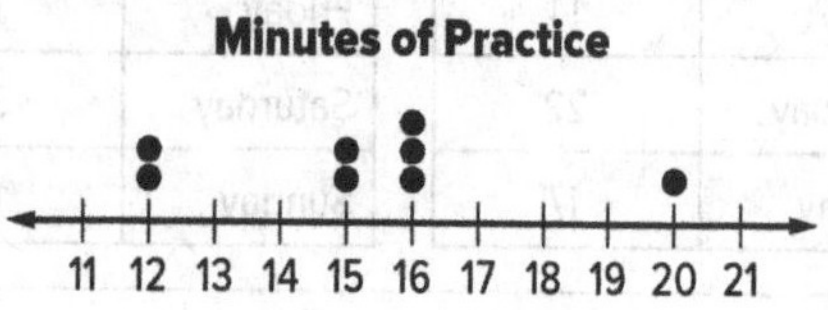

____

2 The graph shows how many rides a group of friends went on at the fair. Each ride costs \$2.75. What was the mean amount of money, in dollars, spent per person to go on the rides? **6.SP.3, MP 2**

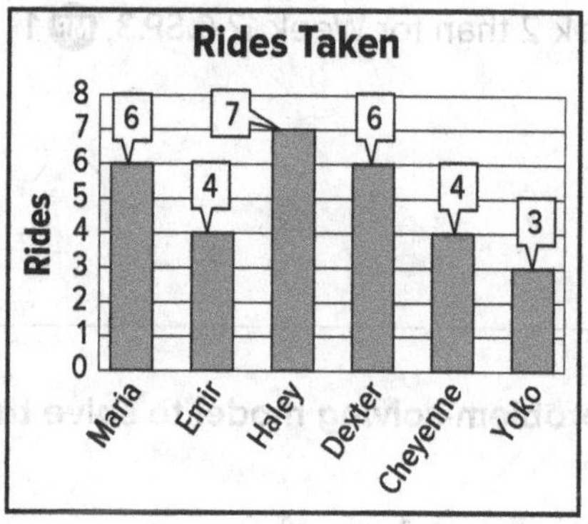

____

3 For Quon's first six quizzes, he had a mean score of 33 points. After the seventh quiz, his mean score was 32 points. After the eighth quiz, the mean was 34. What was the difference in scores between his seventh and eighth quizzes? **6.SP.3, MP 8**

____

4 **H.O.T. Problem** Create a list of 8 values with a mean of 26. Justify your response. **6.SP.3, MP 3**

____

____

____

Lesson 2 Multi-Step Problem Solving

Multi-Step Example

Four students kept track of how long they did homework for five nights. For which student do the mean, median, and mode all have the same value? 6.SP.3, MP 1

Ⓐ Emma
Ⓑ Rosario
Ⓒ Peter
Ⓓ Susana

Day	Emma	Rosario	Peter	Susana
1	1.25	1.25	1	0.75
2	0.75	2.25	1.75	2.5
3	1	1.5	1	1.5
4	1.25	2	0.5	0.75
5	2	0.75	1.5	2

Use a problem-solving model to solve this problem.

1 Understand

Read the problem. Circle the information you know.
Underline what the problem is asking you to find.

2 Plan

What will you need to do to solve the problem? Write your plan in steps.

Step 1 Determine the mean, median, and mode for each student.

Step 2 Compare the measures.

Read to Succeed!
Mean – divide the sum of the values by the number of values
Median – the middle value when the data are ordered
Mode – most occurring number

3 Solve

Use your plan to solve the problem. Show your steps.

	Mean	Median	Mode
Emma			
Rosario			
Peter			
Susana			

Since ________ has all three measures the same, choice ___ is correct. Fill in that answer choice.

4 Check

How do you know your solution is accurate?

____

____

Lesson 2 *(continued)*

Use a problem-solving model to solve each problem.

1 Four drivers recorded the distance they drove each day for a week. Which driver's data set has a mode that is greater than the mean or median AND a median with the lowest value of the three measures? **6.SP.3, MP 1**

Ⓐ Kadisha: 8, 17, 23, 16, 17, 18, 125

Ⓑ Cole: 14, 26, 34, 22, 47, 22, 45

Ⓒ Fabian: 7, 12, 11, 23, 13, 23, 30

Ⓓ Lina: 52, 36, 41, 31, 31, 37, 59

2 The graph shows the number of cell phones per 100 people in certain countries. How much would the difference between the median and the mode change if Finland were not included in the data? **6.SP.5, MP 7**

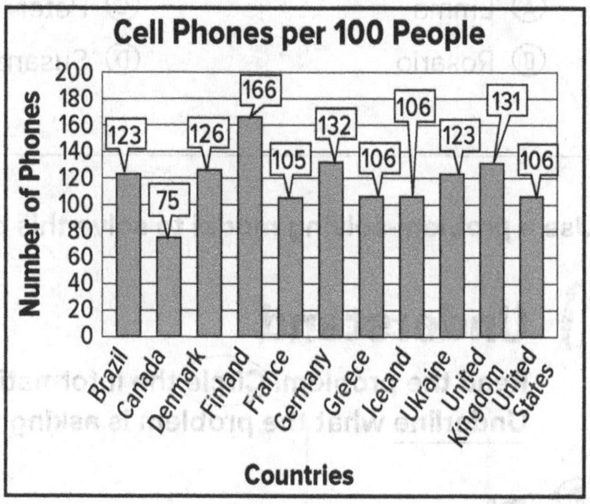

____

3 What is the difference between the medians of Kendra's sprint times and Hakim's sprint times? **6.SP.5, MP 2**

Kendra's Sprint Times (seconds)			
12	14	11	13
15	13	15	14
11	14	17	12

Hakim's Sprint Times (seconds)			
11	11	15	13
14	16	15	14
13	15	12	16

____

4 **H.O.T. Problem** For which data set is the median a better predictor of the rest of the data than the mode? Explain your answer. **6.SP.5b, MP 3**

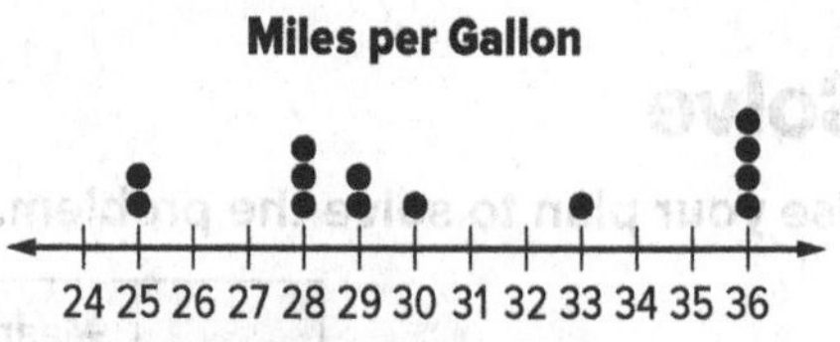

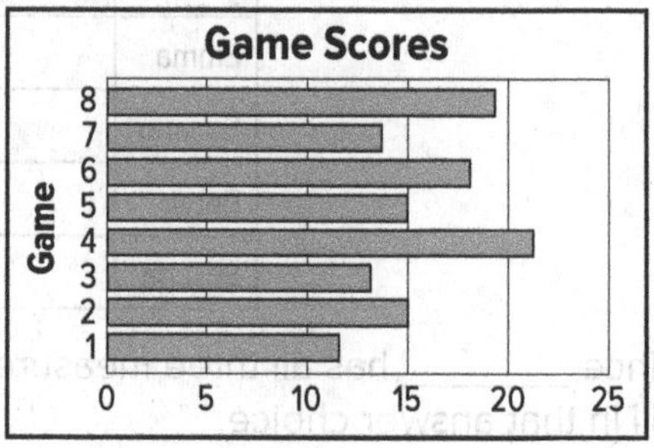

____

____

____

____

NAME ______________________________ DATE ______________ PERIOD ________

Lesson 3 **Multi-Step** Problem Solving

Multi-Step Example

Carmen and Noah are running for president of the middle school student government. Votes are counted by classroom. What is the difference between the interquartile ranges for the two candidates? **6.SP.3, MP 1**

Voting Results					
Room Number	Number of Votes		Room Number	Number of Votes	
	Carmen	Noah		Carmen	Noah
1	12	9	5	8	17
2	6	18	6	2	13
3	14	8	7	18	7
4	20	6	8	12	12

Use a problem-solving model to solve this problem.

1 Understand

Read the problem. Circle the information you know. Underline what the problem is asking you to find.

Read to Succeed!
Interquartile range is the difference between the third quartile and the first quartile.

2 Plan

What will you need to do to solve the problem? Write your plan in steps.

Step 1 Order the values for each person from least to greatest.

Step 2 Determine the IQR for each data set.

Step 3 Subtract to find the difference.

3 Solve

Use your plan to solve the problem. Show your steps.

Carmen: 2, 6, 8, 12, 12, 14, 18, 20 IQR: ____________

Noah: 6, 7, 8, 9, 12, 13, 17, 18 IQR: ____________

The difference between the interquartile ranges is ____________.

4 Check

How do you know your solution is reasonable?

____

____

____

NAME ______________________ DATE ______________ PERIOD ________

Lesson 3 *(continued)*

Use a problem-solving model to solve each problem.

1 Melissa is keeping track of the temperature in her town at noon each day. She has recorded the temperature for six days so far. How much greater will the interquartile range be if Saturday's temperature is 70°F than if it is 58°F? **6.SP.3, MP 1**

Temperature at Noon	
Day	**Temperature (°F)**
Sunday	64
Monday	72
Tuesday	58
Wednesday	54
Thursday	60
Friday	62
Saturday	?

____

2 Jamal cut out these shapes from construction paper. What is the interquartile range for the areas of the shapes? **6.SP.3, MP 7**

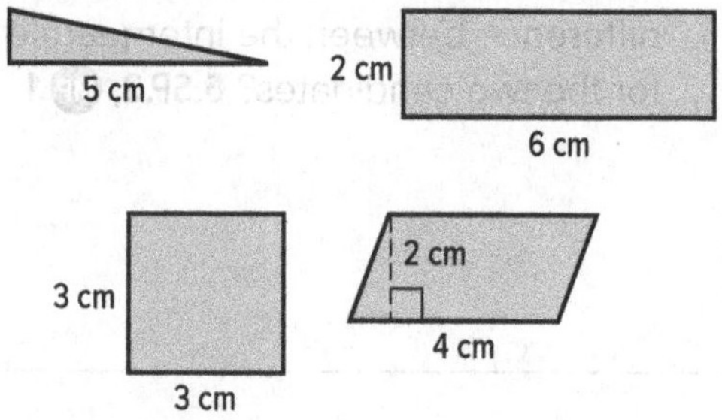

____

3 Tamiko is training for a bicycle race. She made a graph of the number of miles she rode each week for 6 weeks. If the median at Week 7 increases by 1.5, how many miles did she ride in Week 7? **6.SP.5, MP 7**

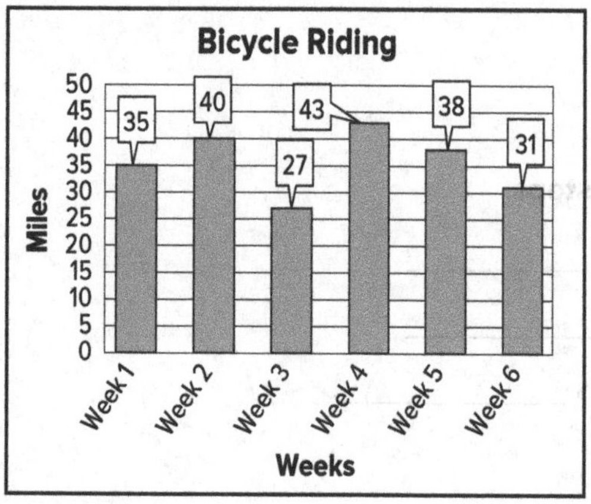

____

4 **H.O.T. Problem** Describe the measures of spread for the data set. Change the data set by adding an outlier. Describe the new measures of spread. **6.SP.5c, MP 3**

1,124	465
650	976
840	711
712	925

____

____

____

____

____

____

____

____

____

NAME_______________________________________ DATE ____________________ PERIOD ___________

Lesson 4 Multi-Step Problem Solving

Multi-Step Example

The table shows the number of hours various students worked on a school project. The students want to compare measures of spread. What is the difference between the interquartile range and mean absolute deviation? **6.SP.5c, MP 1**

Hours Worked			
Luke	0.5	Aria	3.5
Erin	2.5	Donte	1.0
Kim	2.5	Chante	3.5
Sierra	2.0	Kya	0.5

Ⓐ 0.25
Ⓑ 1.25
Ⓒ 2.00
Ⓓ 5.75

Use a problem-solving model to solve this problem.

1 Understand

Read the problem. Circle the information you know.
Underline what the problem is asking you to find.

2 Plan

What will you need to do to solve the problem? Write your plan in steps.

Step 1 Determine the ________________________ and range.

Step 2 Determine the ______________ between the two values.

3 Solve

Use your plan to solve the problem. Show your steps.

The mean absolute deviation is ____. The interquartile range is ____. The difference between the interquartile range and the mean absolute deviation is _____ – _____, or _____.

So, the correct answer is ____. Fill in that answer choice.

Read to Succeed!
Remember to take the absolute value of the differences between each value and the mean when finding the mean absolute deviation.

4 Check

How do you know your solution is accurate?

____

____

____

Lesson 4 *(continued)*

Use a problem-solving model to solve each problem.

1 The table shows the quiz scores of various students. The students want to compare the measures of spread. What is the difference between the interquartile range and mean absolute deviation? **6.SP.5c, MP 1**

Quiz Scores				
95	100	50	75	60
100	100	60	100	60

Ⓐ 20

Ⓑ 21

Ⓒ 31

Ⓓ 40

2 Eight students were asked how many persons live in their home. The results of the survey are shown in the line plot. What is the mean absolute deviation of the data? **6.SP.5c, MP 2**

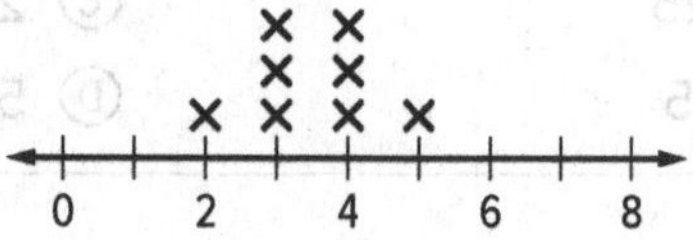

____

3 Some students were asked how much they spent on their last trip to the mall. The results of the survey are shown in the stem-and-leaf plot. What is the mean absolute deviation of the data? **6.SP.5c, MP 7**

Money Spent at Mall

Stem	Leaf
0	6
1	5 9
2	0 5 5
3	0 5 5
4	0

1 | 5 = $15

____

4 **H.O.T. Problem** The mean number of points scored by eight players in a basketball game is 6. Use the table to determine mean absolute deviation of the data. Justify your answer. **6.SP.5b, MP 3**

Points Scored			
Abdi	4	Benito	4
Noah	0	Wen	10
Felix	0	Delon	8
Tyler	2	Sean	?

____

____

____

____

____

____

Lesson 5 Multi-Step Problem Solving

Multi-Step Example

The table shows the ages of the people at a family dinner. Identify the outlier in the data set. Then determine how the outlier affects the mean of the data. 6.SP.5c, MP 1

Age of Family Members (years)						
39	48	47	41	38	84	39

Ⓐ outlier: 48; mean age with the outlier decreased by 48

Ⓑ outlier: 84; mean age with the outlier decreased by 6

Ⓒ outlier: 84; mean age with the outlier increased by 6

Ⓓ outlier: 48; mean age with the outlier increased by 48

Use a problem-solving model to solve this problem.

1 Understand

Read the problem. Circle the information you know.
Underline what the problem is asking you to find.

2 Plan

What will you need to do to solve the problem? Write your plan in steps.

Step 1 Determine the outlier, or deviation from the majority of the data set.

Step 2 Determine the mean for the data set both with and without the outlier.

Step 3 Subtract to compare the mean age with and without the outlier.

Read to Succeed!
Measures of center are used to summarize a data set. Outliers often make one measure more appropriate to use than others.

3 Solve

Use your plan to solve the problem. Show your steps.

Compared to the other ages, 84 is very old. So, ______ is an outlier.

Mean with the outlier: 39 + 47 + 38 + 39 + 48 + 41 + 84 ÷ 7 = ______

Mean without the outliner: 39 + 47 + 38 + 39 + 48 + 41 ÷ 6 = ______

Compare: 48 − 42 = 6. So, the correct answer is _____. Fill in that answer choice.

4 Check

How do you know your solution is reasonable?

____

____

____

Lesson 5 *(continued)*

Use a problem-solving model to solve each problem.

1 The table shows the weekly deposits Malcolm made in his savings account. Identify the outlier in the data set. Then determine how the outlier affects the mean of the data. **6.SP.5c, MP 1**

Deposits in Savings Account ($)				
41	28	26	5	32
41	38	26	36	

Ⓐ outlier: 41; mean with the outlier increased by about 3.2

Ⓑ outlier: 5; mean with the outlier increased by about 3.2

Ⓒ outlier: 41; mean with the outlier decreased by about 3.2

Ⓓ outlier: 5; mean with the outlier decreased by about 3.2

2 The scores Miriam received on the science tests are 95, 80, 95, 85, 45, 95, 75, 85, and 90. Identify the outlier in the data set. Determine the mean, median, and mode without the outlier. Then tell which measure of center best describes the data without the outlier. **6.SP.5d, MP 7**

____

____

____

____

3 List eight data values for which the median is the best measure of center for the data set. Explain. **6.SP.5d, MP 3**

____

____

____

____

____

4 **H.O.T. Problem** The table shows the lengths of some rivers in the United States. Identify the outlier. Find the measures of center with and without the outlier. Tell which measure of center best describes the data with and without the outlier. **6.SP.5c, MP 3**

River	Length (mi)
Columbia	1,243
Mississippi	2,340
Ohio-Allegheny	1,250
Peace	1,210
Red	1,290

____

____

____

____

____

NAME ________________________ DATE ____________ PERIOD ________

Lesson 1 **Multi-Step** Problem Solving

Multi-Step Example

Which description matches the data in the line plot of U.S. presidents' years in office? (Round decimals to the nearest whole number.) **6.SP.4, MP 1**

Presidents' Years in Office 1901–2009

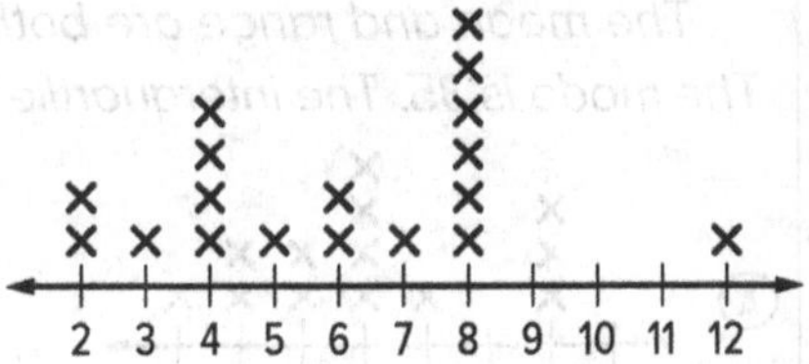

Ⓐ mean: 5, mode: 8, median: 7, range: 10, interquartile range: 4, outlier: 12

Ⓑ mean: 6, mode: 8, median: 6, range: 8, interquartile range: 6, outlier: 12

Ⓒ mean: 6, mode: 8, median: 6, range: 10, interquartile range: 4, no outlier

Ⓓ mean: 7, mode: 8, median: 5, range: 12, interquartile range: 4, no outlier

Use a problem-solving model to solve this problem.

1 Understand

Read the problem. Circle the information you know.
Underline what the problem is asking you to find.

2 Plan

What will you need to do to solve the problem? Write your plan in steps.

Step 1 Determine the measures of center.

Step 2 Determine the measures of spread.

Read to Succeed!
Determine the measures of center and spread and compare your answers to the answer choices, making sure to account for each measure.

3 Solve

Use your plan to solve the problem. Show your steps.

mean: about ☐ range: 12 − 2 or ☐ mode: ☐

IQR: 8 − 4 or ☐ median: ☐ outlier: ☐

Choice ☐ lists the correct measures. Fill in that answer choice.

4 Check

How do you know your solution is reasonable?

____

____

Lesson 1 *(continued)*

Use a problem-solving model to solve each problem.

1 Which line plot matches Noshi's description of his quiz scores? **6.SP.4, MP 1**

The mean and range are both 30.
The mode is 35. The interquartile range is 15.

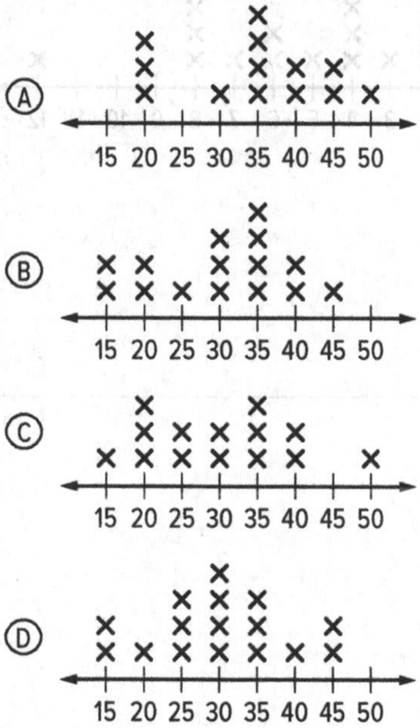

2 Liliana is shopping for pencils that are sold by the dozen (12). She finds out how much one pencil costs at each price and makes a line plot. Show the line plot. What price should have three marks above it? **6.SP.5a, MP 2**

Pencil Prices (per dozen)				
4 for $4.80	4 for $9.60	1 for $2.40	4 for $14.40	2 for $6.00
2 for $3.60	12 for $14.40	4 for $7.20	1 for $3.60	6 for $14.40

Price for One Pencil ($)

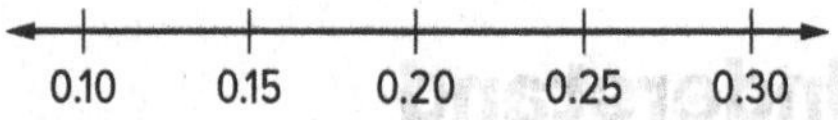

____

3 Joseph asked his friends how many pages of their history book they had read. He made a line plot to show his data and he said that the mean was 6. Then he noticed that he'd forgotten to include one number on the line plot. What number did he forget? **6.SP.5c, MP 4**

Number of Pages Read

____

4 **H.O.T. Problem** The Math Club has been selling cookies during lunch. Make a line plot to show the number of cookies sold each day. Describe the data's measures of spread and center. **6.SP.4, MP 3**

Number of Cookies Sold

12, 16, 10, 12, 15, 13, 14, 20, 15, 15,
13, 11, 13, 14, 15, 14, 11, 12, 13, 15

____

____

____

NAME ______________________ DATE ____________ PERIOD ______

Lesson 2 Multi-Step Problem Solving

Multi-Step Example

The histogram shows the distances a volleyball team travels to their games. What percent of the games did they travel more than 24 miles? Round to the nearest tenth.
6.SP.5a, MP 1

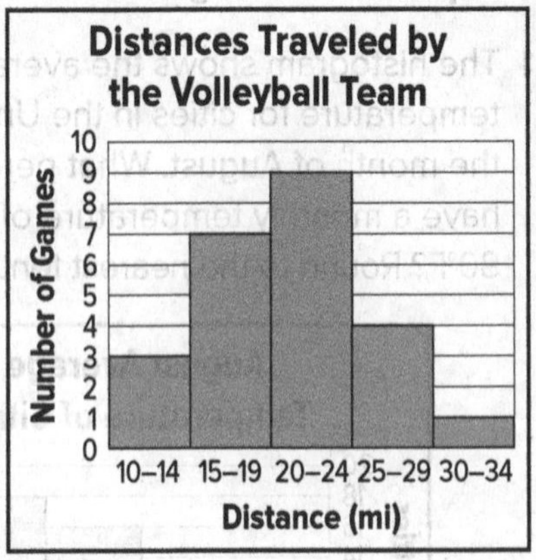

Use a problem-solving model to solve this problem.

1 Understand

Read the problem. Circle the information you know.
Underline what the problem is asking you to find.

2 Plan

What will you need to do to solve the problem?
Write your plan in steps.

Step 1 Determine how many games were more than 24 miles away. Determine the total number of games.

Step 2 Express as a percent.

Read to Succeed!
The percentage is the number of games greater than 24 miles divided by the total number of games. The decimal is then expressed as a percent.

3 Solve

Use your plan to solve the problem. Show your steps.

Number of games greater than 24 miles away: ☐

Total number of games: ☐

So, ☐ out of ☐ games were played greater than 24 miles away.

This is ☐ ÷ ☐ or ☐ or ☐% of the games.

4 Check

How do you know your solution is reasonable?

____

____

____

Lesson 2 *(continued)*

Use a problem-solving model to solve each problem.

1 The histogram shows the average monthly temperature for cities in the United States for the month of August. What percent of cities have a monthly temperature of less than 80°F? Round to the nearest tenth. **6.SP.5a, MP 1**

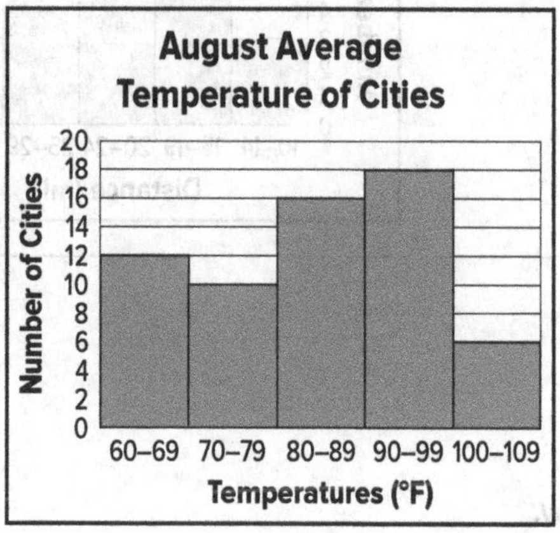

____

2 The students in Mrs. Sanchez's class recorded their heights. The histogram shows the heights of the students. What fraction of the students are taller than 55 inches? Simplify your answer. **6.SP.5a, MP 2**

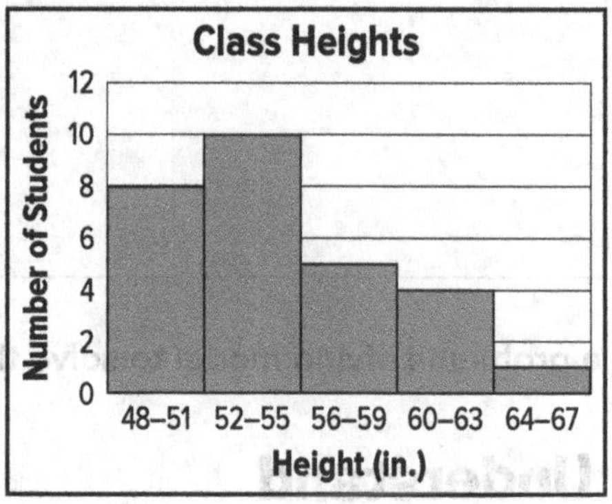

____

3 A government program plants small trees in parks. The histogram shows the number of trees planted in 48 different parks. What is the difference in the number of parks that had the least trees planted and the most trees planted? **6.SP.5a, MP 2**

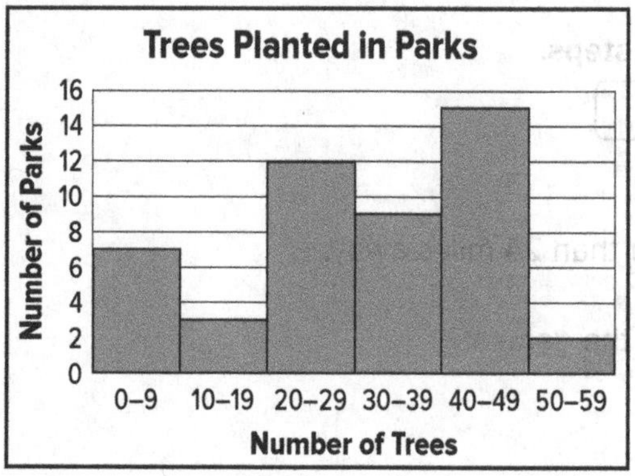

____

4 H.O.T. Problem Valley View Middle School is holding a fundraiser for a local charity. The table shows the number of classes that raised money. Create a histogram for the data. Then find the percent of classes that raised $75 or more. **6.SP.4, MP 4**

Amount ($)	Number of Classes
25–49	|||
50–74	𝍸 ||||
75–99	𝍸 𝍸
100–124	𝍸 |||

____

Lesson 3 Multi-Step Problem Solving

Multi-Step Example

The box plot shows the range of prices of 50 board games available at a local toy store. About how many of the available board games cost less than $10? **6.SP.2, MP 1**

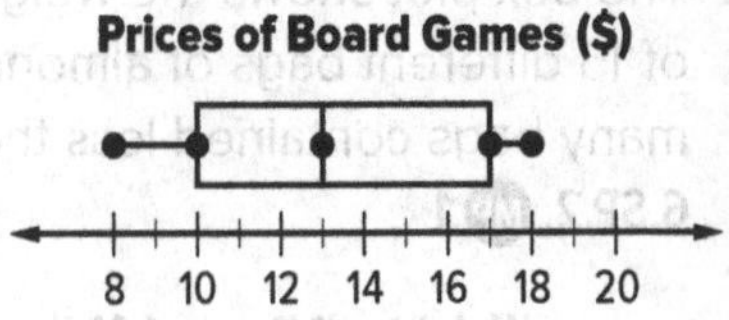

Ⓐ 2 games

Ⓑ 6 games

Ⓒ 13 games

Ⓓ 38 games

Use a problem-solving model to solve this problem.

1 Understand

Read the problem. Circle the information you know.
Underline what the problem is asking you to find.

Read to Succeed!
A box plot shows the data divided into quartiles, or four sections.

2 Plan

What will you need to do to solve the problem? Write your plan in steps.

Step 1 Determine the section of the box plot representing less than $10.

Step 2 Determine the approximate number of board games that cost less than $10.

3 Solve

Use your plan to solve the problem. Show your steps.

The section of the box plot that represents less than $10 is the ___________.

The left whisker represents one-______ of the data.

Since one-fourth of the board games cost less than $10, about _____ ÷ _____ or about _____ games cost less than $10. Choice _____ is correct. Fill in that answer choice.

4 Check

How do you know your solution is reasonable?

____

____

____

Lesson 3 *(continued)*

Use a problem-solving model to solve each problem.

1 The box plot shows the weights, in ounces, of 15 different bags of almonds. About how many bags contained less than 27 ounces? **6.SP.2, MP 1**

Weights of Bags of Almonds (oz)

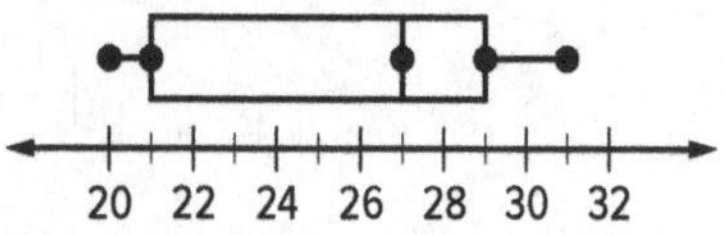

Ⓐ 11 bags

Ⓑ 8 bags

Ⓒ 4 bags

Ⓓ 0 bags

2 The data shows the amounts of flour used in different cookie recipes. What is the third quartile of the data? **6.SP.4, MP 2**

Cups of Flour Used for Cookies						
0.5	1	0.75	1	1.5	1.25	0.75
1.5	0.75	1.75	0.5	0.5	1.5	1

____

3 The line plot shows the ages of dogs at a dog park. What is the first quartile of the data? **6.SP.4, MP 2**

Ages of Dogs at a Dog Park

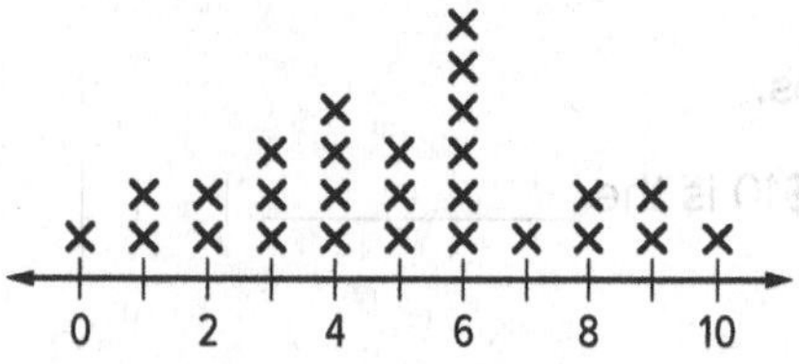

____

4 **H.O.T. Problem** Mr. Jensen assigned a quiz to his math students. The data shows all his students' grades. Create a box plot for the data. Then find the percent of students that scored 82 or higher. **6.SP.4, MP 7**

Math Quiz Grades						
75	81	80	94	77	78	80
74	96	87	84	91	90	83
79	87	100	97	78	76	82

Math Quiz Grades

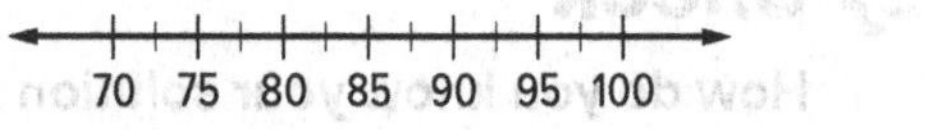

____

Lesson 4 **Multi-Step** Problem Solving

Multi-Step Example

The line plot shows the ages of the students in Session 1 of Mr. Garcia's martial arts class. For Session 2, the students were the same except that a 12-year old dropped out and a 9-year old enrolled. Which shows the best measures of center and spread for the **Session 2** data distribution? **6.SP.5d,** MP 1

Ages in Martial Arts Class
Session 1

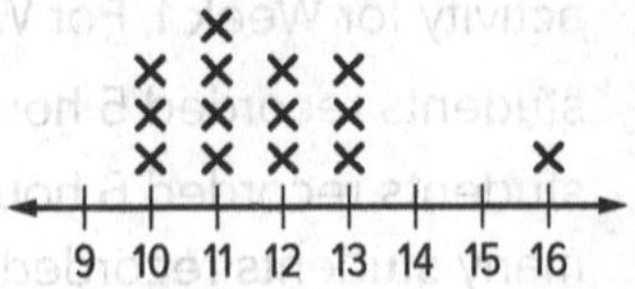

Ⓐ mean ≈ 11.6, range = 7

Ⓑ mean ≈ 11.8, range = 6

Ⓒ median = 11, interquartile range = 3

Ⓓ median = 11.5, interquartile range = 2

Use a problem-solving model to solve this problem.

1 Understand

Read the problem. Circle the information you know.
Underline what the problem is asking you to find.

2 Plan

What will you need to do to solve the problem?
Write your plan in steps.

Step 1 Determine the data set with the values adjusted.

Step 2 Determine the mean, median, range, and interquartile range of the new data set.

Read to Succeed!
List the values in order after replacing the value of 12 with the value of 9.

3 Solve

Use your plan to solve the problem. Show your steps.

The new data set is 9, 10, 10, 10, 11, 11, 11, 11, 12, 12, 13, 13, 13, 16.

The mean is ☐, range is ☐, median is ☐, and interquartile range is ☐. Choice ☐ is correct. Fill in that answer choice.

4 Check

How do you know your solution is reasonable?

____

____

____

Lesson 4 *(continued)*

Use a problem-solving model to solve each problem.

1 The line plot shows the amount of time students recorded engaging in physical activity for Week 1. For Week 2, $\frac{2}{3}$ as many students recorded 5 hours, $\frac{2}{3}$ as many students recorded 6 hours, and twice as many students recorded 10 hours.

Hours of Physical Activity Week 1

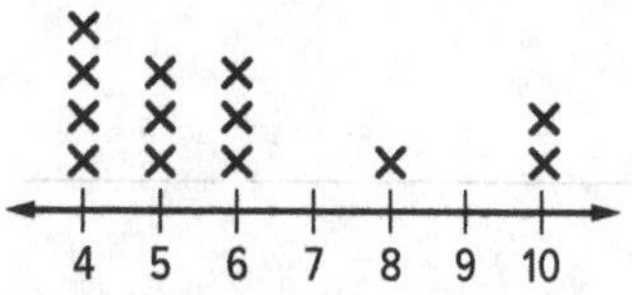

Which option shows the best measures of center and spread for the Week 2 data distribution? **6.SP.5d, MP 1**

Ⓐ mean ≈ 5.9, range = 6

Ⓑ mean ≈ 6.6, range = 6

Ⓒ median = 5, interquartile range = 3

Ⓓ median = 6, interquartile range = 6

2 The stem-and-leaf plot shows students' quiz scores in Ms. Warren's math class.

Quiz Scores

Stem	Leaf
7	5 5 5
8	0 0 0 5 5 5 5
9	0 0 0 0 5 5 5
10	0 0 0

7 | 5 = 75%

Lola found the range and the best measure of center for the data distribution. What two measures did Lola find? What is the sum of the two numbers Lola found? **6.SP.5d, MP 7**

____

3 Donte recorded the number of repairs his bicycle shop completed each day for a month. The box plot shows a summary of the data.

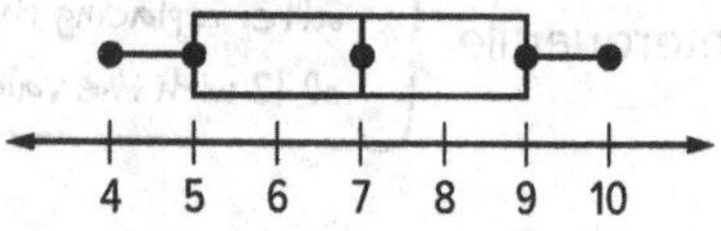

Describe the shape of the distribution **6.SP.2, MP 4**

____

____

____

____

4 **H.O.T. Problem** The histogram summarizes the players' heights on Rachel's volleyball team.

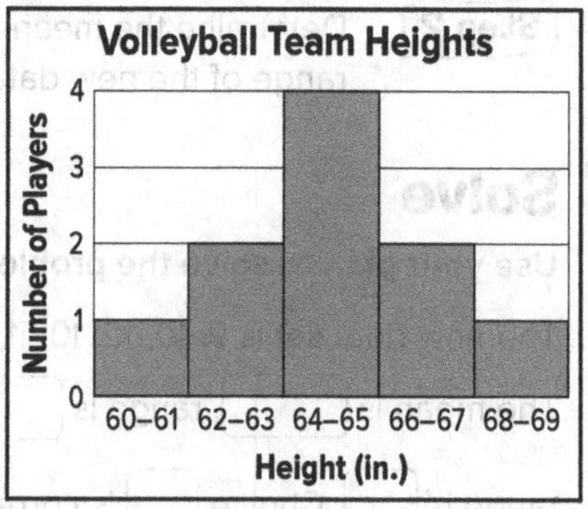

a. What are the range and best measure of center if all the heights are even integers?

b. What are the range and best measure of center if all the heights are odd integers? **6.SP.2, MP 2**

____

____

Lesson 5 **Multi-Step** Problem Solving

Multi-Step Example

The line graph shows the enrollments at Midway Middle School and Oakhill Middle School between the years 1980 and 2010. Based on the graph, predict which school will have a greater enrollment in 2020? Explain.
Extension of **6.SP.4, MP 1**

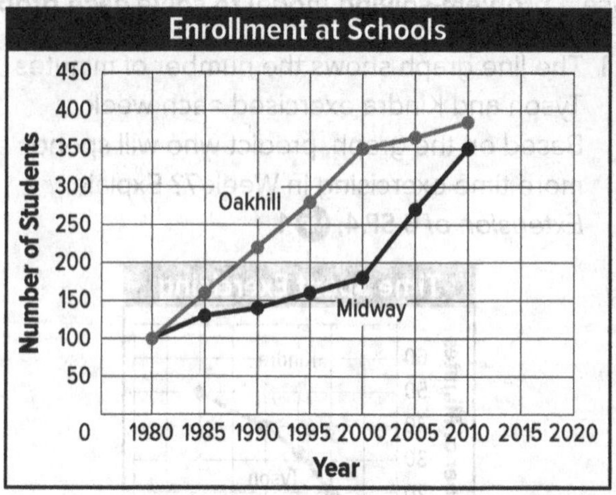

Use a problem-solving model to solve this problem.

1 Understand

Read the problem. Circle the information you know. Underline what the problem is asking you to find.

> **Read to Succeed!**
> Line graphs show change over time. You can look at trends shown in a line graph to predict future data.

2 Plan

What will you need to do to solve the problem? Write your plan in steps.

Step 1 Observe the slant and the steepness of each line.

Step 2 Look for trends and make predictions based on the trends.

3 Solve

Use your plan to solve the problem. Show your steps.

Midway's enrollment increased slowly between 1980 and 2000, and more quickly between ______ and ______. Oakhill's enrollment increased quickly between 1980 and 2000, and more slowly between ______ and ______.

Midway's enrollment in recent years has increased more quickly than Oakhill's enrollment. If the trend continues, the enrollment at ________ will be greater in 2020.

4 Check

How do you know your solution is reasonable?

____

____

Lesson 5 *(continued)*

Use a problem-solving model to solve each problem.

1 The line graph shows the number of minutes Tyson and Kindra exercised each week. Based on the graph, predict who will spend more time exercising in Week 7? Explain.
Extension of **6.SP.4, MP 1**

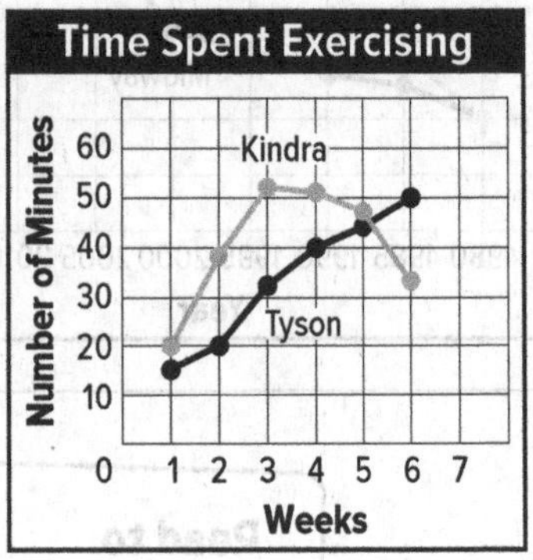

____

____

____

____

2 The line graph shows the circulation of two magazines for ten years. For what percent of these ten years did *The Voice* have a greater circulation than *The Star*?
Extension of **6.SP.4, MP 2**

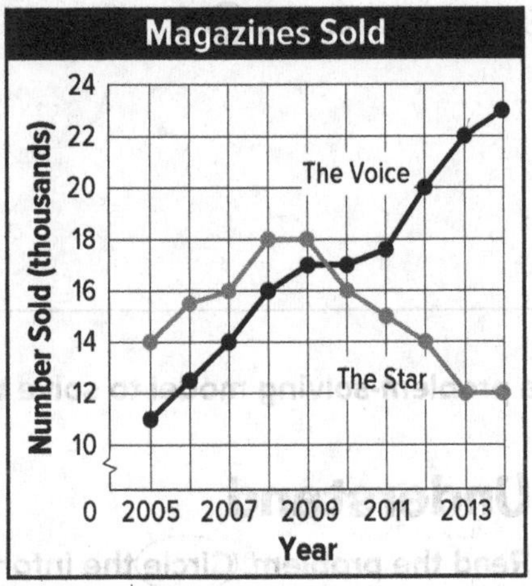

____

3 **H.O.T. Problem** Make a line graph of the data in the table. Suppose the number of books sold in Week 6 was 350, in Week 7 it was 375, and in Week 8 it was 400. Would you predict that the number of books sold in Week 9 will be greater than or less than 400? Explain.
Extension of **6.SP.4, MP 3**

Books Sales at Bob's Book Barn					
Week	1	2	3	4	5
Number Sold	100	125	175	250	350

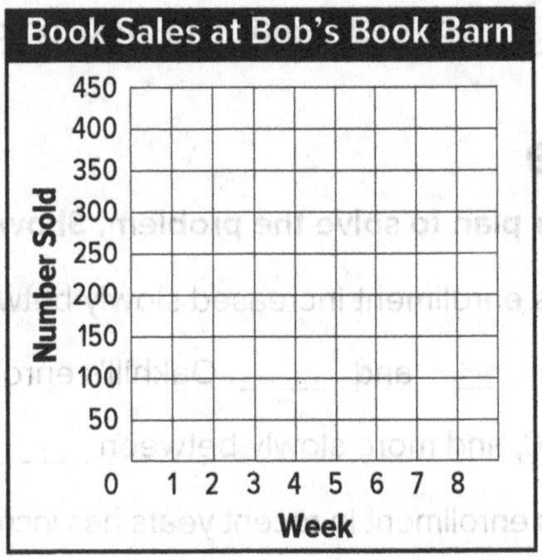

____

____

____

NAME _______________ DATE _______________ PERIOD _______________

Lesson 6 Multi-Step Problem Solving

Multi-Step Example

The table shows the distance 12 students on the track team ran one week. The next week, each student ran exactly twice as far as they did Week 1. Which statement is true about finding the mode of the data for Week 2? *Extension of* **6.SP.4,** MP 1

Number of Miles Run Week 1			
5	7	4	9
10	4	7	4
8	4	6	4

Ⓐ A dot plot will best show that the mode is 8.

Ⓑ A box plot will best show that the mode is 8.

Ⓒ A dot plot will best show that the mode is 11.

Ⓓ A box plot will best show that the mode is 11.

Use a problem-solving model to solve this problem.

1 Understand

Read the problem. Circle the information you know.
Underline what the problem is asking you to find.

2 Plan

What will you need to do to solve the problem? Write your plan in steps.

Step 1 Determine the values for Week 2.

Step 2 Determine the mode of the Week 2.

3 Solve

Use your plan to solve the problem. Show your steps.

The values for Week 2 are 10, 20, 16, 14, 8, 8, 8, 14, 12, 18, 8, 8.

The mode is ☐.

Since a box plot does not show the mode, a dot plot is the best representation to show the mode. Choice ☐ is correct. Fill in that answer choice.

Read to Succeed!
Remember to read each statement before deciding on a choice.

4 Check

How do you know your solution is accurate?

____

____

____

Lesson 6 *(continued)*

Use a problem-solving model to solve each problem.

1 The table shows the number of books each student in an afterschool club read in January. In February, each student met the goal of reading one more book than they did in January. Which statement is true about the mode of the data for February? *Extension of* **6.SP.4, MP 1**

Number of Books Read in January			
7	1	4	4
5	6	8	9
4	5	7	3

Ⓐ A box plot will best show that the number of books read most is 6 books.

Ⓑ A box plot will best show that the number of books read most is between 5 and 8.

Ⓒ A histogram will best show that the number of books read most is between 4 and 7.

Ⓓ A line plot will best show that the number of books read most is 5.

2 Dana created a histogram and a box plot using the data showing the number of minutes she exercised each day for 10 days.

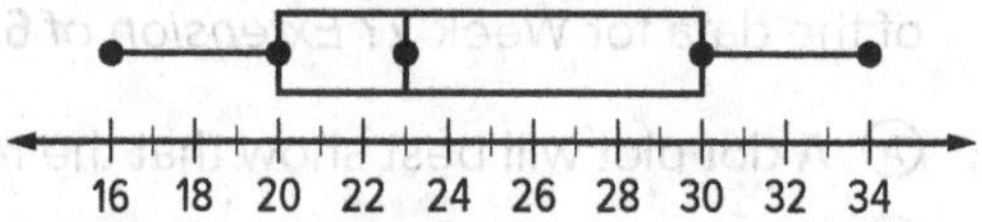

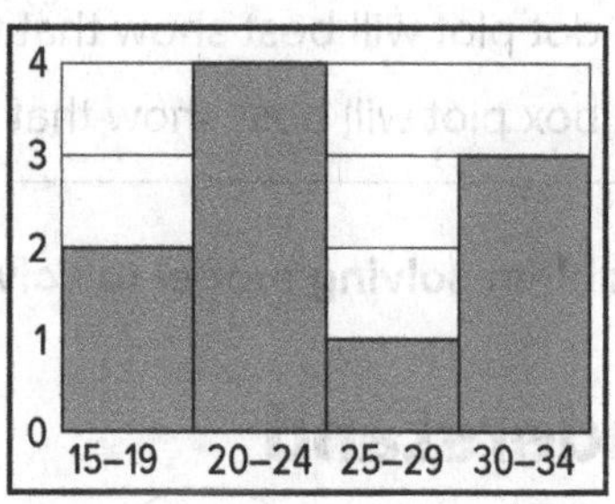

Which representation can she use to determine how many days she exercised 25 minutes or more? How many days did she exercise for 25 minutes or more? *Extension of* **6.SP.4, MP 2**

____

3 **H.O.T. Problem** A real estate agent wants to create a display to show the trend in the median sales of houses over the past 6 months. He has 50 data entries for each month for the past 6 months. Explain what two types of displays he can use, one to show the median and one to show the trend in the median over the past 6 months. *Extension of* **6.SP.4, MP 3**

____

____

____

NAME ________________ DATE ________ PERIOD ________

Lesson 1 **Multi-Step** Problem Solving

Chapter 1

Multi-Step Example

The table shows the school supplies Jin has. He wants to put the pencils, erasers, and notepads in bags, with each item distributed evenly in the bags. If each bag is $2, what is the greatest amount of money he would spend? **6.NS.4, MP 1**

Item	Number
Pencils	48
Pens	32
Erasers	60
Notepads	36

Ⓐ $8 Ⓒ $24 (filled in)
Ⓑ $16 Ⓓ $120

Use a problem-solving model to solve this problem.

1 Understand

Read the problem. Circle the information you know.
Underline what the problem is asking you to find.

2 Plan

What will you need to do to solve the problem? Write your plan in steps.

Step 1 Find the common factors of the number of pencils, the number of erasers, and the number of notepads.

Step 2 Find the greatest common factor of these factors.

Step 3 Determine the greatest amount of money he would spend.

Read to Succeed! Jin is not putting pens in the bags. So, the number of pens is not needed to solve the problem.

3 Solve

Use your plan to solve the problem. Show your steps.

factors of 48: **1, 2, 3, 4, 6, 8, 12, 16, 24, 48**

factors of 60: **1, 2, 3, 4, 5, 6, 10, 12, 15, 20, 30, 60**

factors of 36: **1, 2, 3, 4, 6, 9, 12, 18, 36**

The common factors are **1, 2, 3, 4, 6, and 12**.

The greatest common factor is **12**. The greatest number of bags he can fill is **12**. The most amount of money he would spend is **$24**.

So, the correct answer is **C**. Fill in that answer choice.

4 Check

How do you know your solution is reasonable?

Sample answer: The greatest common factor of 48, 60, and 36 is 12. So, 12 is the greatest number of bags Jin can fill with the with each item distributed evenly in the bags. Each bag is $2, so the greatest amount he can spend is $24.

NAME ________________ DATE ________ PERIOD ________

Lesson 1 *(continued)*

Use a problem-solving model to solve each problem.

1 The table shows the number of muffins Ana baked. She wants to put the oatmeal, banana, and blueberry muffins in rows with the same number of each type of muffin. What is the greatest number of muffins in each row? **6.NS.4, MP 2**

Muffins	
Type	Number
Oatmeal	16
Banana	12
Raisin	10
Blueberry	20

Ⓐ 2 muffins
Ⓑ 4 muffins (filled in)
Ⓒ 12 muffins
Ⓓ 20 muffins

2 Maria will use all the glass, wooden, and pottery beads she buys to make a necklace. The table shows the number of beads that come in each pack. Maria wants to use the same number of each type of bead. What is the least number of each type of bead she will buy? The cost per pack is $1. How much will she spend? **6.NS.4, MP 4**

Beads	
Type	Number per Pack
Glass	6
Wooden	3
Pottery	5

30; $21

3 Carlos has three pieces of fabric. The yellow fabric is 45 inches wide, the blue fabric is 36 inches wide, and the red fabric is 27 inches wide. Carlos wants to cut all three pieces into equal width strips that are as wide as possible. How many strips of each color will he have? **6.NS.4, MP 4**

5 yellow strips; 4 blue strips; 3 red strips

4 **H.O.T. Problem** Suppose the LCM of three numbers is 20, the GCF is 2, and the sum of the three numbers is 16. What are the three numbers? **6.NS.4, MP 7**

2, 4, 10

NAME ____________ DATE ________ PERIOD ________

Lesson 2 **Multi-Step** Problem Solving

Chapter 1

Multi-Step Example

The table shows the types of sandwiches sold on Friday. What is the denominator when the ratio of veggie sandwiches to the total number of sandwiches is written as a fraction in simplest form? **6.RP.1, MP 1**

Sandwich	Number Sold
Turkey	9
Tuna	11
Veggie	6
Chicken	14

Use a problem-solving model to solve this problem.

1 Understand

Read the problem. Circle the information you know.
Underline what the problem is asking you to find.

2 Plan

What will you need to do to solve the problem?
Write your plan in steps.

Step 1 Determine the total number of sandwiches sold.

Step 2 Express the ratio of veggie sandwiches to total sandwiches as a fraction in simplest form.

Read to Succeed!
The order of the words of the ratio gives the order of the values to use as the numerator and denominator.

3 Solve

Use your plan to solve the problem. Show your steps.

The total number of sandwiches sold is **9** + **11** + **6** + **14**

or **40** sandwiches.

The ratio of veggie to total sandwiches is $\frac{6}{40}$ or $\frac{3}{20}$.

So, the simplified denominator is **20**.

4 Check

How do you know your solution is accurate?

Sample answer: Use models to represent veggie sandwiches and the remaining sandwiches.

NAME ____________ DATE ________ PERIOD ________

Lesson 2 *(continued)*

Use a problem-solving model to solve each problem.

1 The table shows the types of breakfasts sold on Thursday. What is the denominator of the simplified ratio of oatmeal orders to total orders? **6.RP.1, MP 1**

Breakfast	Number Sold
Omelets	14
Pancakes	17
Waffles	11
Oatmeal	8

25

2 How many blue counters must be added so that the ratio of yellow counters to total counters is 1:6? **6.RP.1, MP 4**

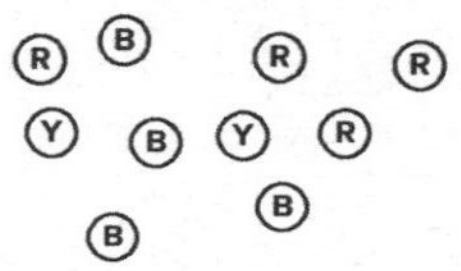

2

3 Cantrise surveyed 100 students about their favorite type of music. After she makes the graph, she receives two more votes for rap. What is the new ratio of rap to total types of music? **6.RP.1, MP 1**

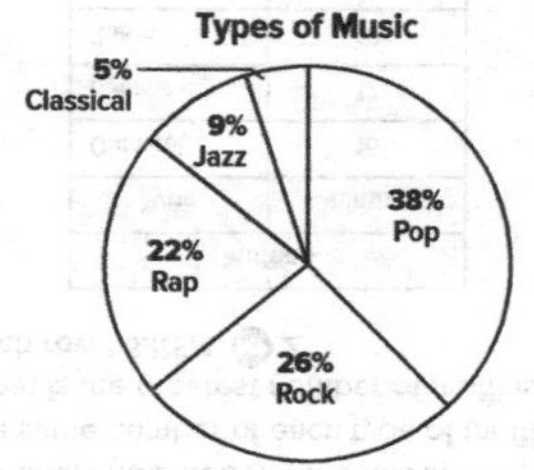

4:17

4 **H.O.T. Problem** The ratio of blue circles to total circles is 4 to 5. There are more than 10 circles. Describe what this group of circles might look like. Explain. **6.RP.1, MP 3**

Sample answer: There could be 4(3) blue circles to 5(3) total circles, or 12 blue circles and 15 circles in all. Multiply 4 and 5 by the same number.

NAME ________ DATE ________ PERIOD ________

Lesson 3 Multi-Step Problem Solving

Multi-Step Example

Zariah is training for a 5-kilometer run, which is about 3 miles. She begins her training by running 1 mile each day for 5 days. She records the number of minutes it takes her to run a mile, as seen in the table. What is her average time in feet per second? **6.RP.2, MP 1**

Day	Time (min)
1	15
2	13
3	16
4	14
5	12

Ⓐ 0.16 ft/sec
Ⓑ 5.87 ft/sec
Ⓒ 6.29 ft/sec
Ⓓ 18.86 ft/sec

Use a problem-solving model to solve this problem.

1 Understand

Read the problem. Circle the information you know. Underline what the problem is asking you to find.

2 Plan

What will you need to do to solve the problem? Write your plan in steps.

Step 1 Determine the average time in minutes. Convert to seconds.

Step 2 Divide the number of feet in a mile by the seconds.

Read to Succeed! There are 5,280 feet in a mile.

3 Solve

Use your plan to solve the problem. Show your steps.

The average time is $\frac{15 + 13 + 16 + 14 + 12}{5}$ or **14** minutes.

14 minutes × 60 = **840** seconds

So, Zariah runs an average of 5,280 ÷ 840 or about **6.29** feet per second.

Choice **C** is correct. Fill in that answer choice.

4 Check

How do you know your solution is accurate?

Sample answer: Convert each time into seconds and find the average number of seconds. Divide the number of feet in a mile, 5,280, by the average number of seconds.

NAME ________ DATE ________ PERIOD ________

Lesson 3 *(continued)*

Use a problem-solving model to solve each problem.

1 Adam records his biking speed for five consecutive days in a table. There are 5,280 feet in a mile. Determine his average rate in feet per minute. Round to the nearest foot per minute. **6.RP.2, MP 1**

Day	Speed (miles per hour)
1	12
2	15
3	13
4	16
5	17

Ⓐ 1,232 ft/min
Ⓑ 1,276 ft/min
Ⓒ 1,285 ft/min
Ⓓ 1,320 ft/min

2 At a grocery store, a 24-pack of 16.9-ounce water bottles is sold for $4.99. There are 128 ounces in a gallon. Determine the price per gallon. Round to the nearest penny. **6.RP.3b, MP 1**

$1.57

3 Carmine is in charge of buying shirts for the senior class. She contacted three companies and recorded their pricing information in the table below. What is the best price per shirt? **6.RP.3b, MP 1**

Company	Cost
A	20 shirts for $38.40
B	25 shirts for $48.75
C	30 shirts for $57.00

Company C; $1.90

4 H.O.T. Problem Using the rate $\frac{y \text{ yards}}{m \text{ minutes}}$, predict what will happen to the value of the ratio for each scenario in the table below. Explain your reasoning. **6.RP.3, MP 7**

y	*m*	value of ratio
increases	unchanged	increases
unchanged	increases	decreases
decreases	unchanged	decreases
unchanged	decreases	increases

Sample answer: When the number of yards increases for the same number of minutes, the value of the ratio is increasing. Likewise, when it takes less time to travel the same distance, the rate is increasing.

When it takes more time to travel the same distance, the rate is decreasing. Likewise, when the number of yards decreases while the time stays the same, the rate is decreasing.

NAME ____________ DATE ____________ PERIOD ____________

Lesson 4 **Multi-Step** Problem Solving

Chapter 1

Multi-Step Example

Marybeth makes money babysitting. She charges a flat rate of \$5 plus \$5 per hour for each child. The rule $5(n) + 5$ can be used to calculate her fee per hour, where n is the number of children. Complete the table to find the ratios of the number of children to the fee. Write each ratio in simplest form. **6.RP.3, MP 1**

Number of children	1	2	3	4	5
Babysitting fee per hour (\$)					

Ⓐ (filled) $\frac{1}{10}, \frac{2}{15}, \frac{3}{20}, \frac{4}{25}, \frac{1}{6}$

Ⓑ $\frac{1}{11}, \frac{1}{6}, \frac{3}{13}, \frac{2}{7}, \frac{1}{3}$

Ⓒ $\frac{10}{1}, \frac{15}{2}, \frac{20}{3}, \frac{25}{4}, \frac{6}{1}$

Ⓓ $\frac{1}{10}, \frac{2}{15}, \frac{3}{20}, \frac{4}{25}, \frac{1}{30}$

Use a problem-solving model to solve this problem.

1 Understand

Read the problem. Circle the information you know.
Underline what the problem is asking you to find.

2 Plan

What will you need to do to solve the problem? Write your plan in steps.

Step 1 Use the rule to determine each fee.

Step 2 Express each ratio in simplest form.

Read to Succeed! Remember to multiply first then add 5.

3 Solve

Use your plan to solve the problem. Show your steps.

Complete the table.

Number of children	1	2	3	4	5
Babysitting fee per hour (\$)	**10**	**15**	**20**	**25**	**30**

The number of children is the numerator and the fee is the denominator. Simplify if necessary.

The ratios are $\frac{1}{10}, \frac{2}{15}, \frac{3}{20}, \frac{4}{25}, \frac{1}{6}$.

Choice **A** is correct. Fill in that answer choice.

4 Check

How do you know your solution is accurate?

Sample answer: The fee increases by \$5 for each child included.

NAME ____________ DATE ____________ PERIOD ____________

Lesson 4 *(continued)*

Use a problem-solving model to solve each problem.

1 A rental car company charges \$0.25 a mile. Use a table like the one below to find the ratios of the cost to the number of miles in simplest form. **6.RP.3, MP 1**

Cost (\$)				
Number of Miles	5	10	15	20

Ⓐ $\frac{5}{4}, \frac{5}{2}, \frac{15}{4}, \frac{5}{1}$

Ⓑ $\frac{4}{1}, \frac{4}{1}, \frac{4}{1}, \frac{4}{1}$

Ⓒ (filled) $\frac{1}{4}, \frac{1}{4}, \frac{1}{4}, \frac{1}{4}$

Ⓓ $\frac{1}{25}, \frac{2}{5}, \frac{3}{75}, \frac{5}{1}$

2 The tables below show the swimming speeds of the King penguin and the Emperor penguin. Determine which table shows an equivalent ratio. Write the equivalent ratio in miles per hour rounded to the hundredths place. **6.RP.3, MP 6**

King Penguin				
Distance (ft)	440	860	1,300	1,740
Time (min)	1	2	3	4

Emperor Penguin				
Distance (ft)	412	824	1,236	1,648
Time (min)	1	2	3	4

Emperor penguin; 4.68 mph

3 Joshua is trying to determine the number of pizzas to order. The table below shows the number of people pizzas will feed. If each pizza costs \$7, how much will it cost to feed 36 people? **6.RP.3, MP 1**

Number of Pizzas	5	7	9
Number of People	15	21	27

\$84

4 H.O.T. Problem On the blueprint below, 1 inch represents 2.5 feet of the house. Complete a table that shows the corresponding number of feet of the house for each length on the blueprint. **6.RP.3a, MP 2**

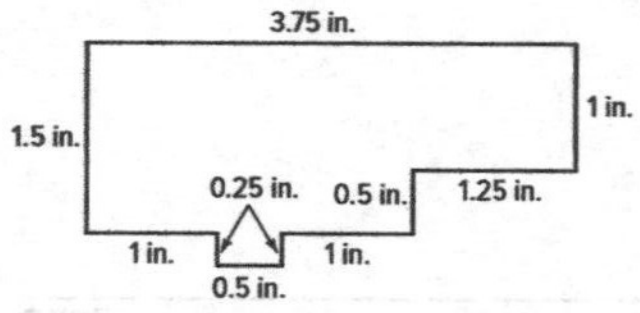

Blueprint (in.)	0.25	0.5	1.5	1.25	3.75
House (ft)	0.625	1.25	3.75	3.125	9.375

NAME________________ DATE ________ PERIOD ________

Lesson 5 Multi-Step Problem Solving

Chapter 1

Multi-Step Example

The height of a tree over time is an equivalent ratio. Which statement best describes the graph of the ratio table? **6.RP.3a, MP 1**

Time (yr)	Height (ft)
3	4.2
5	7
8	11.2

Ⓐ For every one unit right, the line goes up three units.

Ⓑ The line appears to pass through (1, 2).

Ⓒ The line decreases from left to right.

Ⓓ The line appears to pass through the origin.

Use a problem-solving model to solve this problem.

1 Understand

Read the problem. Circle the information you know. Underline what the problem is asking you to find.

2 Plan

What will you need to do to solve the problem? Write your plan in steps.

Step 1 Graph the ordered pairs.

Step 2 Determine which statement is true based on the graph.

Read to Succeed!
When graphing ordered pairs, the x-coordinate tells how far to the right or left. The y-coordinate tells how far up or down.

3 Solve

Use your plan to solve the problem. Show your steps.

Graph the ordered pairs.

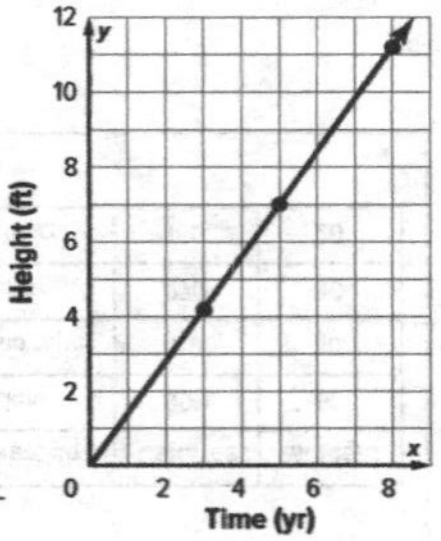

The line appears to increase from left to right, so C is incorrect. The line also does not pass through (1, 2), so B is incorrect. For every unit right, the line goes up 1.4 units, so A is incorrect.

Since the line appears to pass through the origin, choice **D** is correct. Fill in that answer choice.

4 Check

How do you know your solution is accurate?

Sample answer: Determine the unit rate, write a rule, and check each statement based on the rule.

NAME________________ DATE ________ PERIOD ________

Lesson 5 *(continued)*

Use a problem-solving model to solve each problem.

1 The number of bricks is an equivalent ratio to the height of a wall. Which statement describes the pattern in the graph of the ordered pairs? **6.RP.3a, MP 1**

Number of Bricks	Height of Wall (ft)
48	4
84	7
108	9

Ⓐ As the height of the wall increases by 1 foot, there are 12 more bricks in the wall.

Ⓑ For every 48 bricks, the wall increases by 3 feet.

Ⓒ For every 36 bricks, the wall increases by 4 feet.

Ⓓ As the height of the wall increases by 2 feet, there are 48 more bricks in the wall.

2 The perimeter of an equilateral triangle is an equivalent ratio to the length of the sides. An equilateral triangle with a side length of 3 cm has a perimeter of 9 cm, and an equilateral triangle with a side length of 5 cm has a perimeter of 15 cm. What is the perimeter of a triangle with a side length of 7 cm? **6.RP.3, MP 8**

21 centimeters

3 Five yards of material is needed to make 2 curtains. This represents an equivalent ratio. If the equivalent ratio (curtains, material) were graphed, what would the *y*-coordinate be if the *x*-coordinate was 7? **6.RP.3a, MP 4**

17.5

4 **H.O.T. Problem** The table below shows the hours and wages for two friends who babysit. Use a graph to determine if they make the same hourly rate. Explain why or why not. **6.RP.3a, MP 3**

	Zoe	Felix
Hours	3	5
Wages	$21	$30

no; Sample answer: The line drawn showing Zoe's wages is steeper than the line drawn showing Felix's wages.

Answers

NAME________________ DATE ________ PERIOD ________

Lesson 6 **Multi-Step** Problem Solving

Multi-Step Example

Luna wants to burn as many Calories as possible per minute of exercise. Which exercise should Luna choose? **6.RP.3, MP 1**

Exercise	Calories	Minutes
Walking	300	60
Jump rope	110	10
Biking	270	30
Aerobics	160	20

Ⓐ walking
Ⓑ jump rope (filled in)
Ⓒ biking
Ⓓ aerobics

Use a problem-solving model to solve this problem.

1 Understand

Read the problem. Circle the information you know.
Underline what the problem is asking you to find.

2 Plan

What will you need to do to solve the problem? Write your plan in steps.

Step 1 Determine the unit rate for each exercise.

Step 2 Compare the unit rates.

Read to Succeed!
Unit rates have a denominator of 1 when simplified.

3 Solve

Use your plan to solve the problem. Show your steps.

Walking: $\frac{300}{60} = \frac{5}{1}$ Jump rope: $\frac{110}{10} = \frac{11}{1}$

Biking: $\frac{270}{30} = \frac{9}{1}$ Aerobics: $\frac{160}{20} = \frac{8}{1}$

The unit rate **11** Calories per minute is the greatest.

So, Luna should **jump rope**. Choice **B** is correct. Fill in that answer choice.

4 Check

How do you know your solution is accurate?

Sample answer: Divide the number of Calories by the number of minutes using a calculator.

NAME________________ DATE ________ PERIOD ________

Lesson 6 *(continued)*

Use a problem-solving model to solve each problem.

1 Santiago needs to buy apples to make applesauce. He is looking for a better price than $1.29 per pound. Which store has a better price per pound? **6.RP.3, MP 4**

Store	Price ($)	Weight (lb)
Store A	4.08	3
Store B	5.04	4
Store C	12.88	7
Store D	7.35	5

Ⓐ Store A
Ⓑ Store B (filled in)
Ⓒ Store C
Ⓓ Store D

2 Josh is making a scale model of his grandfather's airplane. Given the information in the diagram, how many inches is the wingspan of the model? **6.RP.3, MP 2**

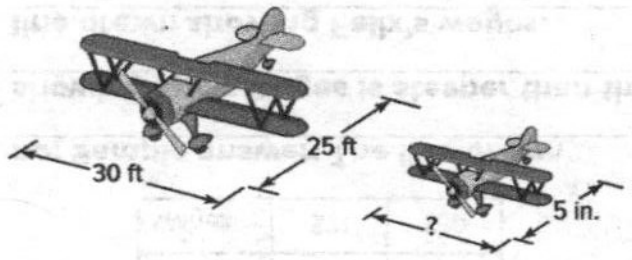

6 in.

3 Riley and Magdalena bought beads to make a necklace. The beads that Riley bought cost 48¢ for 12. Magdalena bought 20 beads for $1.00. How much less do 15 of Riley's beads cost than 15 of Magdalena's beads? **6.RP.3b, MP 4**

15 cents

4 **H.O.T. Problem** Jacob bought 5 pencils for 80¢. At this same rate, how many pencils can he buy for 40¢? Explain why this answer does not make sense. **6.RP.3b, MP 3**

2.5 pencils; This does not make sense because you cannot buy a half of a pencil.

NAME________________ DATE__________ PERIOD__________

Lesson 7 Multi-Step Problem Solving

Chapter 1

Multi-Step Example

The table shows the results of a survey of a group of people about their favorite animal. If 600 people were asked about their favorite animal, predict how many more people would prefer dogs than cats. 6.RP.3, MP 1

Favorite Animal	
Animal	Number of Responses
Bird	5
Cat	13
Dog	19
Iguana	3

Ⓐ 6 Ⓒ 195
Ⓑ 285 Ⓓ 90

Use a problem-solving model to solve this problem.

1 Understand

Read the problem. Circle the information you know.
Underline what the problem is asking you to find.

2 Plan

What will you need to do to solve the problem? Write your plan in steps.

Step 1 Determine the total number of people surveyed.

Step 2 Set up equivalent ratios.

Step 3 Subtract to determine how much more.

Read to Succeed!
When setting up equivalent ratios, remember to keep the same attribute in the numerators and the same in the denominators.

3 Solve

Use your plan to solve the problem. Show your steps.

There were 5 + 13 + 19 + 3 or 40 people surveyed.

Dogs: $\frac{19}{40} = \frac{?}{600} = \boxed{285}$ Cats: $\frac{13}{40} = \frac{?}{600} = \boxed{195}$

So, 285 − 195 or $\boxed{90}$ more people chose dogs than cats. Choice __D__ is correct.
Fill in that answer choice.

4 Check

How do you know your solution is accurate?

Sample answer: There were 6 out of 40 more people that like dogs versus cats. $\frac{6}{40} = \frac{90}{600}$

NAME________________ DATE__________ PERIOD__________

Lesson 7 *(continued)*

Use a problem-solving model to solve each problem.

1 The table shows the results of a survey of a group of people about their favorite sport. If 100 people were asked about their favorite sport, predict how many more people would prefer hockey than volleyball. **6.RP.3,** MP 1

Favorite Sport	
Sport	Number of Responses
Baseball	7
Soccer	10
Volleyball	5
Hockey	8

Ⓐ 3
Ⓑ 10 (filled)
Ⓒ 30
Ⓓ 70

2 Keshia rides her bike 10 miles per hour. At this rate, how many more minutes will it take her to ride 30 miles than 25 miles? **6.RP.3b,** MP 2

30 min

3 Marisol pays $12 for 4 notebooks. How many notebooks could she buy with $62? **6.RP.3b,** MP 2

20

4 **H.O.T. Problem** Jamal is helping a friend with her homework. Look at his friend's work to determine and explain the error she made finding the answer to the problem. **6.RP.3,** MP 3

If 3 dogs eat 4 pounds of food per day, how many dogs eat 15 pounds of food?

$\frac{3}{4} = \frac{15}{?}$

Multiply numerator and denominator by 5. Therefore, 20 dogs eat 15 pounds of food.

Jamal's friend has the second fraction wrong. The 15 should be in the denominator. Both numerators should be the number of dogs and the denominators should be the number of pounds of food.

NAME________________ DATE__________ PERIOD__________

Lesson 1 Multi-Step Problem Solving

Multi-Step Example

The frequency table shows the favorite lunch of some sixth graders. What decimal represents the part of these students that chose pizza or burgers?
Preparation for **6.RP.3c,** MP **1**

Food	Tally	Frequency
Tacos	|||	3
Burgers	𝍸 |	6
Chicken	𝍸	5
Pizza	𝍸 𝍸 |	11

Use a problem-solving model to solve this problem.

1 Understand

Read the problem. Circle the information you know. Underline what the problem is asking you to find.

Read to Succeed!
Use a pencil to write your answer in the boxes at the top of the grid and to fill in the correct bubble(s) of each digit in your answer below.

2 Plan

What will you need to do to solve the problem? Write your plan in steps.

Step 1 Determine the total number of students surveyed.

Step 2 Determine the number of students that chose **pizza** or **burgers**.

Step 3 **Divide** to find the decimal.

3 Solve

Use your plan to solve the problem. Show your steps.

There were 3 + 6 + 5 + 11 or **25** students surveyed.

There were 6 + 11 or **17** students that chose pizza or burgers.

17 ÷ **25** = **0.68**

So, **0.68** of the students surveyed chose pizza or burgers.

4 Check

How do you know your solution is accurate?

Sample answer: Count the tally marks for pizza and burgers and divide by the total number of tallies.

NAME________________ DATE__________ PERIOD__________

Lesson 1 *(continued)*

Use a problem-solving model to solve each problem.

1 The graph shows how Bianca spends her weekly allowance. What decimal represents the part of Bianca's allowance that she spends on entertainment and clothes?
Preparation for **6.RP.3c,** MP **1**

Allowance

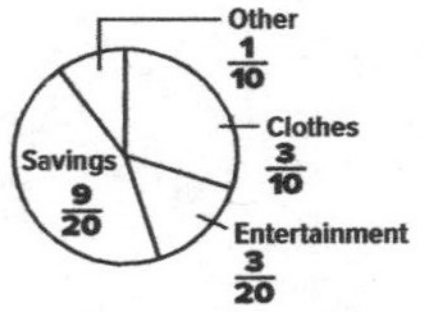

0.45

2 Renee's goal is to run for at least $\frac{1}{4}$ mile more than she ran the previous day. On Day 1, she ran 0.75 mile. The table shows the distances she ran for the next five days. On which of these days did she NOT reach her goal? *Preparation for* **6.RP.3c,** MP **2**

Day	Distance (mi)
2	$1\frac{1}{10}$
3	1.35
4	$1\frac{1}{2}$
5	$1\frac{4}{5}$
6	2.05

4

3 The table shows the number of free throws that Raj successfully made and the number of his attempts over four days.

Day	Shots Made	Attempts
Monday	21	30
Tuesday	18	25
Wednesday	20	32
Thursday	18	24

On which day was the fraction of shots made to attempts the greatest? Express the fraction as a decimal.
Preparation for **6.RP.3c,** MP **2**

Thursday, 0.75

4 **H.O.T. Problem** Write a fraction that is between 0.1 and $\frac{1}{5}$, has a whole-number numerator, and has a denominator of 100. Then express an equivalent decimal form for this fraction. *Preparation for* **6.RP.3c,** MP **6**

Sample answer: $\frac{11}{100}$ and 0.11

NAME________________ DATE ________ PERIOD ________

Lesson 2 Multi-Step Problem Solving

Multi-Step Example

The table shows the percent of time Allison spent studying each of her school subjects last week. What fraction of the subjects studied were math or history?
Preparation for **6.RP.3c,** MP 1

Subject	Time Spent Studying (% of week)
Math	30
Science	10
Language Arts	15
History	10
Reading	20
Music	15

Ⓐ $\frac{1}{10}$ Ⓒ $\frac{3}{10}$

Ⓑ $\frac{2}{5}$ Ⓓ $\frac{4}{5}$

Use a problem-solving model to solve this problem.

1 Understand

Read the problem. Circle the information you know.
Underline what the problem is asking you to find.

2 Plan

What will you need to do to solve the problem? Write your plan in steps.

Step 1 Determine the total percent for both math and history.

Step 2 Express the percent as a fraction in simplest form.

Read to Succeed!
Including Math or History means both subjects need to be included in the studying time.

3 Solve

Use your plan to solve the problem. Show your steps.

Math: **30** %

History: **10** %

Total percentage spent on Math and History: **40** %

So, the simplified fraction is **40** % or $\frac{2}{5}$. So, choice **B** is correct.

Fill in that answer choice.

4 Check

How do you know your solution is accurate?

Sample answer: Convert 30% to a fraction, $\frac{3}{10}$, and 10% to a fraction, $\frac{1}{10}$.
The total is $\frac{3}{10} + \frac{1}{10} = \frac{4}{10}$ or $\frac{2}{5}$.

NAME________________ DATE ________ PERIOD ________

Lesson 2 *(continued)*

Use a problem-solving model to solve each problem.

1 The table shows the percent of each type of car rented last month. What fraction of the rentals were for a sedan or a truck?
Preparation for **6.RP.3c,** MP 1

Type of Car	Percent Rented
Minivan	13
Sport utility	37
Sedan	9
Convertible	4
Sports car	6
Truck	31

Ⓐ $\frac{9}{100}$ Ⓒ $\frac{3}{20}$

Ⓑ $\frac{1}{10}$ **Ⓓ** $\frac{2}{5}$

2 Benito had 10 days of vacation. He spent $\frac{1}{5}$ of his vacation fishing. He spent 30% of his vacation at soccer camp. He spent the rest of the time at the beach. What percent of his vacation did Benito spend at the beach?
Preparation for **6.RP.3c,** MP 2

50%

3 Patricia spent 10 hours at the pool last week. She practiced the butterfly for 2 hours, the breaststroke for 5 hours, and the backstroke for 3 hours. This week she only has 5 hours at the pool. She wants to keep the same percentage of time spent on each stroke. How many more hours will she practice the breaststroke than the backstroke?
Preparation for **6.RP.3c,** MP 2

1 hr

4 **H.O.T. Problem** Ramiro's garden is shown below. What percent of the total area of the garden do the cucumbers cover? Explain.
Preparation for **6.RP.3c,** MP 1

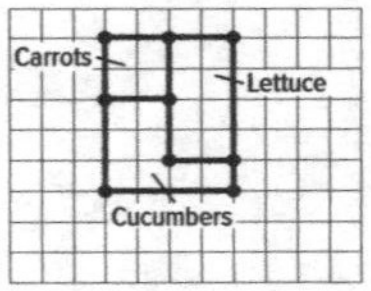

40%; The area of the cucumbers is a 2 × 2 square, which has an area of 4, and a 4 × 1 rectangle, which has an area of 4, for a total of 8 square units. The total area of the 4 × 5 garden is 20 square units.
$\frac{8}{20} = \frac{4}{10} = 40\%$

Answers

NAME ____________ DATE ________ PERIOD ________

Lesson 3 **Multi-Step** Problem Solving

Chapter 2

Multi-Step Example

A yogurt company decreased the amount of yogurt in each container it sells. The new containers contain 15% less yogurt than the original containers. The number line shows the original amount in each container. Which point represents the new amount? *Preparation for* **6.RP.3c,** MP 1

Ⓐ Point *A*
Ⓑ Point *B*
Ⓒ Point *C* (filled)
Ⓓ Point *D*

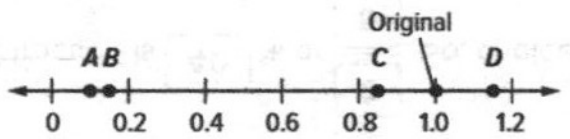

Use a problem-solving model to solve this problem.

1 Understand

Read the problem. Circle the information you know.
Underline what the problem is asking you to find.

2 Plan

What will you need to do to solve the problem? Write your plan in steps.

Step 1 Determine the new percentage of the amount of yogurt.

Step 2 Locate the percent as a decimal on the number line.

3 Solve

Use your plan to solve the problem. Show your steps.

The new percentage of yogurt is 100% − 15% or **85** %.

As a decimal, 85% is **0.85**.

Point **C** is located at 0.85. So, choice **C** is correct. Fill in that answer choice.

Read to Succeed!
To express a percent as a decimal, move the decimal point two places to the left and remove the percent symbol.

4 Check

How do you know your solution is accurate?

Sample answer: Check by using addition. 0.85 + 0.15 = 1.00.

NAME ____________ DATE ________ PERIOD ________

Lesson 3 *(continued)*

Use a problem-solving model to solve each problem.

1 Dexter is tracking his progress in completing math assignments. He has completed 30% of his assignments. What decimal represents the part he has not completed? *Preparation for* **6.RP.3c,** MP 2

Ⓐ 0.03
Ⓑ 0.07
Ⓒ 0.3
Ⓓ 0.7 (filled)

2 Trista deposited money into a savings account and left it there for several years. She earned a total of 12% interest on her deposit. How much did she earn per dollar? *Preparation for* **6.RP.3c,** MP 2

$0.12

3 The table shows the number of students who earned various grades in English. What percent of the students earned an A or a B? *Preparation for* **6.RP.3c,** MP 2

Grade	Tally	Frequency
A	𝍸 ||||	9
B	𝍸 ||	7
C	|||	3
D	|	1

80%

4 **H.O.T. Problem** Autumn used 0.675 of the battery life on her MP3 player. Her brother borrowed it to play a game and used 0.2 of the total battery life. If her MP3 player shows the battery life that remains as a percentage, what does it show after her brother is finished? *Preparation for* **6.RP.3c,** MP 6

12.5%

NAME________________ DATE ________ PERIOD ________

Lesson 4 **Multi-Step** Problem Solving

Multi-Step Example

The table shows a salesperson's commissions during several consecutive years. If the 2010 commission is considered 100% of expected commissions, what percent of expected commissions is the 2011 value?
Preparation for **6.RP.3c, MP 1**

Use a problem-solving model to solve this problem.

1 Understand

Read the problem. Circle the information you know.
Underline what the problem is asking you to find.

2 Plan

What will you need to do to solve the problem? Write your plan in steps.

Step 1 Locate the amounts for 2010 and 2011.

Step 2 Express the amount of 2011 in relation to 2010 as a fraction and convert to a percent.

Read to Succeed!
The amount for 2010 is 100%. So, $10,000 is 100%.

3 Solve

Use your plan to solve the problem. Show your steps.

2011 ⟶ $\frac{\$12,150}{\$10,000}$ = **1.215** = **121.5** %
2010 ⟶

So, the amount for 2011 is **121.5** % of 2010.

4 Check

How do you know your solution is accurate?

Sample answer: Since the amount for 2011 is greater than the amount for 2010, the percent is greater than 100%.

NAME________________ DATE ________ PERIOD ________

Lesson 4 *(continued)*

Use a problem-solving model to solve each problem.

1 It is recommended that 13-year-old girls get 45 mg of Vitamin C each day. The table shows the Vitamin C content of various foods. What percentage of the recommended daily amount will Adelina receive if she eats all of these foods in one day?
Preparation for **6.RP.3c, MP 2**

Food	Approx. Vitamin C Content (mg)
1 orange	70
1 green pepper	100
1 cup cooked broccoli	100

600%

2 Adam has read $\frac{3}{20}$ of a novel in one week. The table shows the fraction of his reading that he completed each day. What percent of the novel did he read on Wednesday?
Preparation for **6.RP.3c, MP 6**

Day	Fraction of Reading Completed
Monday	$\frac{1}{6}$
Tuesday	$\frac{2}{9}$
Wednesday	$\frac{1}{30}$
Thursday	$\frac{4}{9}$
Friday	$\frac{2}{15}$

0.5%

3 If the volume of Rectangular Prism *A* is 0.01% of the volume of Rectangular Prism *B*, what is the ratio of the number of unit cubes it takes to fill Prism *B* to the number it takes to fill Prism *A*? *Preparation for* **6.RP.3c, MP 7**

10,000:1

4 **H.O.T. Problem** Shina wants to plot the values below as decimals on a histogram. Arrange the parts of the same whole shown in increasing order.
Preparation for **6.RP.3c, MP 7**

0.01% 0.05 0.5% 0.500 50 500% 1%

0.01%, 0.5%, 1%, 0.05, 0.500, 500%, 50

NAME ______ DATE ______ PERIOD ______

Lesson 5 **Multi-Step** Problem Solving

Chapter 2

Multi-Step Example

The table shows the portion of an exercise class that each student spent jogging. Who spent the most time jogging?
Preparation for **6.RP.3c, MP 4**

Ⓐ Kyle
Ⓑ Lin (filled in)
Ⓒ Rabi
Ⓓ Sofia

Exercise Class	
Student	**Portion Spent Jogging**
Kyle	$\frac{3}{8}$
Lin	0.43
Rabi	35%
Sofia	$\frac{2}{5}$

Use a problem-solving model to solve this problem.

Read to Succeed!
It is usually easier to compare fractions, decimals, and percents when all the numbers are written as decimals.

1 Understand

Read the problem. Circle the information you know.
Underline what the problem is asking you to find.

2 Plan

What will you need to do to solve the problem? Write your plan in steps.

Step 1 Express each number as a decimal with the same number of places.

Step 2 Locate and compare the numbers on a number line.

3 Solve

Use your plan to solve the problem. Show your steps.

Write each number as a decimal.

$\frac{3}{8}$ = **0.375** 0.43 = **0.430** 35% = **0.350** $\frac{2}{5}$ = **0.400**

Locate the numbers on a number line.

0.300 0.350 0.400 0.450

From least to greatest, the numbers are 35%, $\frac{3}{8}$, $\frac{2}{5}$, and 0.43. Since **0.43** is the greatest number, Lin spent the most time jogging.

So, choice **B** is the correct answer. Fill in that answer choice.

4 Check

How do you know your solution is reasonable?

Sample answer: The number line helps me know that the answer is reasonable, because I know that on a number line the number farthest to the right is the greatest number.

NAME ______ DATE ______ PERIOD ______

Lesson 5 *(continued)*

Use a problem-solving model to solve each problem.

1 The table shows the middle school band concert attendance by class. Which class has the greatest part of the class attending the concert? *Preparation for* **6.RP.3c, MP 1**

Middle School Band Concert	
Class	**Attendance**
Class A	$\frac{4}{5}$
Class B	0.75
Class C	85%
Class D	$\frac{5}{6}$

Ⓐ Class A
Ⓑ Class B
Ⓒ Class C (filled in)
Ⓓ Class D

2 The after-school activities of the students in Caleb's class are shown in the table. Order the activities from least to greatest.
Preparation for **6.RP.3c, MP 2**

After-School Activities	
Activity	**Portion of Students**
Basketball	0.1
Computers	$\frac{3}{10}$
Gymnastics	$\frac{1}{5}$
Soccer	25%
Other	$\frac{3}{20}$

Basketball, Other, Gymnastics, Soccer, Computers

3 The table shows the portion of math homework that students have completed. Which students have completed more than 75% of their math homework?
Preparation for **6.RP.3c, MP 2**

Math Homework	
Student	**Portion Completed**
Clara	$\frac{5}{8}$
Kyra	68%
Levon	0.85
Noah	$\frac{8}{10}$
Soto	$\frac{3}{4}$

Levon and Noah

4 H.O.T. Problem Write a fraction that is between 0.4 and 60%. Then write a fraction that is between 0.4 and the fraction you wrote. Write all four numbers in order from least to greatest. Explain how you know your answer is correct.
Preparation for **6.RP.3c, MP 3**

Sample answer: $\frac{1}{2}$; $\frac{9}{20}$; 0.4, $\frac{9}{20}$, $\frac{1}{2}$, 60%;

Sample explanation: I can write all the numbers as decimals: 0.4 = 0.40, $\frac{9}{20}$ = 0.45, $\frac{1}{2}$ = 0.50, and 60% = 0.60. Then I can compare the four decimals.

Copyright © McGraw-Hill Education. Permission is granted to reproduce for classroom use.

NAME_______________ DATE _________ PERIOD _________

Lesson 6 **Multi-Step** Problem Solving

Chapter 2

Multi-Step Example

Sabrina takes her car to the car wash and gets the Gold Star service that includes a wash, wax, and interior cleaning. This service costs $51.99, but she must also pay a 6% sales tax. Estimate the total amount Sabrina paid at the car wash. **6.RP.3, MP 1**

Ⓐ $3.00
Ⓑ $47.50
Ⓒ $53.00
Ⓓ $75.00

Use a problem-solving model to solve this problem.

1 Understand

Read the problem. Circle the information you know.
Underline what the problem is asking you to find.

2 Plan

What will you need to do to solve the problem? Write your plan in steps.

Step 1 Determine the benchmark percent and use it to determine the tax.

Step 2 Add to determine the total spent.

3 Solve

Use your plan to solve the problem. Show your steps.

6% is a multiple of **1**%. $51.99 is about $**50**.

Using the benchmark percent, 1% of $50 is $**0.50**.

The total tax is 6 × $0.50 or $**3.00**.

The total spent is about $**50** + $**3** or $**53**.

So, Sabrina will spend about $**53**. Choice **C** is correct. Fill in that answer choice.

Read to Succeed!
By using a benchmark percent, you can mentally estimate the total.

4 Check

How do you know your solution is accurate?

Sample answer: Use a 10 by 10 grid to determine the tax per 1% and multiply by 6 to find the estimated tax. Add to find the total spent.

NAME_______________ DATE _________ PERIOD _________

Lesson 6 *(continued)*

Use a problem-solving model to solve each problem.

1 A sporting goods store purchases a skateboard for $100 and marks the price up by 40%. The store is having a sale where everything is 15% off the sticker price. Estimate the final price of a skateboard. **6.RP.3, MP 1**

Ⓐ $119
Ⓑ $125
Ⓒ $140
Ⓓ $155

2 Emilio buys 3 pizzas, 4 subs, and 8 sodas for a party. The sales tax is 7.5%. What is the minimum number of $20 bills Emilio should pay with? **6.RP.3, MP 6**

Item	Cost ($)
Pizza	10
Sub	5
Soda	1

four $20 bills

3 There were 485 people who went to an amusement park on Monday. Sixty percent of the people wanted to ride the new roller coaster. Twenty-three percent of those people decided not to ride the coaster because the line was too long. About how many people waited in line? **6.RP.3, MP 1**

about 225

4 **H.O.T. Problem** Suppose the area of the rectangle below was increased by 20%. What would be the perimeter of the larger rectangle if the length of 15 feet stayed the same? Explain the steps you used to solve this problem. **6.RP.3, MP 1**

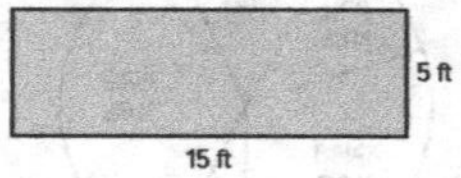

The area of the smaller rectangle is 75 square feet. A 20% increase would make the area 90 square feet. Use the area formula to determine the width, which is 6, since 6 × 15 is 90. Then use the perimeter formula: 2(6) + 2(15) = 42. The perimeter of the larger rectangle is 42 feet.

Answers

NAME______________ DATE ________ PERIOD ________

Lesson 7 **Multi-Step** Problem Solving

Chapter 2

Multi-Step Example

Students were asked which night they planned on attending the book fair. The results of a survey are shown in the table. If 25% of the people who responded with Thursday did not go to the book fair that night, how many people did go on Thursday? **6.RP.3, MP 1**

Day	Number of People
Monday	55
Tuesday	80
Wednesday	70
Thursday	112
Friday	65

Ⓐ 112
Ⓑ 100
Ⓒ 84 (filled in)
Ⓓ 28

Use a problem-solving model to solve this problem.

1 Understand

Read the problem. Circle the information you know. Underline what the problem is asking you to find.

2 Plan

What will you need to do to solve the problem? Write your plan in steps.

Step 1 Determine the number of people that did not go to the book fair on Thursday.

Step 2 Subtract to determine the number of people that did go on Thursday.

3 Solve

Use your plan to solve the problem. Show your steps.

The number of people that did not go to the book fair on Thursday is 25% of 112 or **28**.

The number of people that did go on Thursday is 112 − **28** or **84**.

So, choice **C** is correct. Fill in that answer choice.

Read to Succeed! When finding the percent of a number, rename the percent as a decimal and multiply.

4 Check

How do you know your solution is accurate?

Sample answer: If 25% of the people did not go to the book fair on Thursday, then 75% did go. 75% of 112 is 84.

NAME______________ DATE ________ PERIOD ________

Lesson 7 *(continued)*

Use a problem-solving model to solve each problem.

1 Students were surveyed about their summer plans. Of the people that stated they were traveling abroad, 30% did not actually travel abroad. Of the people that stated they were going to summer camp, 25% did not actually go. How many more students went to summer camp than traveled abroad? **6.RP.3, MP 1**

Summer Plans	Number of People
Summer camp	252
Traveling abroad	180
Visiting grandparent	327

Ⓐ 315
Ⓑ 189
Ⓒ 126
Ⓓ 63 (filled in)

2 Five hundred students were asked what color they prefer for the new school colors. The results are shown in the circle graph. How many students prefer red or black? **6.RP.3, MP 4**

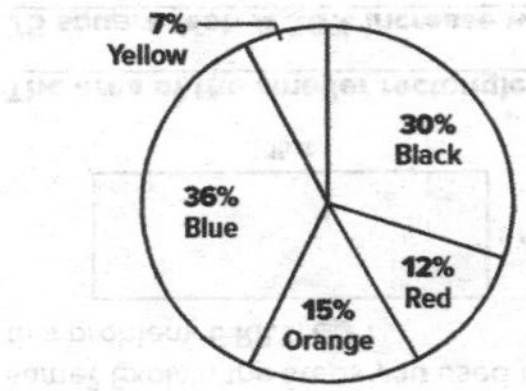

210

3 Carmen is going to buy a pair of sneakers that cost $63. The sales tax rate is 7.5%. What is the total cost of the sneakers to the nearest cent? **6.RP.3, MP 1**

$67.73

4 **H.O.T. Problem** Aaron is estimating the growth of his puppy over time. If the puppy continues to grow at the same rate, how old will the puppy be when it is 250% of its 2-month weight? **6.RP.3, MP 8**

Age (months)	Weight (lb)
2	4
3	5.5

6 months

NAME______________________ DATE ____________ PERIOD ________

Lesson 8 **Multi-Step** Problem Solving

Multi-Step Example

The table shows the percentage of each type of popcorn flavor at a specialty food store. A store clerk put all of the bags of cinnamon popcorn and cheese popcorn in a display in the front of the store. If the clerk put 60 bags up front, how many bags of popcorn does the store have in all? **6.RP.3,** MP 1

Popcorn Flavor	
Kettle corn	60%
Cinnamon	15%
Caramel	10%
Cheese	15%

Ⓐ 18
Ⓑ 100
Ⓒ 200
Ⓓ 400

Use a problem-solving model to solve this problem.

1 Understand

Read the problem. Circle the information you know.
Underline what the problem is asking you to find.

2 Plan

What will you need to do to solve the problem? Write your plan in steps.

Step 1 Determine the total percentage of bags displayed.

Step 2 Use the percent proportion to determine the whole.

3 Solve

Use your plan to solve the problem. Show your steps.

The percentage of bags displayed is 15% + 15% or **30** %.

The situation can be represented by the proportion: $\frac{60}{■} = \frac{30}{100}$.

Since 30 × **2** is 60, multiply 100 × **2**.

So, the total number of bags the store has is 100 × 2 or **200**.

Choice **C** is correct. Fill in that answer choice.

Read to Succeed!
The percent proportion is $\frac{part}{whole} = \frac{percent}{100}$.

4 Check

How do you know your solution is accurate?

Sample answer: 15% of 200 is 30; The store clerk put out 30 + 30 or 60 bags of popcorn.

NAME______________________ DATE ____________ PERIOD ________

Lesson 8 *(continued)*

Use a problem-solving model to solve each problem.

1 The table shows the percentage of each type of puzzle in a toy store. During a sale, the store sold all of the 300-piece and 500-piece puzzles. If they sold 120 puzzles, how many puzzles did the store have before the sale? **6.RP.3,** MP 2

Jigsaw Puzzles	
300-piece	50%
500-piece	30%
750-piece	15%
1,000-piece	5%

Ⓐ 150
Ⓑ 240
Ⓒ 400
Ⓓ 600

2 Miguel surveyed 150 students about their favorite sport. The results are shown in the circle graph. How many more students chose basketball than tennis? **6.RP.3,** MP 4

Favorite Sport

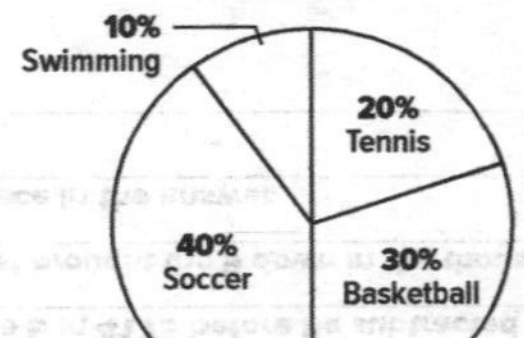

15 students

3 A tile wall is shown below. Each square tile has a side length of 4 inches. What percent of the wall is gray? **6.RP.3,** MP 4

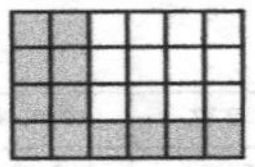

50%

4 **H.O.T. Problem** The perimeter of a square is 30% of the perimeter of a larger square. If the perimeter of the smaller square is 4.8 inches, what is the side length of the larger square? Explain how you found your answer. **6.RP. 3,** MP 7

4 in.; I can write and solve the proportion $\frac{4.8}{■} = \frac{30}{100}$ to determine the perimeter of the larger square is 16 inches. If the perimeter is 16, then a side length is 16 ÷ 4 or 4 inches.

NAME ______ DATE ______ PERIOD ______

Lesson 1 Multi-Step Problem Solving

Multi-Step Example

The table shows the amount of time Jessica spent practicing different strokes in swim class. If the class is 1 hour long, how many minutes did Jessica have left to practice freestyle? **6.NS.3, MP 1**

Jessica's Swim Times	
Swimming Stroke	**Time (min)**
Backstroke	12.5
Breaststroke	13.75
Butterfly	18.1
Freestyle	?

Ⓐ 15.65 min Ⓒ 26.65 min
Ⓑ 24.35 min Ⓓ 44.35 min

Use a problem-solving model to solve this problem.

1 Understand

Read the problem. Circle the information you know.
Underline what the problem is asking you to find.

2 Plan

What will you need to do to solve the problem? Write your plan in steps.

Step 1 Estimate.

Step 2 Line up the decimal points and add.

Step 3 Subtract the sum from the number of minutes in an hour.

Read to Succeed!
When adding and subtracting decimals, use zeros to write equivalent decimals. This way all the decimals have the same number of places and are aligned properly.

3 Solve

Use your plan to solve the problem. Show your steps.

Estimate: 12.5 + 13.75 + 18.1 ≈ 13 + 14 + 18 or **45** and 60 − 45 = **15**

Line up the decimal points and add.

```
  12.50
  13.75
+ 18.10
  44.35
```

Subtract.

```
  60.00
− 44.35
  15.65
```

So, **A** is the correct answer. Fill in that answer choice.

4 Check

How do you know your solution is reasonable?

Sample answer: 15.65 is close to the estimate. So, the answer is reasonable.

NAME ______ DATE ______ PERIOD ______

Lesson 1 *(continued)*

Use a problem-solving model to solve each problem.

1 The table shows the weight of three different-size bags of peanuts sold at the store. How many more ounces of peanuts will you get if you buy 2 large bags than if you buy 2 small bags and 1 medium bag? **6.NS.3, MP 1**

Bags of Peanuts	
Bag Size	**Weight (oz)**
Small bag	12.5
Medium bag	18.25
Large bag	22.8

Ⓐ 2.25 oz
Ⓑ 2.35 oz
Ⓒ 12.35 oz
Ⓓ 18.85 oz

2 Brandon has a board that is 10 feet long. He wants to cut the board into 4 shelves. The lengths of 3 shelves are shown in the table. How long will the fourth shelf be? **6.NS.3, MP 6**

Brandon's Shelves	
Shelf	**Length (ft)**
Shelf A	1.5
Shelf B	2.35
Shelf C	3.85
Shelf D	?

2.3 ft

3 Tia bought crackers for $3.68, two loaves of bread for $3.29 each, a bag of apples for $5.99, and three dozen eggs for $1.65 per dozen. How much more than $20 do these items cost? **6.NS.3, MP 1**

$1.20

4 **H.O.T. Problem** Logan says that 41.75 − 24.689 = 17.079. What might he have done wrong? **6.NS.3, MP 3**

Sample answer: He did not write a 0 after the 5 in 41.75 before he subtracted and just brought the 9 down in the thousandths place in the answer.

NAME ______________ DATE ________ PERIOD ________

Lesson 2 Multi-Step Problem Solving

Chapter 3

Multi-Step Example

David, Nicole, and Tyrell each earn $19.15 an hour working at a science museum. The table shows how many hours they work each week. About how much more does Tyrell earn than Nicole in a week? Estimate to solve the problem. 6.NS.3, MP 1

Employee	Hours Worked
David	18
Nicole	26.5
Tyrell	37.5

Use a problem-solving model to solve this problem.

Read to Succeed! *There are different ways to round numbers. It is easier to multiply numbers that have been rounded to the greatest place value.*

1 Understand

Read the problem. Circle the information you know.
Underline what the problem is asking you to find.

2 Plan

What will you need to do to solve the problem? Write your plan in steps.

Step 1 Determine the numbers needed to solve the problem. Round each number to the greatest place value to make it easier to compute.

Step 2 Multiply the rounded numbers to find about how much each employee earns.

Step 3 Subtract to find the difference.

3 Solve

Use your plan to solve the problem. Show your steps.

Round to the greatest place value. 19.15 ≈ 20 26.5 ≈ **30** 37.5 ≈ **40**

Estimate what Nicole earns. $20 \times 30 =$ **600**
Estimate what Tyrell earns. $20 \times 40 =$ **800**
Find the difference. $800 - 600 =$ **200**

Nicole makes about $600 a week. Tyrell makes about $800 a week. So Tyrell makes about $200 more than Nicole in a week.

4 Check

How do you know your solution is accurate?

Sample answer: I can solve the problem a different way. Tyrell works about 10 hours a week more than Nicole. 10 × 20 = 200, so Tyrell earns about $200 more each week than Nicole.

NAME ______________ DATE ________ PERIOD ________

Lesson 2 *(continued)*

Use a problem-solving model to solve each problem.

1 Irene, Pablo, and Kari each earn $8.85 an hour working at an animal shelter. The table shows how many hours they work each week. About how much less does Kari earn than Irene in a week? Estimate to solve the problem. 6.NS.3, MP 1

Employee	Hours Worked
Irene	32.5
Pablo	15.75
Kari	21.5

Sample answer: about $90

2 The table shows the price per pound of some items at the deli in the grocery store. About how much would 3 pounds of Swiss cheese and 4 pounds of baked ham cost? 6.NS.3, MP 4

Item	Cost per Pound
Baked ham	$6.25
Roast beef	$7.99
Swiss cheese	$4.95
Smoked turkey	$6.49

Sample answer: about $39

3 LaToya wants to buy a backpack for $15.39 and 5 notebooks for $4.25 each. She estimates that the items will cost $35. Is the actual cost more than or less than her estimate? Explain your reasoning. 6.NS.3, MP 3

More than; Sample answer: LaToya rounded the costs down for both the backpack and the notebook. So, the items will actually cost more than her estimate.

4 **H.O.T. Problem** Jason says that if both factors are rounded up when estimating the product of two factors, the estimated product will always be greater than the actual product. Is he correct? Explain your reasoning. 6.NS.3, MP 3

Yes; Sample answer: If each factor is rounded up, each factor will be greater; the product of the two greater factors will also be greater.

Answers

NAME ______ DATE ______ PERIOD ______

Lesson 3 Multi-Step Problem Solving

Multi-Step Example

The table shows the ingredients in a fruit punch recipe. Luis has 3.75 quarts of cranberry juice. How many cups of cranberry juice will he have left after he makes 3 batches of the punch? 6.NS.3, MP 1

Ⓐ 4.75 c
Ⓑ 5.25 c (filled in)
Ⓒ 9.75 c
Ⓓ 15 c

Fruit Punch Recipe

Ingredient	Amount (c)
Cranberry juice	3.25
Lemonade	1.75
Orange juice	2.25
Sparkling water	4.0

Use a problem-solving model to solve this problem.

Read to Succeed!
Hint: There are 4 cups in 1 quart.

1 Understand

Read the problem. Circle the information you know. Underline what the problem is asking you to find.

2 Plan

What will you need to do to solve the problem? Write your plan in steps.

Step 1 Multiply to determine the number of cups in 3 batches of punch.

Step 2 Convert 3.75 quarts to cups.

Step 3 Subtract to determine the number of cups left.

3 Solve

Use your plan to solve the problem. Show your steps.

Multiply. $3.25 \times 3 = \boxed{9.75}$

There are 4 cups in 1 quart. $3.75 \times 4 = \boxed{15.00}$

So, there are 15 cups in 3.75 quarts.

Subtract. $15.00 - 9.75 = \boxed{5.25}$

So, Luis will have 5.25 cups of cranberry juice left.

So, __B__ is the correct answer. Fill in that answer choice.

4 Check

How do you know your solution is reasonable?

Sample answer: Estimate the number of cups of juice Luis needs: 3 × 3 = 9 and the number of cups he has: 4 × 4 = 16. Subtract: 16 − 9 = 7. Compare the actual answer to the estimate.

NAME ______ DATE ______ PERIOD ______

Lesson 3 *(continued)*

Use a problem-solving model to solve each problem.

1 The table shows the ingredients in 1 batch of snack mix. Kenji has 1.5 pounds of raisins. How many more ounces of raisins does he need to make 5 batches of snack mix? (*Hint*: There are 16 ounces in 1 pound.). 6.NS.3, MP 1

Snack Mix Recipe

Ingredient	Weight (oz)
Flaked coconut	3.5
Peanuts	6.8
Raisins	7.5
Toasted oats	4.25

Ⓐ 13.5 oz (filled in)
Ⓑ 21.1 oz
Ⓒ 24 oz
Ⓓ 37.5 oz

2 Rosa had $35.20 in her checking account. She wrote 4 checks for $7.59 each. How much did she have left in her checking account after she wrote the 4 checks? 6.NS.3, MP 6

$4.84

3 The library is having a book sale. All hardcover books cost $3.75 and all paperback books cost $2.15. Shari has $20. How much more money does she need to buy 4 hardcover books and 4 paperback books? 6.NS.3, MP 5

$3.60

4 **H.O.T. Problem** Who solved the problem correctly? Explain your answer. 6.NS.3, MP 3

Matthew	Kendra
175 × 0.05 = 8.75	175 × 0.05 = 0.875

Matthew; Sample answer: There are two decimal places in 0.05, so there should be two decimal places in the product.

NAME_______________ DATE ________ PERIOD ________

Chapter 3

Lesson 4 Multi-Step Problem Solving

Multi-Step Example

Cheyenne buys the amounts of fruit shown in the table. She uses 1.05 pounds of grapes in a salad. How many times more pounds of grapes does she have now than apples? Round to the nearest hundredth. **6.NS.3, MP 1**

Fruit	Pounds
Cherries	3.2
Grapes	4.8
Bananas	3.8
Apples	2.25

Ⓐ 1.47
Ⓑ 1.50
Ⓒ 1.67
Ⓓ 2.13

Use a problem-solving model to solve this problem.

1 Understand

Read the problem. Circle the information you know.
Underline what the problem is asking you to find.

2 Plan

What will you need to do to solve the problem? Write your plan in steps.

Step 1 **Subtract** to find the number of pounds of grapes she has after making the salad.

Step 2 **Divide** to find how many times more pounds of grapes she has than apples.

Read to Succeed!
Divide to determine how many times greater one quantity is of another.

3 Solve

Use your plan to solve the problem. Show your steps.

She has 4.8 – 1.05 or **3.75** pounds of grapes left.

She has **3.75** ÷ 2.25 or **$1.6\overline{6}$** times more pounds of grapes than apples.

Since $1.6\overline{6}$ is about **1.67**, the correct choice is **C**. Fill in that answer choice.

4 Check

How do you know your solution is reasonable?

Sample answer: Use estimation. 5 lb of grapes – 1 lb = 4 lb;

4 lb grapes ÷ 2 lb apples = 2; Since 1.67 is about 2, the answer is reasonable.

NAME_______________ DATE ________ PERIOD ________

Lesson 4 *(continued)*

Use a problem-solving model to solve each problem.

1 The lengths of two wires are shown below. A project uses 32.6 centimeters of Wire B. How many times longer is Wire B than Wire A after part of Wire B is used? (Wires not drawn to scale.) **6.NS.3, MP 2**

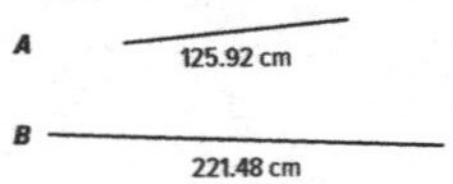

Ⓐ 1.5 (filled)
Ⓑ 2.4
Ⓒ 3.9
Ⓓ 6.8

2 Mika has $9.50. She buys as many bumper stickers as she can afford. Then she earns $2.50 for her allowance. How much money, in dollars, does she have now? **6.NS.3, MP 1**

Item	Cost
Bumper sticker	$1.25
Hat	$6.00
Mug	$5.50

$3.25

3 Elijah has a red ribbon that is 12.2 inches long. He has a green ribbon that is 3.4 inches longer than the red ribbon. He has a yellow ribbon that is 28.08 inches long. How many times longer is the yellow ribbon than the green ribbon? **6.NS.3, MP 4**

1.8

4 **H.O.T. Problem** Determine which quotient is greater, without performing the division. Explain your reasoning. **6.NS.3, MP 7**

8.54 ÷ 5.23 8.54 ÷ 5.32

The quotient of 8.54 ÷ 5.23 is greater.

If you divide by a lesser number, the quotient will be greater.

Answers

NAME ______ DATE ______ PERIOD ______

Lesson 5 Multi-Step Problem Solving

Multi-Step Example

The table shows the number of cookies made for the bake sale. The cookies were put into bags with a dozen cookies in each bag. How many bags of a dozen cookies were there? **6.NS.2, MP 1**

Ⓐ 60 Ⓑ 55 Ⓒ 50 Ⓓ 45

Bake Sale Cookies	
Type	**Number**
Chocolate chip	125
Oatmeal	60
Peanut butter	245
Sugar	116

Use a problem-solving model to solve this problem.

1 Understand

Read the problem. Circle the information you know.
Underline what the problem is asking you to find.

Read to Succeed!
When you divide, use estimation to help you place the first digit in the quotient.

2 Plan

What will you need to do to solve the problem? Write your plan in steps.

Step 1 Add to determine the total number of cookies.

Step 2 Write the division problem. Estimate the quotient.

Step 3 Divide and interpret the quotient.

3 Solve

Use your plan to solve the problem. Show your steps.

Add 125 + 60 + 245 + 116 = **546**

Divide 546 ÷ 12

Estimate 500 ÷ 10 = **50**

Divide 12)546 = **45** R **6**
−48
66
−60
6

There are 45 bags with a dozen cookies in each bag and 6 cookies left over.

So, **D** is the correct answer. Fill in that answer choice.

4 Check

How do you know your solution is reasonable?

Sample answer: I can compare the exact answer with the estimate. Since 45 R6 is close to the estimate of 50, the quotient is reasonable.

NAME ______ DATE ______ PERIOD ______

Lesson 5 *(continued)*

Use a problem-solving model to solve each problem.

1 There are 24 seats in each row of the middle school auditorium. The table shows the number of students from each grade who attended a concert in each grade of the middle school. If the students fill each row in the auditorium, how many rows are needed for all the students? **6.NS.2, MP 1**

Grade	Number of Students
Sixth	310
Seventh	256
Eighth	272

Ⓐ 25 rows
Ⓑ 32 rows
Ⓒ 35 rows
Ⓓ 38 rows

2 Ryland and Charlotte each drove their own cars on 4–day vacations. The table shows the number of miles each person drove each day. Ryland used 46 gallons of gas and Charlotte used 49 gallons of gas. Who got more miles per gallon on his or her trip? How much more? **6.NS.2, MP 6**

Day	Miles Ryland Drove	Miles Charlotte Drove
Thursday	425	157
Friday	312	253
Saturday	175	279
Sunday	330	536

Ryland; 2 miles per gallon more

3 A stadium holds 37,402 people. There are 18,682 empty seats in the stadium. Suppose, the stadium has 36 sections and each section has the same number of people. How many people are in each section of the stadium? **6.NS.2, MP 4**

520 people

4 **H.O.T. Problem** Write and solve a real-world problem that uses division. Make the solution to the problem the remainder in the division problem. **6.NS.2, MP 2**

Sample answer: Chali has 100 apples to divide into 12 baskets. If she divides them evenly, how many apples will she have left over? 4 apples

NAME_______________ DATE _______ PERIOD _______

Lesson 6 **Multi-Step** Problem Solving

Chapter 3

Multi-Step Example

The table shows the average monthly precipitation in Phoenix, Arizona and Atlanta, Georgia during the months of January, February, and March. About how many times more inches of precipitation does Atlanta have than Phoenix during the three months? **6.NS.2, MP 1**

Average Precipitation (in.)		
Month	Phoenix, AZ	Atlanta, GA
January	0.8	5.0
February	0.8	4.7
March	1.1	5.4

Ⓐ about 3 times
Ⓑ about 4 times
Ⓒ about 5 times
Ⓓ about 7 times

Use a problem-solving model to solve this problem.

Read to Succeed! You can use rounding and compatible numbers to help you estimate quotients.

1 Understand

Read the problem. Circle the information you know. Underline what the problem is asking you to find.

2 Plan

What will you need to do to solve the problem? Write your plan in steps.

Step 1 Add to determine the precipitation for each city.

Step 2 Write the division problem.

Step 3 Use rounding to estimate the quotient.

3 Solve

Use your plan to solve the problem. Show your steps.

Add $0.8 + 0.8 + 1.1 = \boxed{2.7}$ $\quad 5.0 + 4.7 + 5.4 = \boxed{15.1}$ $\quad 2.7\overline{)15.1} \rightarrow 3\overline{)15}$, quotient 5

Round 2.7 to 3 and 15.1 to 15 to write compatible numbers.

So, **C** is the correct answer. Fill in that answer choice.

4 Check

How do you know your solution is reasonable?

Sample answer: Using compatible numbers is helpful because I can divide mentally. 15 ÷ 3 = 5, so 15.1 ÷ 2.7 must be about 5.

NAME_______________ DATE _______ PERIOD _______

Lesson 6 *(continued)*

Use a problem-solving model to solve each problem.

1 The table shows the average monthly snowfall on the north rim of the Grand Canyon and Zion National Park during the months of January, February, and March. About how many times more inches of snow fall on the north rim of the Grand Canyon than in Zion National Park during the three months? **6.NS.2, MP 1**

Average Snowfall (in.)		
Month	Grand Canyon	Zion National Park
January	34.8	3.3
February	25.2	1.7
March	28.2	1.1

Ⓐ about 6
Ⓑ about 10
Ⓒ about 12
Ⓓ about 15

2 The table shows the weight of the peanuts, walnuts, pecans, and almonds that Mario has to make snack bags of mixed nuts. He wants to put 8 ounces of mixed nuts in each snack bag. Estimate to determine whether or not Mario has enough mixed nuts for 25 snack bags? Explain why your answer is reasonable. [*Hint:* There are 16 ounces in 1 pound.] **6.NS.2, MP 2**

Type of Nut	Weight (lb)
Peanuts	4.2
Walnuts	2.4
Pecans	0.8
Almonds	2.7

No; 4.2 + 2.4 + 0.8 + 2.7 ≈ 4 + 2 + 1 + 3 = 10; 10 × 16 = 160; 160 ÷ 8 = 20; 20 < 25, so Mario does not have enough mixed nuts for 25 bags.

3 Noor wants to buy a laptop that costs \$698.99. The tax on the laptop will be \$48.93. Noor has saved \$252. She will use the money she has saved to help pay for the laptop. Noor estimates that if she saves \$70 per month for 6 months, she will have enough money to buy the computer. Is her estimate reasonable? Explain your answer. **6.NS.2, MP 3**

Sample answer: No; using compatible numbers: \$700 + \$50 = \$750; \$750 − \$250 = \$500; so, Noor needs about \$500 more. \$70 × 6 = \$420. \$500 > \$420, so she will not have enough money.

4 **H.O.T. Problem** Show two different ways you can use compatible numbers to estimate the quotient of the problem shown. Which pair of compatible numbers do you think has a quotient closer to the actual quotient? Explain why your answer is reasonable. **6.NS.2, MP 7**

377.5 ÷ 23.15

Sample answer: 375 ÷ 25 = 15 and 400 ÷ 20 = 20; the quotient for 375 ÷ 25 is closer to the actual quotient because the compatible numbers are closer to the actual numbers.

Answers

NAME ____________ DATE ____________ PERIOD ____________

Lesson 7 **Multi-Step** Problem Solving

Multi-Step Example

The table shows the price for cheese pizzas at the Pizza Parlor. Each added topping is $0.75. Eight friends are sharing 2 small pizzas with mushrooms and 1 large pizza with sausage and peppers. If the friends share the cost equally, how much should each friend pay for the pizzas? **6.NS.3, MP 1**

Pizza Parlor Prices	
Size	**Price**
Small	$8.50
Medium	$10.25
Large	$11.60

Ⓐ $3.45
Ⓑ $3.95 (filled in)
Ⓒ $4.45
Ⓓ $4.95

Use a problem-solving model to solve this problem.

Read to Succeed! When dividing with decimals, you can use the estimate to help you place the decimal point in the correct location in the quotient.

1 Understand

Read the problem. Circle the information you know. Underline what the problem is asking you to find.

2 Plan

What will you need to do to solve the problem? Write your plan in steps.

Step 1 Find the total cost of the pizza.

Step 2 Write the division problem. Estimate the quotient.

Step 3 Divide to find how much each friend should pay.

3 Solve

Use your plan to solve the problem. Show your steps.

Multiply and add.

2 small pizzas → 2($8.50 + $0.75) = **$18.50**

1 large pizza → $11.60 + 2($0.75) = **$13.10**

Total cost → $18.50 + $13.10 = **$31.60**

Divide.
$31.60 ÷ 8
Estimate.
$32 ÷ 8 = **$4**

Divide.

$$\begin{array}{r} 3.95 \\ 8\overline{)31.60} \\ -24 \\ \hline 76 \\ -72 \\ \hline 40 \\ -40 \\ \hline 0 \end{array}$$

So, **B** is the correct answer. Fill in that answer choice.

4 Check

How do you know your solution is reasonable?

Sample answer: The estimate is about $4. The quotient is $3.95. Compared to the estimate, the quotient is reasonable.

NAME ____________ DATE ____________ PERIOD ____________

Lesson 7 *(continued)*

Use a problem-solving model to solve each problem.

1 The table shows the price for some bracelets at Jewelry Gems. Charms can be added to a bracelet for an additional $2.50 each. Maggie buys 2 silver bracelets with 3 charms each and 1 gold bracelet with 4 charms. What is the average price of the bracelets? **6.NS.3, MP 1**

Bracelet Prices	
Type	**Price**
Bronze	$15.95
Silver	$21.75
Gold	$28.25

Ⓐ $12.75
Ⓑ $29.75
Ⓒ $31.75
Ⓓ $32.25 (filled in)

2 The table shows the weight of a Labrador retriever puppy. Was the puppy's average monthly weight gain greater between the ages 2 months and 4 months or between 8 months and 10 months? How much greater? **6.NS.3, MP 1**

Labrador Retriever Puppy	
Age (months)	**Weight (lb)**
2	15.4
4	30.8
6	44.8
8	52.9
10	57.3

between 2 months and 4 months; 5.5 lb greater

3 Waylen painted a house in 15 hours. He was paid $35.75 per hour for painting it. He used some of his earnings to buy four new paintbrushes that each cost the same amount. If Waylen had $468.69 of his earnings left, how much did each paintbrush cost? **6.NS.3, MP 2**

$16.89

4 H.O.T. Problem Find the quotient for 124.66 ÷ 23. Then explain how you can use a pattern to find the quotients for these problems. Write the quotients. **6.NS.3, MP 7**

1246.6 ÷ 23 12.466 ÷ 23 1.2466 ÷ 23

5.42; Sample answer: The quotient has the same number of decimal places as the dividend. 54.2, 0.542, 0.0542

NAME________________ DATE__________ PERIOD__________

Lesson 8 Multi-Step Problem Solving

Multi-Step Example

The table shows the cost of produce per pound at a farmer's market. Mr. Gonzalez bought 0.75 pound of pears and 3.5 pounds of plums. What was his change from $10? **6.NS.3, MP 1**

Produce	Cost per Pound
Pears	$0.98
Oranges	$1.29
Carrots	$1.18
Plums	$1.49

Ⓐ $4.05

Ⓑ $5.95

Ⓒ $6.33

Ⓓ $10.50

Use a problem-solving model to solve this problem.

1 Understand

Read the problem. Circle the information you know.
Underline what the problem is asking you to find.

2 Plan

What will you need to do to solve the problem? Write your plan in steps.

Step 1 Determine the total amount spent on pears and the total amount spent on plums.

Step 2 Subtract the amounts spent on pears and plums from $10.00.

3 Solve

Use your plan to solve the problem. Show your steps.

Pears: 0.75 × $ **0.98** = $ **0.735**

Plums: 3.5 × $ **1.49** = $ **5.215**

Total spent: $ **0.735** + $ **5.215** = $ **5.95**

$10.00 − $ **5.95** = $ **4.05**

So, Mr. Gonzalez would receive $ **4.05** in change. The correct answer is **A**.

Read to Succeed! When rounding, wait to round until the end of the problem.

4 Check

How do you know your solution is accurate?

Sample answer: Estimate the cost of pears and the cost of plums. Subtract the estimate from $10.

NAME________________ DATE__________ PERIOD__________

Lesson 8 *(continued)*

Use a problem-solving model to solve each problem.

1 The table shows the cost per yard of different types of fabric. Sierra bought 2.5 yards of nylon and 4.5 yards of cotton. What is the change from $5? Round to the nearest cent. **6.NS.3, MP 1**

Fabric	Cost per Yard
Cotton	$0.50
Linen	$0.25
Rayon	$0.125
Nylon	$0.75

Ⓐ $0.87

Ⓑ $2.00

Ⓒ $2.25

Ⓓ $4.13

2 The rectangle represents Demarco's garden. For each square foot, he needs to use 2 scoops of fertilizer. How many scoops does he use? **6.NS.3, MP 2**

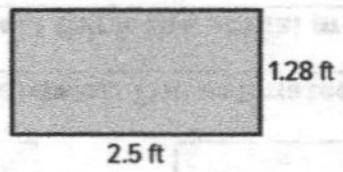

6.4

3 Abigail runs at a constant rate of 7.057 miles per hour. Jamal runs at a constant rate of 6.4 miles per hour. At these rates, how much farther did Abigail run than Jamal after the first 0.5 hour? Round your answer to the nearest hundredth. (*Hint*: Distance = rate × time) **6.NS.3, MP 1**

0.33

4 H.O.T. Problem A name brand cereal is on sale and costs $4.50 for an 18.2-oz box. The grocery store version of the cereal costs $4.03 for a 13.1-oz box. Which cereal costs less per ounce? **6.NS.3, MP 4**

Name brand cereal: $0.25 per ounce; Grocery store cereal: $0.31 per ounce; The name brand cereal is a better buy.

NAME ________ DATE ________ PERIOD ________

Lesson 1 Multi-Step Problem Solving

Multi-Step Example

Jake made a drawing of the vegetable garden he wants to plant. Estimate the area of the vegetable garden.
Preparation for **6.NS.1**, MP 1

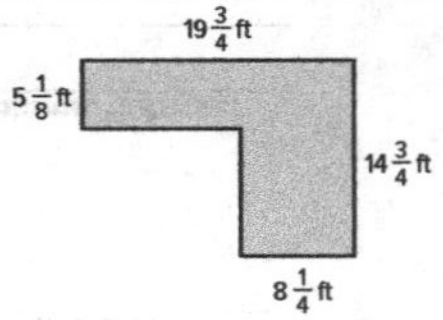

Ⓐ about 120 ft² Ⓒ about 200 ft²
Ⓑ about 180 ft² Ⓓ about 300 ft²

Use a problem-solving model to solve this problem.

1 Understand

Read the problem. Circle the information you know.
Underline what the problem is asking you to find.

Read to Succeed! To find the area of a rectangle, multiply length by width.

2 Plan

What will you need to do to solve the problem? Write your plan in steps.

Step 1 Separate the garden into two rectangles. Round each mixed number to the nearest whole number. Subtract to find the estimated width of the smaller rectangle.

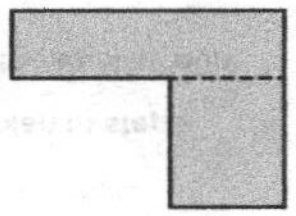

Step 2 Multiply to find the area of each rectangle.

Step 3 Add the two areas to find the estimated area of the garden.

3 Solve

Use your plan to solve the problem. Show your steps.

Round: $19\frac{3}{4} \rightarrow$ **20** $5\frac{1}{8} \rightarrow$ **5** $14\frac{3}{4} - 5 \rightarrow$ **10** $8\frac{1}{4} \rightarrow$ **8**

Multiply: $20 \times 5 = 100$ and $10 \times 8 = 80$ Add: $100 + 80 = 180$
The garden is about 180 square feet.

So, **B** is the correct answer. Fill in that answer choice.

4 Check

How do you know your solution is reasonable?

Sample answer: Solve it a different way. Find the area of the rectangle that is 20 by 15 and then subtract the area of the rectangle that is 12 by 10.

$20 \times 15 = 300$ and $12 \times 10 = 120$. $300 - 120 = 180$.

NAME ________ DATE ________ PERIOD ________

Lesson 1 *(continued)*

Use a problem-solving model to solve each problem.

1 Carolina drew a sketch of the patio she wants to build. Estimate the area of the patio.
Preparation for **6.NS.1**, MP 1

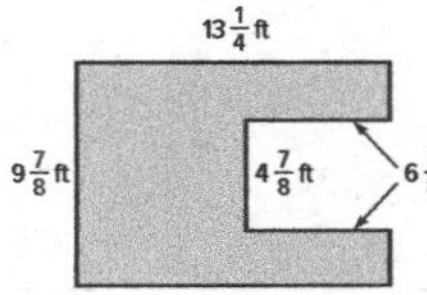

Ⓐ about 100 ft²
Ⓑ about 130 ft²
Ⓒ about 160 ft²
Ⓓ about 190 ft²

2 Layton bought $3\frac{1}{4}$ pounds of grapes. He and his friends ate $\frac{5}{6}$ of the grapes he bought. Jaylinn bought $5\frac{7}{8}$ pounds of grapes. She and her friends ate $\frac{2}{5}$ of the grapes she bought. Estimate to determine who has more grapes left? Explain your answer.
Preparation for **6.NS.1**, MP 4

Jaylinn has more grapes left; Sample answer: Estimate by rounding: $3 \times 1 = 3$ and $6 \times \frac{1}{2} = 3$. They ate about the same number of pounds of grapes. Jaylinn started with more grapes so she has more grapes left.

3 Kuni has a piece of ribbon that is $45\frac{1}{2}$ yards long. Estimate to determine how many more pieces she will have if she cuts the ribbon into $2\frac{7}{8}$-yard strips than if she cuts the ribbon into $4\frac{1}{4}$-yard strips.
Preparation for **6.NS.1**, MP 2

Sample answer: 4 pieces

4 **H.O.T. Problem** Drawings of a square room and a rectangular room are shown. Estimate to determine how the areas of the floors compare. Which room do you think has the greater actual area? Explain how you know. *Preparation for* **6.NS.1**, MP 3

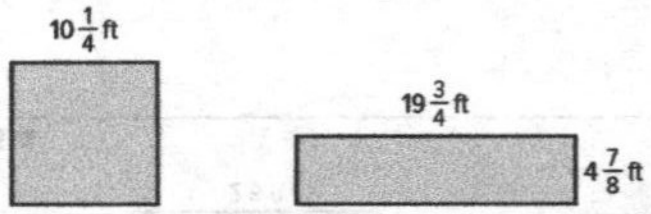

Sample answer: The estimated areas are the same. I think the actual area of the square floor is greater than the actual area of the rectangular floor because I rounded the length down to find the area of the square and I rounded the length and the width up to find the area of the rectangle.

NAME_______________ DATE __________ PERIOD ________

Lesson 2 Multi-Step Problem Solving

Multi-Step Example

Sophia is $\frac{3}{4}$ as tall as Mandy. Alexis is $\frac{5}{6}$ as tall as Mandy. What is the difference in height between Sophia and Alexis?
Preparation for **6.NS.1,** MP 1

Girl	Height (ft)
Sophia	
Mandy	5
Alexis	

Ⓐ $\frac{1}{12}$ foot
Ⓑ $\frac{5}{12}$ foot (filled)
Ⓒ $4\frac{1}{6}$ feet
Ⓓ $3\frac{3}{4}$ feet

Use a problem-solving model to solve this problem.

1 Understand

Read the problem. Circle the information you know.
Underline what the problem is asking you to find.

2 Plan

What will you need to do to solve the problem? Write your plan in steps.

Step 1 Determine the heights of Sophia and Alexis.

Step 2 Subtract to determine the difference between Sophia's height and Alexis's height.

Read to Succeed!
When subtracting mixed numbers, remember to regroup when necessary.
$4\frac{2}{12} = 3\frac{14}{12}$

3 Solve

Use your plan to solve the problem. Show your steps.

Sophia: $\frac{3}{4} \times 5 = \frac{15}{4} = 3\frac{3}{4}$ feet tall

Alexis: $\frac{5}{6} \times 5 = \frac{25}{6} = 4\frac{1}{6}$ feet tall

$4\frac{1}{6} - 3\frac{3}{4} = 4\frac{2}{12} - 3\frac{9}{12}$

$= 3\frac{14}{12} - 3\frac{9}{12}$ or $\frac{5}{12}$

So, Alexis is $\frac{5}{12}$ feet taller than Sophia. The correct answer is **B**.

Fill in that answer choice.

4 Check

How do you know your solution is accurate?

Sample answer: Divide Mandy's height by Sophia's height then by Alexis's height to check your solution to their heights.

Chapter 4

NAME_______________ DATE __________ PERIOD ________

Lesson 2 *(continued)*

Use a problem-solving model to solve each problem.

1 In June, Arturo spent 20 hours walking dogs in his neighborhood. In July, he spent $\frac{4}{5}$ as much time walking dogs. In August, he spent $\frac{9}{10}$ as much time walking dogs as he did in June. How many more hours did Arturo walk dogs in August than July?
Preparation for **6.NS.1,** MP 1

Ⓐ $\frac{1}{10}$ hour
Ⓑ $\frac{18}{25}$ hour
Ⓒ 2 hours (filled)
Ⓓ 16 hours

2 The table shows the number of students in three classes at Hammond Middle School. Of all these students, $\frac{3}{8}$ plan to play in the school band and $\frac{1}{4}$ plan to play sports. How many more students plan to play in the band than play sports?
Preparation for **6.NS.1,** MP 2

Class	Total Students
Ms. Chen	33
Mr. Rice	28
Ms. Lang	35

12

3 During a read-a-thon, 40 students read as many books as they could in one month. The circle graph below shows the fraction of students that read less than 10 books, 10 to 20 books, and more than 20 books. How many more students read more than 20 books versus less than 10 books?
Preparation for **6.NS.1,** MP 4

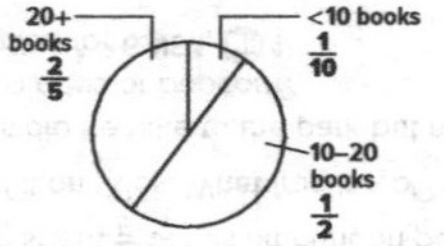

12

4 **H.O.T. Problem** Kisho wants to make ten dozen chocolate chip cookies. He needs $\frac{3}{4}$ cup granulated sugar for the recipe for one dozen cookies. He only has $\frac{1}{4}$ cup of sugar. How many cookies can he make with this amount of sugar? *Preparation for* **6.NS.1,** MP 7

4

Answers

NAME ____________ DATE ________ PERIOD ________

Lesson 3 Multi-Step Problem Solving

Chapter 4

Multi-Step Example

Ella had $\frac{1}{3}$ left of a wall to paint in her bedroom. She painted dots on $\frac{1}{4}$ of what was left to paint and stripes on $\frac{1}{2}$ of it. What is the area of the region that Ella has not yet painted? *Preparation for* **6.NS.1, MP 1**

Ⓐ 88 ft² Ⓒ 8 ft² (filled in)

Ⓑ 16 ft² Ⓓ 1 ft²

Use a problem-solving model to solve this problem.

1 Understand

Read the problem. Circle the information you know.
Underline what the problem is asking you to find.

2 Plan

What will you need to do to solve the problem? Write your plan in steps.

Step 1 Divide the width into thirds. Divide one of the thirds into fourths. Determine the length of the area not painted.

Step 2 Multiply to find the area.

3 Solve

Use your plan to solve the problem. Show your steps.

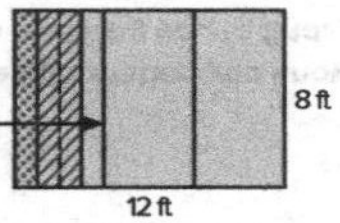

Read to Succeed! The area of a rectangle is found by using the formula $A = l \cdot w$.

So, Ella has [8] square feet left to paint. The correct choice is [C].

Fill in that answer choice.

4 Check

How do you know your solution is accurate?

Sample answer: By dividing the wall into twelfths, I can see that each section is 1 foot wide. The area she has yet to paint is 1 ft × 8 ft or 8 ft².

NAME ____________ DATE ________ PERIOD ________

Lesson 3 *(continued)*

Use a problem-solving model to solve each problem.

1 Denzel earned money after school. He put $\frac{1}{2}$ of this month's earnings into savings. He took the rest to spend at the amusement park. He spent $\frac{1}{5}$ of this amount on popcorn and $\frac{3}{4}$ of it on rides. What fraction of his earnings did he take to the park but not spend on rides or popcorn? *Preparation for* **6.NS.1, MP 1**

Ⓐ $\frac{1}{40}$ (filled in)

Ⓑ $\frac{11}{20}$

Ⓒ $\frac{1}{10}$

Ⓓ $\frac{3}{8}$

2 The table shows how Mura spends her free time on a typical Saturday. If she has 6 hours of free time, how many hours does she spend playing board games or going to the park? *Preparation for* **6.NS.1, MP 6**

Activity	Fraction of Free Time
Board games	$\frac{1}{10}$
Park	$\frac{2}{5}$
Piano	$\frac{3}{7}$
Reading	$\frac{1}{14}$

3

3 Ricardo needs to pave the two rectangular sections shown. Determine the total area that Ricardo needs to pave. *Preparation for* **6.NS.1, MP 4**

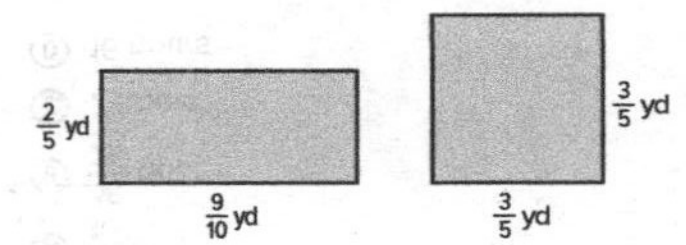

$\frac{18}{25}$ **yd²**

4 **H.O.T. Problem** Without multiplying, determine where the product of $2 \times \frac{12}{7} \times \frac{1}{7}$ is located on the number line. Choose *A*, *B*, or *C*. Justify your reasoning. *Preparation for* **6.NS.1, MP 3**

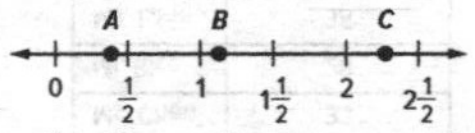

A; $2 \times \frac{1}{7}$ is $\frac{2}{7}$, and $\frac{2}{7}$ is less than $\frac{1}{2}$. $\frac{12}{7}$ is less than 2. So, the product of these two numbers would be less than $\frac{1}{2} \times 2$.

NAME______________ DATE__________ PERIOD__________

Lesson 4 Multi-Step Problem Solving

Multi-Step Example

On Saturday, Ishan rode his bike $5\frac{1}{2}$ miles. Ama rode her bike $1\frac{1}{4}$ times as far as Ishan. Joseph rode his bike $1\frac{2}{5}$ times as far as Ama. How many more miles did Joseph ride than Ama? *Preparation for* **6.NS.1, MP 1**

Ⓐ $1\frac{1}{10}$ mi
Ⓒ $6\frac{7}{8}$ mi
Ⓑ $2\frac{3}{4}$ mi
Ⓓ $9\frac{5}{8}$ mi

Use a problem-solving model to solve this problem.

1 Understand

Read the problem. Circle the information you know.
Underline what the problem is asking you to find.

2 Plan

What will you need to do to solve the problem? Write your plan in steps.

Step 1 Multiply to find the number of miles Ama rode. Then multiply to find the number of miles Joseph rode.

Step 2 Subtract to find how many more miles Joseph rode than Ama.

Read to Succeed!
When you multiply fractions and mixed numbers, remember to simplify your answer.

3 Solve

Use your plan to solve the problem. Show your steps.

Miles Ama rode: $5\frac{1}{2} \times 1\frac{1}{4} = \frac{11}{2} \times \frac{5}{4} = \frac{55}{8} = 6\frac{7}{8}$

Miles Joseph rode: $6\frac{7}{8} \times 1\frac{2}{5} = \frac{55}{8} \times \frac{7}{5} = \frac{77}{8} = 9\frac{5}{8}$

Find the difference: $9\frac{5}{8} - 6\frac{7}{8} = 8\frac{13}{8} - 6\frac{7}{8} = 2\frac{6}{8} = 2\frac{3}{4}$

So, **B** is the correct answer. Fill in that answer choice.

4 Check

How do you know your solution is reasonable?

Sample answer: Estimate the number of miles Ama and Joseph rode and subtract to find the difference. Compare the estimate to the solution.

NAME______________ DATE__________ PERIOD__________

Lesson 4 *(continued)*

Use a problem-solving model to solve each problem.

1 It took Everett $4\frac{3}{4}$ hours to write his science report. Hudson took $2\frac{2}{3}$ times as long as Everett to write his report. Brannon took $1\frac{1}{2}$ times as long as Hudson to write his report. How many more hours did it take Brannon to write his report than Hudson? *Preparation for* **6.NS.1, MP 1**

Ⓐ $1\frac{1}{6}$ hr
Ⓑ $6\frac{1}{3}$ hr
Ⓒ $12\frac{2}{3}$ hr
Ⓓ 19 hr

2 Horaclo's garden is shown below. He needs $1\frac{1}{3}$ scoops of fertilizer for each square foot of the garden. How many scoops of fertilizer does Horaclo need for the entire garden? *Preparation for* **6.NS.1, MP 4**

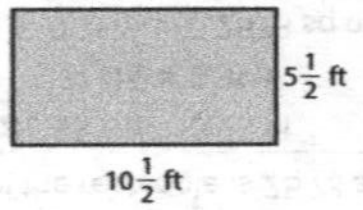

77

3 Jan and Dan work part time. Jan earns \$9.50 an hour. Dan earns \$8.25 an hour. The table shows how many hours they worked on Monday, Wednesday, and Friday. Who earned more money? How much more? *Preparation for* **6.NS.1, MP 2**

Day	Jan's Time (hr)	Dan's Time (hr)
Monday	$3\frac{1}{3}$	$4\frac{2}{5}$
Wednesday	$4\frac{1}{2}$	$2\frac{3}{4}$
Friday	$2\frac{2}{3}$	$4\frac{1}{4}$

Jan; \$5.70

4 H.O.T. Problem Without multiplying, explain which expression has the greater product. *Preparation for* **6.NS.1, MP 3**

$8\frac{1}{2} \times 7\frac{7}{8}$ or $8\frac{1}{2} \times 7\frac{2}{9}$

$\frac{7}{8}$ is greater than $\frac{2}{9}$. So $7\frac{7}{8}$ is greater than $7\frac{2}{9}$. The first factor in each expression is the same, so the expression with the greater second factor will have the greater product. $8\frac{1}{2} \times 7\frac{7}{8} > 8\frac{1}{2} \times 7\frac{2}{9}$

Answers

NAME______________________ DATE ____________ PERIOD ________

Lesson 5 **Multi-Step** Problem Solving

Multi-Step Example

The table shows the amount of water each athlete drinks during soccer practice. How many quarts of water are needed for these five athletes during practice? **6.RP.3, 6.RP.3d, MP 1**

Athlete	Amount (c)
Deon	2
Sierra	1.5
Carmen	3.5
Mia	3
Ella	2

Use a problem-solving model to solve this problem.

1 Understand

Read the problem. Circle the information you know.
Underline what the problem is asking you to find.

2 Plan

What will you need to do to solve the problem? Write your plan in steps.

Step 1 Determine the total number of cups drank during practice.

Step 2 Convert cups to quarts.

Read to Succeed!
There are 2 cups in a pint and 2 pints in a quart.

3 Solve

Use your plan to solve the problem. Show your steps.

The total amount drank is 2 + 1.5 + 3.5 + 3 + 2 or __12__ cups.

__12__ cups = __6__ pints = __3__ quarts

So, the team drank __3__ quarts of water.

4 Check

How do you know your solution is accurate?

Sample answer: Write equivalent ratios to determine the number of quarts.

NAME______________________ DATE ____________ PERIOD ________

Lesson 5 *(continued)*

Use a problem-solving model to solve each problem.

1 William collects metal to sell to a recycling plant. The table shows the amount of metal he has collected over several days. He needs to collect 4 tons before he can take the load to the recycling plant. How many more pounds does he need to reach 4 tons? **6.RP.3, 6.RP.3d, MP 1**

Day	Metal (lb)
Monday	2,500
Tuesday	1,375
Wednesday	2,550
Thursday	1,075

500

2 Joaquin drank 6 glasses of water each containing 10 fluid ounces. His goal was to drink 2 quarts. How many more fluid ounces does he have to drink to reach his goal? **6.RP.3, 6.RP.3d, MP 6**

4

3 A football team needs to travel 80 yards from their current location to their opponent's end zone to score a touchdown. The team is now 6 feet away from their opponent's end zone, ready to score the touchdown. How many feet have they already traveled down the field? **6.RP.3, 6.RP.3d, MP 2**

234

4 **H.O.T. Problem** The dimensions of a rectangle are given in the diagram. Jane wanted to know the area in square meters. She used two different methods to determine the area. Which method is correct, and why? **6.RP.3, 6.RP.3d, MP 3**

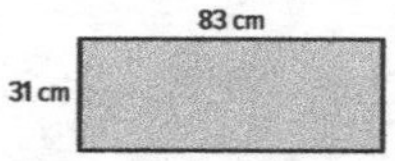

Method 1 83 cm × 31 cm = 2,573 sq cm
2,573 cm = 25.73 m
The area of the rectangle is 25.73 sq m.

Method 2 83 cm = 0.83 m
31 cm = 0.31 m
0.83 m × 0.31 m = 0.2573 sq m
The area of the rectangle is 0.2573 sq m.

Method 2 is correct. In Method 1, the correct number of sq cm was found, but the conversion to sq m is incorrect. When converting square units, the conversion involves two dimensions. So, instead of dividing by 100, you must divide by 10,000 to convert sq cm to sq m.

NAME ______________ DATE ________ PERIOD ________

Lesson 6 Multi-Step Problem Solving

Multi-Step Example

The table shows the ingredients needed to make one batch of salad dressing. A chef has 3 tablespoons of minced garlic. She made the greatest number of batches possible. How many tablespoons of garlic were left? **6.NS.1, MP 1**

Ingredient	Amount
Oil	1 cup
Vinegar	$\frac{3}{4}$ cup
Minced garlic	$\frac{2}{3}$ tbsp

Ⓐ $\frac{1}{2}$ tablespoon
Ⓑ $\frac{1}{3}$ tablespoon (filled)
Ⓒ $\frac{2}{3}$ tablespoon
Ⓓ $\frac{5}{6}$ tablespoon

Use a problem-solving model to solve this problem.

1 Understand

Read the problem. Circle the information you know.
Underline what the problem is asking you to find.

2 Plan

What will you need to do to solve the problem? Write your plan in steps.

Step 1 Divide to determine the number of batches made.

Step 2 Subtract to determine the amount remaining.

Read to Succeed! *When dividing by fractions, multiply by the reciprocal.*

3 Solve

Use your plan to solve the problem. Show your steps.

The chef could make $3 \div \frac{2}{3}$ or $4\frac{1}{2}$ batches.

Since she made 4 full batches, there is $\frac{1}{2}$ batch left.

One-half of a batch uses $\frac{1}{2} \times \frac{2}{3}$ or $\frac{1}{3}$ tablespoon. The correct choice is B.
Fill in that answer choice.

4 Check

How do you know your solution is accurate?

Sample answer: Multiply to find the total used in 4 batches then subtract to determine how much garlic is left.

NAME ______________ DATE ________ PERIOD ________

Lesson 6 *(continued)*

Use a problem-solving model to solve each problem.

1 The table shows the time it takes each person to build a house of cards. If there are 2 hours available to make houses of cards, how many more houses can Fina make than Logan? **6.NS.1, MP 1**

Person	Time (hours)
Jenna	$\frac{1}{4}$
Logan	$\frac{1}{3}$
Fina	$\frac{1}{5}$

Ⓐ 3
Ⓑ 4 (filled)
Ⓒ 5
Ⓓ 6

2 Aria made 9 pounds of fudge. She separates the fudge into $\frac{3}{4}$-pound portions. She sells each portion for $6.50. If she sells all the fudge, how much money will she make? **6.NS.1, MP 2**

$78

3 Robert and Judi are ordering lasagnas for a party. Robert ordered 10 large lasagnas. Each lasagna is cut into tenths. Judi ordered 6 smaller lasagnas. Each lasagna was cut into eighths. How many pieces did they order in all? **6.NS.1, MP 2**

148

4 **H.O.T. Problem** Cadence is making gift bags filled with different colored beads for her jewelry party. She fills the bags using a mixture of $1\frac{1}{2}$ pounds pink beads, $\frac{3}{4}$ pound purple beads, and $1\frac{1}{4}$ pounds green beads. She divides the mixture into 8 packages. How much is in each package? **6.NS.1, MP 4**

$\frac{7}{16}$ pound

Answers

NAME ______ DATE ______ PERIOD ______

Lesson 7 **Multi-Step** Problem Solving

Chapter 4

Multi-Step Example

Alfonso is making snack bags with different types of nuts as shown in the table. Each snack bag contains $\frac{1}{8}$ pound of one type of nut. How many more whole servings of walnuts can he make than peanuts? **6.NS.1, MP 1**

Type of Nut	Weight (lb)
Almonds	$\frac{1}{2}$
Cashews	$\frac{1}{4}$
Peanuts	$\frac{2}{5}$
Walnuts	$\frac{3}{4}$

Ⓐ 1
Ⓑ 2
Ⓒ 3 (shaded)
Ⓓ 6

Use a problem-solving model to solve this problem.

1 Understand

Read the problem. Circle the information you know.
Underline what the problem is asking you to find.

2 Plan

What will you need to do to solve the problem? Write your plan in steps.

Step 1 Divide to determine the number of servings of walnuts and peanuts.

Step 2 Subtract to determine how many more servings of walnuts than peanuts.

3 Solve

Use your plan to solve the problem. Show your steps.

Walnuts: $\frac{3}{4} \div \frac{1}{8} = \frac{3}{4} \cdot \frac{8}{1}$ or **6** servings

Peanuts: $\frac{2}{5} \div \frac{1}{8} = \frac{2}{5} \cdot \frac{8}{1}$ or **3** $\frac{1}{5}$ servings

So, Alfonso made 6 − 3 or **3** more whole servings of walnuts than peanuts. The correct choice is **C**.

Read to Succeed!
The number of whole servings of peanuts is 3 because $\frac{1}{5}$ is not a whole serving.

4 Check

How do you know your solution is accurate?

Sample answer: Estimate the number of servings of walnuts and the number of servings of peanuts. Subtract to find an approximate difference.

NAME ______ DATE ______ PERIOD ______

Lesson 7 *(continued)*

Use a problem-solving model to solve each problem.

1 Anabella is using ribbon to decorate the edge of a picture frame with a length of $\frac{1}{4}$ yard and a width of $\frac{1}{6}$ yard. She will only use one color to decorate the frame. Each color of ribbon is available in lengths as shown in the table. How many more strips of green ribbon than blue ribbon would she need for the frame? **6.NS.1, MP 1**

Color	Strip Length (yd)
Black	$\frac{1}{2}$
Blue	$\frac{2}{3}$
Green	$\frac{1}{6}$

Ⓐ 5
Ⓑ 3 (shaded)
Ⓒ 2
Ⓓ 1

2 Camillo is decorating birthday cards with glitter to send to his friends. The table shows the different colors of glitter that he has. He will mix all these colors together, and then use $\frac{1}{4}$ tube of glitter on each card. How many birthday cards can he decorate? **6.NS.1, MP 4**

Color	Tubes
Red	2
Yellow	$\frac{3}{5}$
Purple	$\frac{3}{8}$
Pink	$\frac{1}{2}$

13

3 Stephanie usually jogs $\frac{3}{4}$ mile every day. She decides that she wants to sprint for a part of this distance. She will jog for $\frac{1}{2}$ of $\frac{3}{4}$ mile and will sprint the rest, but she only sprints $\frac{1}{8}$ mile at a time before resting. How many sprints will Stephanie do each day? **6.NS.1, MP 2**

3

4 **H.O.T. Problem** Without doing any calculations, which expression does not have the same value as $\frac{1}{2} \div \frac{2}{3}$? Explain. **6.NS.1, MP 7**

A	$\frac{1}{2} \div \frac{4}{6}$
B	$\frac{1}{2} \times \frac{3}{2}$
C	$\frac{3}{6} \times \frac{3}{2}$
D	$\frac{3}{6} \div \frac{6}{4}$

D; Sample answer: Dividing by a fraction is the same as multiplying by the reciprocal, so B has the same value. C and A are both the same, just written with equivalent fractions.

NAME_______________ DATE ___________ PERIOD _________

Lesson 8 Multi-Step Problem Solving

Multi-Step Example

The table shows the side lengths of four square mirrors. How many times greater is the area of mirror B than the area of mirror C? **6.NS.1, MP 1**

Mirror	Side Length (ft)
A	$1\frac{1}{4}$
B	$2\frac{1}{2}$
C	$1\frac{3}{4}$
D	$3\frac{1}{6}$

Ⓐ $2\frac{2}{49}$ (filled)
Ⓑ $3\frac{1}{16}$
Ⓒ $3\frac{3}{16}$
Ⓓ $6\frac{1}{4}$

Use a problem-solving model to solve this problem.

1 Understand

Read the problem. Circle the information you know.
Underline what the problem is asking you to find.

2 Plan

What will you need to do to solve the problem? Write your plan in steps.

Step 1 Use the formula $A = \ell \cdot w$ to determine the area of each mirror.

Step 2 Divide to determine how many times greater mirror B is than mirror C.

3 Solve

Use your plan to solve the problem. Show your steps.

Mirror B: $2\frac{1}{2} \cdot 2\frac{1}{2} = \frac{5}{2} \cdot \frac{5}{2} = \frac{25}{4}$ or $6\frac{1}{4}$ square feet

Mirror C: $1\frac{3}{4} \cdot 1\frac{3}{4} = \frac{7}{4} \cdot \frac{7}{4} = \frac{49}{16}$ or $3\frac{1}{16}$ square feet

So, Mirror B is $6\frac{1}{4} \div 3\frac{1}{16}$ or **$2\frac{2}{49}$** times larger than Mirror C.

Choice **A** is correct. Fill in that answer choice.

Read to Succeed!
To determine the area of a square, multiply the length times the width.

4 Check

How do you know your solution is accurate?

Sample answer: Estimate the area of each mirror and divide to find how many times greater.

Chapter 4

NAME_______________ DATE ___________ PERIOD _________

Lesson 8 *(continued)*

Use a problem-solving model to solve each problem.

1 The table shows the dimensions of two fenced-in areas at a dog park. How many times greater is the area enclosed by the wood fence than the area enclosed by the metal fence? **6.NS.1, MP 1**

Fence	Length (yd)	Width (yd)
Wood	$6\frac{1}{2}$	$2\frac{1}{4}$
Metal	$2\frac{3}{4}$	$2\frac{1}{2}$

Ⓐ $1\frac{1}{9}$
Ⓑ $2\frac{7}{55}$ (filled)
Ⓒ $2\frac{4}{11}$
Ⓓ $7\frac{3}{4}$

2 Mylie has $35\frac{3}{4}$ yards of red ribbon and $30\frac{1}{3}$ yards of green ribbon. She cuts the red ribbon into strips that are each $3\frac{1}{4}$ yards long and the green ribbon into strips that are each $2\frac{1}{6}$ yards long. How many more green strips than red strips does she have? **6.NS.1, MP 2**

3

3 On Saturday, Justine studied $1\frac{1}{4}$ times as long as Shantel and $1\frac{3}{4}$ times as long as Nicole. If Justine studied $3\frac{1}{2}$ hours on Saturday, how much longer did Shantel study than Nicole on Saturday? Express your answer as a number of hours in decimal notation. **6.NS.1, MP 2**

0.8

4 **H.O.T. Problem** Without dividing, explain whether $1\frac{1}{2} \div 3\frac{1}{4} \div 2\frac{5}{6}$ is greater or less than $2\frac{5}{6} \div 3\frac{1}{4} \div 1\frac{1}{2}$. **6.NS.1, MP 7**

less than; The first number, $1\frac{1}{2}$, is smaller than the last number, $2\frac{5}{6}$, and therefore is divided into a greater number of parts.

Answers

NAME ______ DATE ______ PERIOD ______

Lesson 1 Multi-Step Problem Solving

Multi-Step Example

Golf scores are measured by the number of strokes over or under par. Scores over par can be represented by a positive integer. Scores under par can be represented by a negative integer. The table shows the golf scores of four players in a golf tournament. Which golfer's score is represented by point *B* on the number line below? **6.NS.6c, MP 1**

Golfer	Score
Chase	2 under par
Augustus	3 over par
Etu	1 under par
Miles	1 over par

Ⓐ Chase
Ⓑ Augustus
Ⓒ Etu (filled)
Ⓓ Miles

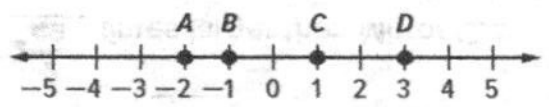

Chapter 5

Use a problem-solving model to solve this problem.

1 Understand

Read the problem. Circle the information you know.
Underline what the problem is asking you to find.

2 Plan

What will you need to do to solve the problem? Write your plan in steps.

Step 1 Determine the number located at point __B__.

Step 2 Determine the corresponding value in the table.

3 Solve

Use your plan to solve the problem. Show your steps.

Point B is located at __−1__. −1 is 1 __under__ par.

The golfer that scored −1 is __Etu__.

So, the correct answer is __C__. Fill in that answer choice.

Read to Succeed!
Numbers to the left of zero on a number line are negative. Numbers to the right are positive.

4 Check

How do you know your solution is accurate?

Sample answer: Looking at Etu's golf score, 1 under par can be represented as −1. On the number line, −1 is located at point B.

NAME ______ DATE ______ PERIOD ______

Lesson 1 *(continued)*

Use a problem-solving model to solve each problem.

1 The table shows the changes in the value of four stocks over one day. Which point on the number line represents the change in value of Stock *R*? **6.NS.6c, MP 1**

Stock	Change in Value
Stock *Q*	Up $2
Stock *R*	Down $3
Stock *S*	Down $1
Stock *T*	Up $3

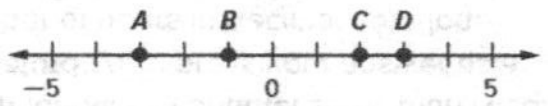

Ⓐ point *A* (filled)
Ⓑ point *B*
Ⓒ point *C*
Ⓓ point *D*

2 Monique is playing a board game where players move about the board using numbered cards. If the card is green, you move forward the number of spaces indicated on the card. If the card is red, you move backward the number of spaces indicated on the card. The number line shows the number of spaces Monique moved in her first five turns. How many red cards did Monique get in her first five turns? **6.NS.6c, MP 3**

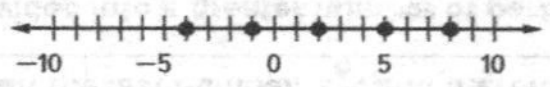

2

3 A football team has four chances, called downs, to gain at least 10 yards. The number line shows the number of yards gained and lost in four downs of a football game.

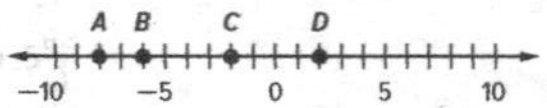

The table shows which downs correspond to the points on the graph. During which down did the team gain 2 yards? **6.NS.6c, MP 3**

Down	Point
1	*B*
2	*C*
3	*A*
4	*D*

4

4 **H.O.T. Problem** The temperature outside is 4°F. The temperature drops 6°F. Between which two points is the location of the temperature after the change? **6.NS.6c, MP 2**

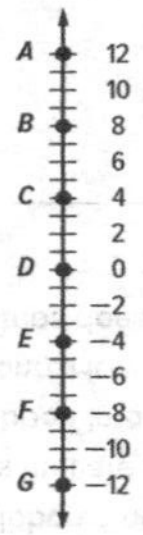

D and E

NAME________________ DATE ________ PERIOD ______

Lesson 2 Multi-Step Problem Solving

Multi-Step Example

The graph shows the freezing and boiling points of water in degrees Fahrenheit. How much greater is the absolute value of the boiling point of water than the absolute value of the freezing point of water? **6.NS.7c, MP 1**

Ⓐ −32
Ⓑ 0
Ⓒ 180 (filled in)
Ⓓ 244

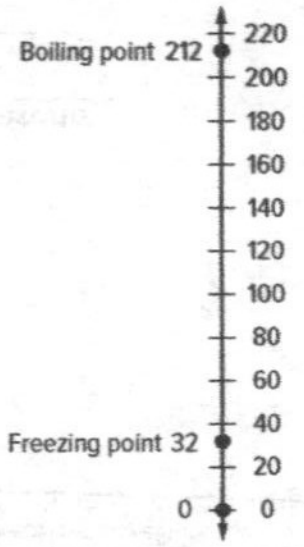

Use a problem-solving model to solve this problem.

1 Understand

Read the problem. Circle the information you know.
Underline what the problem is asking you to find.

2 Plan

What will you need to do to solve the problem? Write your plan in steps.

Step 1 Determine the absolute value of the boiling point of water and the absolute value of the freezing point of water.

Step 2 Subtract to find how much greater.

3 Solve

Use your plan to solve the problem. Show your steps.

Boiling point of water $|212| =$ **212**

Freezing point of water $|32| =$ **32**

So, the absolute value of the boiling point of water is **212** − **32**

or **180** degrees greater than the absolute value of the freezing point of water. Choice C is correct. Fill in that answer choice.

Read to Succeed!
Absolute value is the distance a number is from zero and is always positive.

4 Check

How do you know your solution is accurate?

Sample answer: The distance from the freezing point of water to zero is 32. The distance from zero to the boiling point of water is 212. The difference in distances is 212 − 32 or 180.

NAME________________ DATE ________ PERIOD ______

Lesson 2 *(continued)*

Use a problem-solving model to solve each problem.

1 What is the difference between the absolute value of point *B* and the absolute value of point *D*? **6.NS.7c, MP 1**

Ⓐ 8
Ⓑ 4
Ⓒ 0 (filled in)
Ⓓ −4

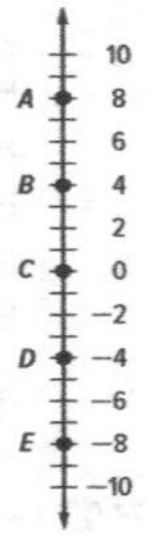

2 The table shows the account balances of five students.

Student	Balance ($)
Yen	−9
Mark	11
Aisha	−3
Wendy	10
Ross	6

What is the difference between the absolute value of Wendy's balance and the absolute value of Yen's balance, in dollars? 6.NS.7c, MP 2

1

3 The graph shows the path Chante walked, beginning at point *A*. The distance between two tick marks represents 1 meter. Chante walked 3 meters east to point *B* and then 5 meters east to point *C*. How many meters west must Chante walk from point *C* to be at the point represented by the opposite of *C*? **6.NS.6a, MP 4**

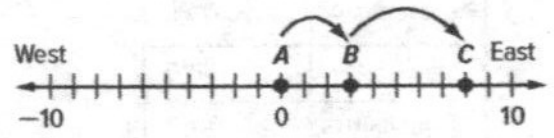

16

4 H.O.T. Problem Is the opposite of the absolute value of a number *always*, *sometimes*, or *never* equal to the absolute value of the opposite of a number? Explain your response and give examples. **6.NS.7c, MP 3**

sometimes; Sample answer: If the number is 0, then $-|0| = |-0|$. If the number is not 0, for example: −2, then $-|-2| \neq |-(-2)|$.

NAME ________ DATE ________ PERIOD ________

Lesson 3 **Multi-Step** Problem Solving

Multi-Step Example

The table shows the freezing points in degrees Celsius of four substances. Which substance(s) have greater freezing points than aniline? **6.NS.7, MP 1**

Substance	Freezing Point (°C)
Aniline	−6
Acetic acid	17
Acetone	−95
Water	0

Ⓐ water only
Ⓑ acetic acid only
Ⓒ acetic acid, acetone, and water
Ⓓ acetic acid, water (filled in)

Use a problem-solving model to solve this problem.

1 Understand

Read the problem. Circle the information you know.
Underline what the problem is asking you to find.

2 Plan

What will you need to do to solve the problem? Write your plan in steps.

Step 1 Locate the numbers on a **number line**.

Step 2 Compare the numbers based on their location on the number line.

3 Solve

Use your plan to solve the problem. Show your steps.

Locate the values on a number line.

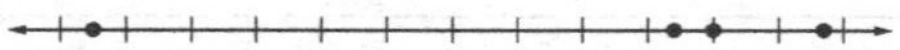

Read to Succeed! Make sure to read the directions carefully.

Compare the numbers.
The numbers greater than −6 are 0 and 17, which correspond to water and acetic acid. Choice **D** is correct. Fill in that answer choice.

4 Check

How do you know your solution is accurate?

Sample answer: The only substance with a lower freezing point is acetone because −95 < −6.

NAME ________ DATE ________ PERIOD ________

Lesson 3 *(continued)*

Use a problem-solving model to solve each problem.

1 Golf scores are measured as over or under par. The winner has the least score. The table shows the golf scores of five players.

Golfer	Score
Jamal	2 under par
Zaire	1 over par
Dante	even
Alexandra	3 under par
Ajay	4 over par

Which lists the players in order from first place to fifth place? **6.NS.7, MP 1**

Ⓐ Dante, Zaire, Jamal, Alexandra, Ajay
Ⓑ Alexandra, Jamal, Dante, Zaire, Ajay (filled in)
Ⓒ Ajay, Zaire, Dante, Jamal, Alexandra
Ⓓ Dante, Alexandra, Jamal, Zaire, Ajay

2 The table shows the rise and fall in the value of a certain stock over five days. Which day shows the greatest fall in stock value? **6.NS.7, MP 2**

Day	Change in Stock Value ($)
1	$-1\frac{1}{8}$
2	$4\frac{3}{8}$
3	$6\frac{1}{2}$
4	$-3\frac{1}{4}$
5	$1\frac{3}{4}$

Day 4

3 When a football player causes a penalty during a game, the team can lose 5, 10, or 15 yards on the play. The table shows the players, by jersey number, and the number of penalty yards the team was given based on each player's penalties. How many players caused more penalty yards than the player with jersey number 10? **6.NS.7, MP 2**

Player (jersey number)	Penalty Yards
12	−15
8	−25
28	−30
17	−10
10	−20
48	−5

2

4 **H.O.T. Problem** Order the numbers from greatest to least. Explain how you know which number is the greatest. **6.NS.7, MP 3**

$\frac{1}{2}$, $-|-3|$, -0.5, $|-2|$, -1

$|-2|$, $\frac{1}{2}$, -0.5, -1, $-|-3|$; Sample answer: $|-2| = 2$ which is greater than $\frac{1}{2}$. The other numbers are less than zero, which are also less than 2.

Copyright © McGraw-Hill Education. Permission is granted to reproduce for classroom use.

NAME_______________ DATE_______ PERIOD_______

Lesson 4 Multi-Step Problem Solving

Chapter 5

Multi-Step Example

There are 84 chairs that need to be set up in the school's auditorium. Students were asked to write an expression to show how to find the number of chairs needed in each of 12 rows. The table shows samples of expressions given by students. Which sample(s) result in the correct solution? *Preparation for* **6.NS.6c,** MP 1

Sample	Expression
A	$12 \div 84$
B	$84 \div 12$
C	$\frac{12}{84}$
D	$\frac{84}{12}$

Ⓐ Sample A
Ⓑ Sample B
Ⓒ Samples A and C
Ⓓ Samples B and D

Use a problem-solving model to solve this problem.

1 Understand

Read the problem. Circle the information you know.
Underline what the problem is asking you to find.

2 Plan

What will you need to do to solve the problem? Write your plan in steps.

Step 1 Determine the number of chairs needed in each row.

Step 2 Determine whether each sample's expression results in the correct solution.

Read to Succeed! Make sure to read all choices given when answering multiple choice questions.

3 Solve

Use your plan to solve the problem. Show your steps.

There are 84 ÷ **12**, or **7** chairs needed in each row.

Sample A $12 \div 84 \approx$ **0.14** Sample C $\frac{12}{84} \approx$ **0.14**

Sample B $84 \div 12 =$ **7** Sample D $\frac{84}{12} =$ **7**

Samples **B** and **D** result in the correct solution. So, choice **D** is correct.
Fill in that answer choice.

4 Check

How do you know your solution is accurate?

Sample answer: I know that Samples A and C are not correct because the dividend or numerator should be 84.

NAME_______________ DATE_______ PERIOD_______

Lesson 4 *(continued)*

Use a problem-solving model to solve each problem.

1 Selina has 8 cups of sugar to use for baking pies. She would like to bake 12 pies. She knows she can divide to find how much sugar to use for each pie. Which expression is equivalent to 8 ÷ 12?
Preparation for **6.NS.6c,** MP 1

Ⓐ 2×3
Ⓑ 3×2
Ⓒ $\frac{2}{3}$
Ⓓ $\frac{3}{2}$

2 Alivia needs to classify the numbers in the table using the Venn diagram shown.

Numbers
$\lvert -6 \rvert, \frac{3}{4}, 1.55, 11 \div 6, -2, 0.\overline{7}, \frac{2}{3}, 6 \div 2$

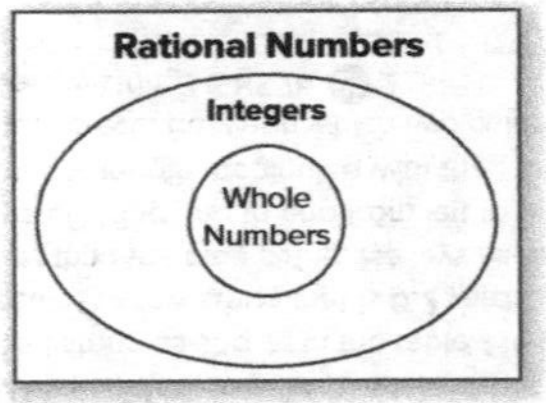

She decides to first color-code the numbers by highlighting the numbers that represent *terminating* decimals in yellow and *repeating* decimals in red. How many numbers should Alivia highlight in red?
Preparation for **6.NS.6c,** MP 8

3

3 The table shows the perimeters of different equilateral triangles. How many of the triangles have side lengths that are terminating decimals?
Preparation for **6.NS.6c,** MP 8

Perimeter (in.)
11
$15.\overline{6}$
18
21
$21.\overline{33}$
23.3

2

4 **H.O.T. Problem** Which number is greater, $-\frac{1}{3}$ or -0.3? Use the number line to help explain your answer. **6.NS.7a,** MP 3

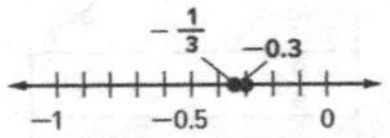

−0.3; Possible answer: First write the fraction as a decimal. $-\frac{1}{3} = -0.333...$ and $-0.\overline{3} < -0.3$ because $-0.\overline{3}$ is to the left of −0.3 on the number line. So, $-0.3 > -\frac{1}{3}$.

NAME________________ DATE ________ PERIOD ________

Lesson 5 **Multi-Step** Problem Solving

Multi-Step Example

The table shows the difference between the actual amount of rainfall, in inches, that a city received over four weeks and the average amount that it usually receives during those weeks. Which shows the weeks in order of the differences from least to greatest? **6.NS.7b, MP 1**

Week	Difference (in.)
1	$\frac{1}{3}$
2	−1.6
3	0.3
4	$-1\frac{1}{2}$

Ⓐ 3, 1, 4, 2
Ⓑ 2, 4, 3, 1 (filled in)
Ⓒ 1, 3, 4, 2
Ⓓ 2, 4, 1, 3

Use a problem-solving model to solve this problem.

1 Understand

Read the problem. Circle the information you know.
Underline what the problem is asking you to find.

2 Plan

What will you need to do to solve the problem? Write your plan in steps.

Step 1 Compare the negative rational numbers.
Step 2 Compare the positive rational numbers.
Step 3 Order the rational numbers.

Read to Succeed!
When ordering from least to greatest, remember negative numbers are less than positive numbers.

3 Solve

Use your plan to solve the problem. Show your steps.

$-1.6 < -1\frac{1}{2}$ and $\frac{1}{3} > 0.3$

So, the correct order of weeks is 2, 4, 3, 1. Choice **B** is correct. Fill in that answer choice.

4 Check

How do you know your solution is accurate?

Sample answer: I can graph each rational number on a number line and determine the order by writing the numbers from left to right.

NAME________________ DATE ________ PERIOD ________

Lesson 5 *(continued)*

Use a problem-solving model to solve each problem.

1 In last year's diving competition, Stefani's average score per dive was 9.55 points. The table shows the difference between her average score and her actual scores for her first four dives from this year's competition. Which of the following lists the dives in order of the differences from greatest to least? **6.NS.7b, MP 1**

Dive	Difference (points)
1	$\frac{1}{4}$
2	−0.35
3	$-\frac{3}{10}$
4	0.4

Ⓐ 4, 1, 3, 2 (filled in)
Ⓑ 1, 3, 2, 4
Ⓒ 4, 1, 2, 3
Ⓓ 2, 3, 1, 4

2 The table shows the heights, in feet, of five classmates. How many of these classmates are taller than $5\frac{1}{2}$ feet? **6.NS.7b, MP 2**

Student	Height (ft)
Mario	5.6
Phong	$5\frac{1}{3}$
Travis	$5\frac{5}{6}$
Zack	5.45
Tavon	$5\frac{3}{5}$

3

3 The table shows the dimensions, in centimeters, of four rectangles. How many centimeters wider is the rectangle with the greatest perimeter than the rectangle with the least perimeter? **6.NS.7b, MP 2**

Rectangle	Width (cm)	Length (cm)
A	6.25	8.9
B	6.3	8.73
C	6.5	8.7
D	6.6	8.5

0.2

4 **H.O.T. Problem** Jaquan has lengths of colored string as shown in the table. He finds a piece of yellow string that is 0.2 yard shorter than the blue string. He lays all four pieces of string end to end from left to right in order of length, beginning with the shortest piece. Between which two colors is the yellow string? **6.NS.7b, MP 2**

Color	Length (yd)
Red	$3\frac{2}{5}$
Blue	$3\frac{5}{8}$
Green	3.5

red and green

NAME _______________ DATE _______ PERIOD _______

Lesson 6 **Multi-Step** Problem Solving

Multi-Step Example

Samantha drew a map of the park in her neighborhood. She put a point on the grid for the playground. The fountain is located at the reflection of the location of the playground across the *y*-axis. The picnic tables are located at the reflection of the location of the fountain across the *x*-axis. What ordered pair describes the location of the picnic tables? **6.NS.6b, MP 1**

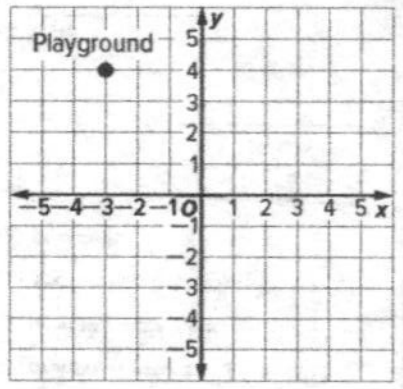

Use a problem-solving model to solve this problem.

1 Understand

Read the problem. Circle the information you know.
Underline what the problem is asking you to find.

2 Plan

What will you need to do to solve the problem? Write your plan in steps.

Step 1 Identify the ordered pair for the playground.

Step 2 Identify the reflection across the *y*-axis for the location of the fountain.

Step 3 Identify the reflection across the *x*-axis for the location of the picnic tables.

Read to Succeed! When locating points on a grid, begin at the origin and move horizontally along the x-axis and then move vertically along or parallel to the y-axis.

3 Solve

Use your plan to solve the problem. Show your steps.

The ordered pair for the playground is **(−3, 4)**.

The reflection across the *y*-axis for the location of the fountain is **(3, 4)**.

The reflection across the *x*-axis for the location of the picnic tables is **(3, −4)**.

So, the location of the picnic tables is at **(3, −4)**.

4 Check

How do you know your solution is reasonable?

Sample answer: I can start at the location of the picnic tables, reflect this location across the *x*-axis to the fountain, and then reflect the location of the fountain across the *y*-axis to see if I end up at (−3, 4), the location of the playground.

Chapter 5

NAME _______________ DATE _______ PERIOD _______

Lesson 6 *(continued)*

Use a problem-solving model to solve each problem.

1 Emily drew a map of her backyard. She put a point on the grid for the flower garden. The bench is located at the reflection of the location of the flower garden across the *x*-axis. The swing set is located at the reflection of the location of the bench across the *y*-axis. What ordered pair describes the location of the swing set? **6.NS.6b, MP 1**

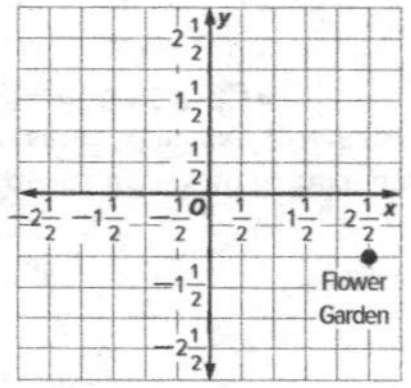

$\left(-2\frac{1}{2}, 1\right)$

2 Jorge identified the ordered pair that is a reflection of (3, −2) across the *y*-axis. Juliana identified the ordered pair that is a reflection of (−3, 2) across the *x*-axis. Camila identified the ordered pair that is a reflection of (−3, −2) across the *y*-axis. Who identified a point inside the square? **6.NS.6b, MP 1**

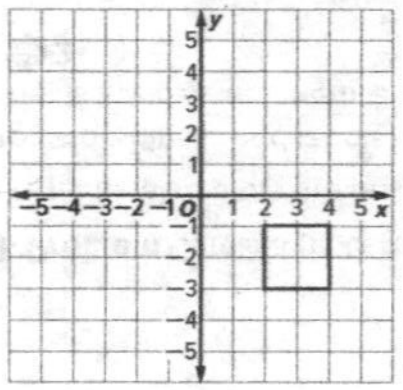

Camila

3 Carlos says all ordered pairs that have a 0 as either the *x*-coordinate or the *y*-coordinate are on the *x*-axis. Is he correct? Explain. **6.NS.6c, MP 3**

No; ordered pairs with 0 as the *y*-coordinate are on the *x*-axis, but ordered pairs with 0 as the *x*-coordinate are on the *y*-axis.

4 H.O.T. Problem Write each ordered pair from the box in the correct column of the table shown. Explain how you knew where to write the ordered pairs. **6.NS.6b, MP 7**

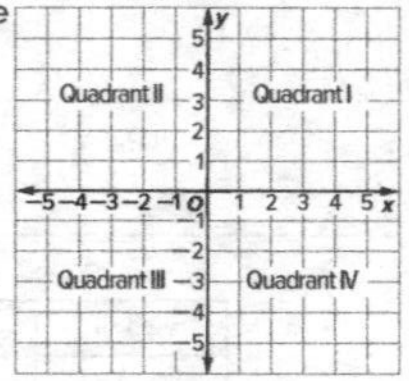

(2, 6) (−4, −3) (1, −7) (−8, 9)
$\left(5\frac{1}{3}, -3\frac{5}{6}\right)$ (−1.5, 2.2) $\left(-6\frac{2}{5}, -3\right)$ $\left(1\frac{1}{2}, 2\right)$

Quadrant I	Quadrant II	Quadrant III	Quadrant IV
(2, 6); $\left(1\frac{1}{2}, 2\right)$	**(−8, 9); (−1.5, 2.2)**	**(−4, −3); $\left(-6\frac{2}{5}, -3\right)$**	**(1, −7); $\left(5\frac{1}{3}, -3\frac{5}{6}\right)$**

The signs of the *x*-coordinate and the *y*-coordinate indicate the quadrant for each ordered pair.

Answers

NAME ______ DATE ______ PERIOD ______

Lesson 7 **Multi-Step** Problem Solving

Chapter 5

Multi-Step Example

The table shows the locations for several different places around town. The grid shows a map of the town, and each square on the grid represents one city block. Ben needs to go to the dry cleaner, which is 5 blocks north of the library. Where on the grid must he go? **6.NS.8, MP 1**

Place	Location
Bank	(5, −4)
Grocery	(−3, 0)
Library	(0, −3)
Post Office	(−4, 5)

Ⓐ (0, −8)
Ⓑ (5, −3)
● (0, 2)
Ⓓ (0, 5)

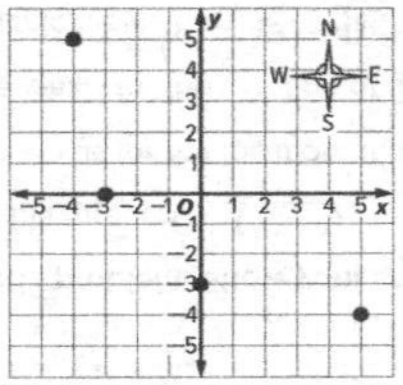

Use a problem-solving model to solve this problem.

1 Understand

Read the problem. Circle the information you know.
Underline what the problem is asking you to find.

2 Plan

What will you need to do to solve the problem? Write your plan in steps.

Step 1 Determine the dot on the grid that corresponds to the **library**.

Step 2 Determine the ordered pair of the location **5** blocks **north** of the library.

3 Solve

Use your plan to solve the problem. Show your steps.

The library is located at (**0** , **−3**). Five blocks north would be the ordered pair (**0** , **2**). Choice **C** is correct. Fill in that answer choice.

Read to Succeed!
The x-coordinate of an ordered pair tells you left or right and the y-coordinate tells you up or down.

4 Check

How do you know your solution is accurate?

Sample answer: Moving 5 blocks south from Choice C, I end up at the library.

NAME ______ DATE ______ PERIOD ______

Lesson 7 *(continued)*

Use a problem-solving model to solve each problem.

1 Rosa is currently at (4, −2) on the map. To which place in the table is she the closest? **6.NS.8, MP 1**

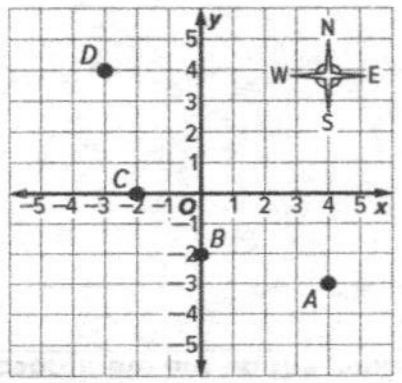

Place	Location
Stadium	*A*
Playground	*B*
Mall	*C*
Hospital	*D*

● Stadium
Ⓑ Playground
Ⓒ Mall
Ⓓ Hospital

2 Josie posts stakes at the following locations. She ties rope to the stakes to section off a rectangle. Each unit represents 1 foot. What is the perimeter of the rectangle in feet? **6.NS.8, MP 2**

Stake	Location
Stake 1	(2, 0)
Stake 2	(2, −4)
Stake 3	(−1, −4)
Stake 4	(−1, 0)

14

3 Eduardo drives from *A* to *B*. Each unit on the map represents 10 miles. How many miles does he drive? **6.NS.8, MP 4**

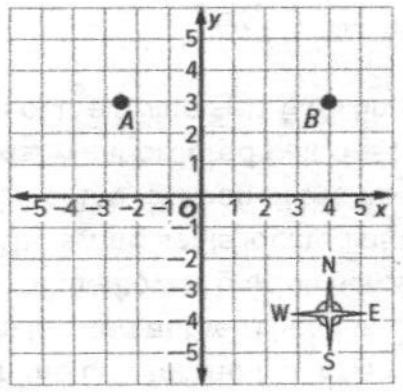

65

4 **H.O.T. Problem** Catalina plots point *C* at $\left(4, -1\frac{1}{2}\right)$. She also plots the point with the opposite *y*-coordinate and labels the point as *D*. What is the distance from *C* to *D*? **6.NS.8, MP 2**

3 units

NAME ______________ DATE ______ PERIOD ______

Lesson 1 **Multi-Step** Problem Solving

Chapter 6

Multi-Step Example

Delmar is studying the reproduction rate of a specific type of bacteria. He places 3 cells in a Petri dish and records the number of bacteria over time. He notices a pattern, which is shown in the table. Predict the number of bacteria in the Petri dish after 25 hours. Express the answer using exponents. Then evaluate to determine the number of bacteria. **6.EE.1, MP 1**

Number of Hours	Number of Bacteria
5	3 × 3
10	3 × 3 × 3
15	3 × 3 × 3 × 3
20	3 × 3 × 3 × 3 × 3

Ⓐ 3^6; 729 Ⓑ 6^3; 216 Ⓒ 3^5; 243 Ⓓ 3^6; 18

Use a problem-solving model to solve this problem.

1 Understand

Read the problem. Circle the information you know.
Underline what the problem is asking you to find.

2 Plan

What will you need to do to solve the problem? Write your plan in steps.

Step 1 Determine the pattern in the table.

Step 2 Express the answer using an exponent based on the pattern and evaluate.

3 Solve

Use your plan to solve the problem. Show your steps.

Every 5 hours, the number of bacteria triples. So, the number of bacteria after 25 hours is 3 × 3 × 3 × 3 × 3 × 3 or 3^6.

Since 3^6 = **729**, choice **A** is correct. Fill in that answer choice.

Read to Succeed! The exponent tells you how many times a base is used as a factor.

4 Check

How do you know your solution is accurate?

Sample answer: After 5 hours, there are 9 bacteria. After 10 hours, there are 27, then 81, then 243. After 25 hours, there are 243 × 3 or 729 bacteria.

NAME ______________ DATE ______ PERIOD ______

Lesson 1 *(continued)*

Use a problem-solving model to solve each problem.

1 Sonia is studying the reproduction rate of fleas. She places 2 fleas in an enclosed habitat and records the number of eggs each day. She notices a pattern, which is shown in the table. Predict the number of eggs in the habitat on the 5th day. Express the answer using exponents. Then evaluate to determine the number of eggs. **6.EE.1, MP 1**

Number of Days	Number of Eggs
1	5 × 5
2	5 × 5 × 5
3	5 × 5 × 5 × 5
4	5 × 5 × 5 × 5 × 5

Ⓐ 6^5; 7,776
Ⓑ 5^5; 3,125
Ⓒ 5^6; 15,625
Ⓓ 6^6; 46,656

2 Elena has a fish tank that holds 2^5 gallons of water. How many fluid ounces of water does the fish tank hold? (*Hint:* 1 gal = 128 fl oz) **6.EE.1, MP 2**

4,096 fl oz

3 Faith is turning 12 this year. She asks her parents to give her $1 on her birthday and to double that amount for her next birthday. If she continues with this pattern, how much money will Faith get on her 20th birthday? **6.EE.1, MP 7**

Birthday	Amount ($)
12th	2^0
13th	2^1
14th	2^2
15th	2^3

$256

4 **H.O.T. Problem** In any cube, the length, width, and height each have the same measure. The volume of the cube below can be found by calculating a^3, where a is the length of a side. Suppose a is 8 inches. How many gallons of water would the cube hold if 231 cubic inches is equal to 1 gallon? Round the answer to the nearest tenth of a gallon. List the steps you used to find your answer. **6.EE.1, MP 1**

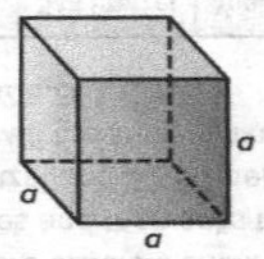

about 2.2 gallons; Find the volume of the cube: $V = 8^3 = 512$ in³. Then use a unit rate to find the number of gallons: $512 \text{ in}^3 \cdot \frac{1 \text{ gal}}{231 \text{ in}^3} \approx 2.2$ gal

NAME ____________ DATE ________ PERIOD ________

Lesson 2 **Multi-Step** Problem Solving

Multi-Step Example

An art store sells art kits that include crayons and a sketch pad. The table shows the number of crayons and sketch pad pages in each art kit size. A school buys 30 small, 10 large, and 24 medium art kits. Then they return 18 medium art kits. How many crayons do they have in all? **6.EE.1, MP 1**

Art Kit Size	Number of Crayons	Sketch Pad Pages
Small	16	20
Medium	24	40
Large	68	100

Ⓐ 3,240　Ⓑ 2,560　Ⓒ 1,736　**Ⓓ 1,304**

Use a problem-solving model to solve this problem.

1 Understand

Read the problem. Circle the information you know.
Underline what the problem is asking you to find.

2 Plan

What will you need to do to solve the problem? Write your plan in steps.

Step 1 Write an expression to represent the situation.

Step 2 Evaluate the expression.

Read to Succeed!
When evaluating the expression, remember to follow the order of operations. Perform the operations in the parentheses first then add.

3 Solve

Use your plan to solve the problem. Show your steps.

The school bought a total of 30 small, 6 medium, and 10 large kits.

$(30 \times$ **16** $) + (6 \times$ **24** $) + (10 \times$ **68** $) =$ **1,304**

So, the school has **1,304** crayons. Choice **D** is correct.

Fill in that answer choice.

4 Check

How do you know your solution is accurate?

Sample answer: Estimate the total number of crayons. 30 × 20 = 600, 10 × 70 = 700, and 10 × 24 = 240; 600 + 700 + 240 = 1,540

NAME ____________ DATE ________ PERIOD ________

Lesson 2 *(continued)*

Use a problem-solving model to solve each problem.

1 The table shows the chocolate chip cookies in each container type. Yesterday, the bakery sold 4 tubs, 10 baskets, and 12 boxes. However, 3 of the boxes were returned. How many total cookies were sold by the end of the day? **6.EE.1, MP 1**

Container Type	Chocolate Chip Cookies
Basket	16
Box	36
Tub	48

Ⓐ 160　Ⓑ 192　Ⓒ 432　**Ⓓ 676**

2 The table shows the number of salt and yogurt pretzels that come in different sized boxes. A store orders 3 small boxes and 5 large boxes and then sells them at \$2.00 per salt pretzel and \$2.50 per yogurt pretzel. How much will the store make if they sell all the pretzels? **6.EE.1, MP 2**

Box Size	Salt Pretzel	Yogurt Pretzel
Small	36	12
Large	48	24

\$1,086

3 A luxury line of furniture at a store sells couches for \$4,000, reclining chairs for \$2,040, and loveseats for \$2,800. This luxury line went on sale where the price of each piece of furniture was divided by 4. During the sale, how much would 2 couches, 1 reclining chair, and 3 loveseats cost? **6.EE.1, MP 2**

\$4,610

4 **H.O.T. Problem** Determine the value of the expression below. Explain your answer. **6.EE.1, MP 3**

$$\frac{120 - (3^3 - 36 \div 2)^2}{10 + 3(10 - 6) - 27 \div 3}$$

3; Sample answer: Use the order of operations to determine the numerator and denominator. The value of the numerator is 39. The value of the denominator is 13. Divide 39 ÷ 13 = 3.

NAME ______ DATE ______ PERIOD ______

Lesson 3 **Multi-Step** Problem Solving

Chapter 6

Multi-Step Example

The table shows the dimensions of three picture frame sizes available at a framing shop. What is the total perimeter, in inches, of two small frames and three large frames? The perimeter of a rectangle is $2\ell + 2w$, where ℓ is the length and w is the width. **6.EE.2c, MP 1**

Picture Frame Size	Length (in.)	Width (in.)
Small	3	5
Medium	5	7
Large	8	10

Use a problem-solving model to solve this problem.

1 Understand

Read the problem. Circle the information you know.
Underline what the problem is asking you to find.

2 Plan

What will you need to do to solve the problem? Write your plan in steps.

Step 1 Determine the perimeter of a small frame and a large frame.

Step 2 Determine the total perimeter for the five frames.

Read to Succeed!
Remember that 2ℓ is the same as $2 \times \ell$ and $2w$ is $2 \times w$.

3 Solve

Use your plan to solve the problem. Show your steps.

The perimeter of a small frame is 2(**3**) + 2(**5**) or **16** inches.

The perimeter of a large frame is 2(**8**) + 2(**10**) or **36** inches.

The total perimeter of the five frames is 2(16) + 3(36) or **140** inches.

4 Check

How do you know your solution is accurate?

Sample answer: Draw a diagram for each frame, label the lengths and widths, and add to find the perimeters.

NAME ______ DATE ______ PERIOD ______

Lesson 3 *(continued)*

Use a problem-solving model to solve each problem.

1 The table shows different dog carrier sizes available at a pet store. What is the total perimeter, in feet, of one large, two extra-small, and three medium dog carriers? Represent the situation with an expression. **6.EE.2c, MP 1**

Carrier Size	Length (in.)	Width (in.)
Extra-Small	19	13
Small	24	18
Medium	30	19
Large	36	23

45 ft; $\dfrac{(2 \cdot 36 + 2 \cdot 23) + 2(2 \cdot 19 + 2 \cdot 13) + 3(2 \cdot 30 + 2 \cdot 19)}{12}$

2 Gabby is going to cover her ruler shown below with construction paper for an art project. The area of the ruler can be found using the expression ℓw, where ℓ is the length and w is the width of the rectangle. What is the area, in square inches, of the construction paper needed if 4 rulers are used in her art project? Represent the situation with an expression. **6.EE.2c, MP 4**

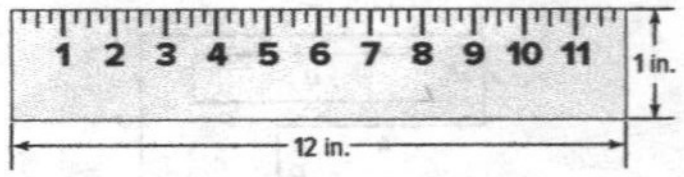

48 in²; 4(12 • 1)

3 Calvin is filling a sandbox with sand. The volume of the sandbox can be found using the expression ℓwh where ℓ is the length, w is the width, and h is the height. What is the volume, in cubic inches, of two of these sandboxes? Represent the situation with an expression. (*Hint: There are 12 × 12 × 12 cubic inches in a cubic foot.*) **6.EE.2c, MP 4**

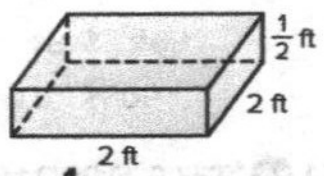

6,912 in³; $2[(\frac{1}{2} \cdot 2 \cdot 2)] \cdot 12 \cdot 12 \cdot 12$

4 H.O.T. Problem Write an expression for the perimeter of the irregular-shaped figure shown below. Explain. **6.EE.6, MP 4**

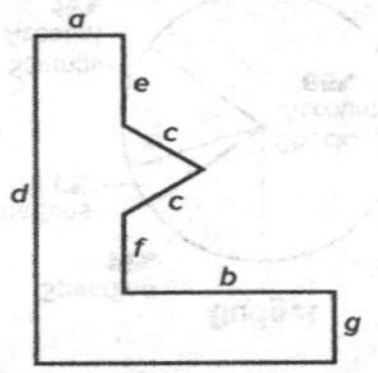

The perimeter would be $d + a + e + c + c + f + b + g + (a + b)$ **or** $2a + 2b + 2c + d + e + f + g$**. The missing side length was found using parallel side lengths of** a **and** b **in the figure.**

Answers

NAME ______________________ DATE ____________ PERIOD ________

Lesson 4 **Multi-Step** Problem Solving

Chapter 6

Multi-Step Example

The cost of tickets at a movie theater is shown in the table. Write an expression to represent the total cost of tickets using the variables in the table. Then use the expression to find the total ticket cost in dollars for 2 adults, 3 children, and 1 senior. **6.EE.6,** MP 1

Type of Ticket	Number of Tickets	Cost ($)
Adult	a	8
Child	c	5
Senior	s	6

Ⓐ $a + c + s$; $6
Ⓑ $(8 + 5 + 6)(a + c + s)$; $114
Ⓒ $8a + 5c + 6s$; $37
Ⓓ $8a + 5c + 6s$; $19

Use a problem-solving model to solve this problem.

1 Understand
Read the problem. Circle the information you know.
Underline what the problem is asking you to find.

2 Plan
What will you need to do to solve the problem? Write your plan in steps.

Step 1 Represent the situation with an expression.
Step 2 Substitute the values given and evaluate.

Read to Succeed! Follow the order of operations when evaluating expressions. Multiply first then add.

3 Solve
Use your plan to solve the problem. Show your steps.

The cost of a adult tickets is **8a**, c child tickets is **5c**, and s senior tickets is **6s**.

The total for any number of tickets is **8a** + **5c** + **6s**.

So, the total for 2 adults, 3 children, and 1 senior is 8(2) + 5(3) + 6(1) or **$37**.

Choice **C** is correct. Fill in that answer choice.

4 Check
How do you know your solution is accurate?
Sample answer: Substitute the values in the table with the number of tickets purchased. Determine the total cost to check your solution.

NAME ______________________ DATE ____________ PERIOD ________

Lesson 4 *(continued)*

Use a problem-solving model to solve each problem.

1 Shelby is attending classes at a university that charges $175 per credit, plus an application fee of $45. Write an expression that represents the total cost of tuition based on the number of credits c. Then use the expression to determine the tuition amount if Shelby plans on taking 14 credits. **6.EE.6,** MP 1

Ⓐ 45c + 175; $805
Ⓑ 175c + 45; $2,495
Ⓒ (175 + 45)c; $3,080
Ⓓ $\frac{175}{c}$ + 45; $57.50

2 Michael is working on his budget and decides to allocate a percentage of each paycheck as described in the circle graph. Write an expression to determine the total amount he would deposit in the bank b. How much does he deposit if his check is $800? **6.EE.3,** MP 2

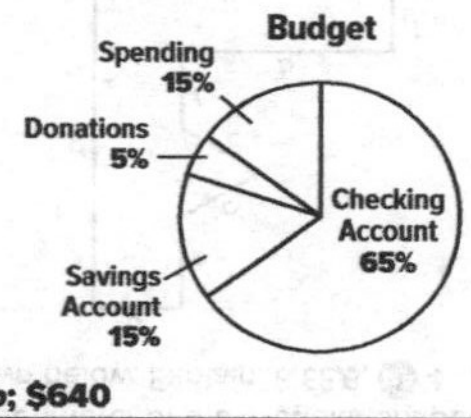

0.8b; $640

3 Brad is designing an A-frame house and needs to include angle measures on the diagram below. The sum of angles 1, 2, and 3 is 180°. The sum of angles 3 and 4 is 180°. The measure of angle 1 is 50° and the measure of angle 2 is 20°. Write an equation that can be used to find the measure of angle 4. **6.EE.6,** MP 4

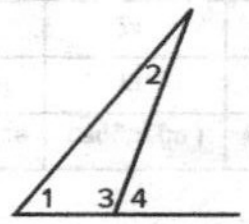

Let a = the measure of angle 4;
$[180 - (50 + 20)] + a = 180$

4 **H.O.T. Problem** Write an expression to find the area of the shaded region. **6.EE.6,** MP 4

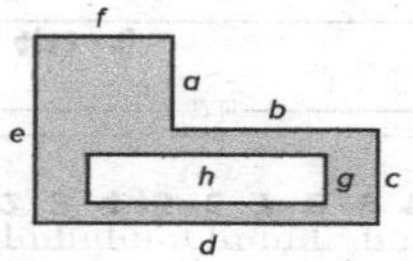

$fa + cd - gh$

NAME ______________________ DATE ____________ PERIOD ________

Lesson 5 Multi-Step Problem Solving

Multi-Step Example

Marla calculated the sum 1 + 2 + 3 + 4 + 5 + … + 12 + 13 + 14 by performing the additions shown in the table and then multiplying 15 × 7. What property did Marla use to make the addition problem easier to compute? 6.EE.3, MP 1

Addends	Sum
1 + 14	15
2 + 13	15
3 + 12	15
4 + 11	15
5 + 10	15
6 + 9	15
7 + 8	15

Ⓐ Commutative Property of Addition (filled in)
Ⓑ Commutative Property of Multiplication
Ⓒ Associative Property of Addition
Ⓓ Associative Property of Multiplication

Use a problem-solving model to solve this problem.

1 Understand

Read the problem. Circle the information you know.
Underline what the problem is asking you to find.

2 Plan

What will you need to do to solve the problem? Write your plan in steps.

Step 1 Determine the pattern shown in the table.

Step 2 Determine the property used.

Read to Succeed!
The Commutative Property says the order in which two numbers are added or multiplied does not matter.
The Associative Property says the grouping of the numbers when added or multiplied does not matter.

3 Solve

Use your plan to solve the problem. Show your steps.

When Marla reordered the addends, she used the first and last, second and second-to-last addend, and so on. When reordering the addends, she used the Commutative Property of Addition.

So, choice __A__ is correct. Fill in that answer choice.

4 Check

How do you know your solution is accurate?

Sample answer: Since the situation uses addition, choices B and D are incorrect. Since the ordering of the addends changed, choice C is incorrect.

NAME ______________________ DATE ____________ PERIOD ________

Lesson 5 *(continued)*

Use a problem-solving model to solve each problem.

1 The table shows the number of students that chose different sports based on gender. Kevin wants to determine the total number of boys surveyed. Which of the following expressions uses the Associative Property to help him mentally determine the sum? 6.EE.3, MP 1

Sport	Number of Girls	Number of Boys
Basketball	14	11
Football	2	19
Lacrosse	7	5
Soccer	13	12
Swimming	12	13

Ⓐ 14 + 2 + 7 + 13 + 12
Ⓑ (11 + 19) + [5 + (12 + 13)] (filled in)
Ⓒ (11 + 5) + (19 + 12 + 13)
Ⓓ 2(13) + 2(12)

2 The table shows pairs of equivalent expressions. Which expression does not model one of the different properties of number operations? 6.EE.3, MP 7

9 + 7 = 7 + 9	9 + 0 = 9
10 + 10 = 5 + 15	(8 + 6) + 0 = 8 + (6 + 0)
0 × 8 = 8 × 0	8 × 1 = 8

10 + 10 = 5 + 15

3 Four students were each given 48 blocks to arrange to make a rectangular prism. For two of these students, the length and width model the Commutative Property. What is the product of the length and width that they used? 6.EE.3, MP 7

Student	Length	Width	Height
Lucy	12	4	1
Jose	4	4	3
Maria	3	4	4
Tony	4	12	1

48

4 **H.O.T. Problem** How could you use the Commutative and Associative Properties to quickly find the value of the expression $28 + (9^2 + 72)$? 6.EE.3, MP 3

Sample answer: Find the value of 9^2, which is 81. Then, using both the Commutative and Associative properties, rewrite the expression as 28 + (81 + 72) = 28 + (72 + 81) = (28 + 72) + 81 = 100 + 81 = 181.

NAME ______________________ DATE ____________ PERIOD ________

Lesson 6 **Multi-Step** Problem Solving

Chapter 6

Multi-Step Example

Wen is buying bottles of apple juice and wants to mentally calculate how much they will cost. He buys 5 bottles of juice at $2.15 each. Which of the following shows equivalent expressions using the Distributive Property? How much change will he receive from $20? **6.EE.3, MP 1**

Ⓐ 5(2.15) = 5(2) + 0.15; $10.75

Ⓑ 5(2.15) = 5(2) − 0.15; $9.25

● 5(2.15) = 5(2) + 5(0.15); $9.25

Ⓓ 5(2.15) = 5(2) − 5(0.15); $10.75

Use a problem-solving model to solve this problem.

1 Understand

Read the problem. Circle the information you know.
Underline what the problem is asking you to find.

2 Plan

What will you need to do to solve the problem? Write your plan in steps.

Step 1 Generate equivalent expressions using the Distributive Property.

Step 2 Determine the total spent and subtract from $20 to find the change he will receive.

Read to Succeed!
The number 2.15 can be expressed as 2 + 0.15.

3 Solve

Use your plan to solve the problem. Show your steps.

Five bottles of apple juice will cost 5(**2.15**). An equivalent expression, using the Distributive Property, is 5(2) + 5(0.15).

The total spent would be $ **10.75** . Wen will receive $20.00 − $ **10.75** or $ **9.25** in change.

So, choice **C** is correct. Fill in that answer choice.

4 Check

How do you know your solution is accurate?

Sample answer: Evaluate 5(2.15) and 5(2) + 5(0.15) to determine if they are equal.

NAME ______________________ DATE ____________ PERIOD ________

Lesson 6 *(continued)*

Use a problem-solving model to solve each problem.

1 Cole is connecting 3 benches to make one long bench. Each bench has a length of $5\frac{1}{2}$ feet. Which of the following show equivalent expressions using the Distributive Property? **6.EE.3, MP 1**

● $3\left(5\frac{1}{2}\right) = 3(5) + 3\left(\frac{1}{2}\right)$

Ⓑ $3\left(5\frac{1}{2}\right) = 3(5) - 3\left(\frac{1}{2}\right)$

Ⓒ $3\left(5\frac{1}{2}\right) = 3(5) + \left(\frac{1}{2}\right)$

Ⓓ $3\left(5\frac{1}{2}\right) = 5(3) + 5\left(\frac{1}{2}\right)$

2 The total area of two rectangles can be calculated using different methods. What number can be substituted for *x* so that each expression will show the total area of the two rectangles? **6.EE.3, MP 4**

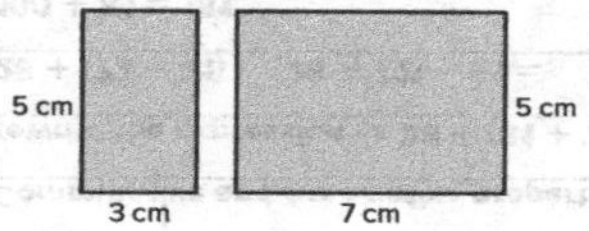

$x(3 + 7) = x(3) + x(7)$

5

3 Elijah bought 5 more candles, *c*, than Bianca. The candles cost $2 each. What number should be replaced for *x* to give an equivalent expression for the total amount Elijah and Bianca spent? **6.EE.3, MP 2**

$2c + 2(c + 5) = xc + 10$

4

4 **H.O.T. Problem** Marta needs to buy 8 notebooks that cost $3.75 each. Two friends show a shortcut for mentally calculating the total price. Evaluate each student's method. **6.EE.3, MP 3**

Juanita	8(3 + 0.75)
Joe	8(4 − 0.25)

Both are correct. When the expression is simplified, the result is the same. Both expressions in the parentheses equal 3.75, the cost of 1 notebook.

NAME ____________ DATE ________ PERIOD ________

Lesson 7 Multi-Step Problem Solving

Chapter 6

Multi-Step Example

When buying frozen yogurt, there are many choices for the toppings. Which simplified expression represents the price of 2 cones with fruit and 3 cones with candy and syrup? **6.EE.2, MP 1**

Frozen Yogurt	Price ($)
Cone	x
Candy Topping	add $0.75
Syrup Topping	add $0.50
Fruit Topping	add $1.00

Ⓐ $5x + 2.25$
Ⓑ $5x + 5.75$ (filled in)
Ⓒ $2x + 2$
Ⓓ $2x + 4.50$

Use a problem-solving model to solve this problem.

1 Understand

Read the problem. Circle the information you know.
Underline what the problem is asking you to find.

2 Plan

What will you need to do to solve the problem? Write your plan in steps.

Step 1 Determine the expressions for each type of cone.

Step 2 Combine like terms to simplify the expressions.

Read to Succeed!
Remember to combine like terms when simplifying expressions. Like terms have the same variable.

3 Solve

Use your plan to solve the problem. Show your steps.

Two cones with fruit: $2(x + 1.00) =$ **2x** + **2**

Three cones with candy and syrup: $3(x + 0.75 + 0.50) =$ **3x** + **3.75**

Adding the two expressions and simplifying results in **5x** + **5.75**.

Choice **B** is correct. Fill in that answer choice.

4 Check

How do you know your solution is accurate?

Sample answer: Draw a bar diagram to represent the addition of each cone and type of topping.

NAME ____________ DATE ________ PERIOD ________

Lesson 7 *(continued)*

Use a problem-solving model to solve each problem.

1 At a taco stand, chicken tacos are available with a choice of 3 different toppings. Which simplified expression represents the price of 2 tacos with cheese and lettuce, and 1 taco with cheese, lettuce, and sour cream? **6.EE.2, MP 1**

Item	Price ($)
Chicken taco	x
Cheese	$0.25
Lettuce	$0.15
Sour cream	$0.45

Ⓐ $2x + 0.40$
Ⓑ $3x + 1.25$
Ⓒ $3x + 1.65$ (filled in)
Ⓓ $3x + 2.55$

2 Treven and Dylan each bought fruit at the market. Treven bought 3 pears and 7 apples. Dylan bought 4 pears and 9 apples and spent $3.00 on oranges. If p represents the cost of each pear and a represents the cost of each apple, the expression below represents the total cost. What is the value of x? **6.EE.2b, MP 2**

$$xp + 16a + 3$$

7

3 The length of a rectangle is x feet more than its width. The area of the rectangle is $5x + 25$. What is the width of the rectangle? **6.EE.3, MP 2**

5 ft

4 H.O.T. Problem Julian asked x students about their favorite color, and displayed the results in the circle graph. Claire thinks that the same number of people voted for orange and red as did green and purple. Write three equal expressions that show that she is correct. **6.EE.3, MP 4**

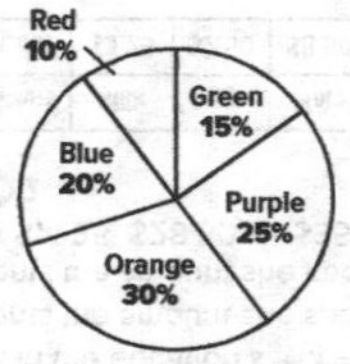

0.30x + 0.10x = 0.15x + 0.25x = 0.40x

Answers

NAME ____________ DATE ________ PERIOD ________

Lesson 1 Multi-Step Problem Solving

Multi-Step Example

Sandi bought a sandwich and a milkshake. She spent $12 in all. The equation $s + 4.25 = 12$ can be used to determine the cost of the sandwich. Which of the following is the solution to the equation? **6.EE.5, MP 1**

Item	Cost ($)
Sandwich	s
Milkshake	4.25

Ⓐ $7.75 (filled)
Ⓑ $8.75
Ⓒ $16
Ⓒ $16.25

Use a problem-solving model to solve this problem.

1 Understand

Read the problem. Circle the information you know.
Underline what the problem is asking you to find.

2 Plan

What will you need to do to solve the problem? Write your plan in steps.

Step 1 Test each answer choice to determine if a true number sentence is created.

3 Solve

Use your plan to solve the problem. Show your steps.

$7.75 + 4.25 = 12$ ✓

$8.75 + 4.25 \neq 12$ X

$16 + 4.25 \neq 12$ X

$16.25 + 4.25 \neq 12$ X

Since **7.75** + 4.25 = 12, choice **A** is correct. Fill in that answer choice.

Read to Succeed!
Substitute each answer choice for s and add to determine if the number sentence is true.

4 Check

How do you know your solution is accurate?

Sample answer: Since the left side of the equal sign is the same as the right side, the answer is correct.

NAME ____________ DATE ________ PERIOD ________

Lesson 1 *(continued)*

Use a problem-solving model to solve each problem.

1 Kaleena paid $26.25 to rent a kayak for 3 hours. The equation $3x = 26.25$ can be used to determine the amount she paid per hour. Which of the following is the solution to the equation? **6.EE.5, MP 1**

Ⓐ $8.66
Ⓑ $8.75 (filled)
Ⓒ $23.25
Ⓓ $29.25

2 Shalah went to the grocery store with $50 in cash and bought the items shown in the table. In the equation $s + x = 50$, s represents the amount she spent, and x represents the amount she had left. Does she have $14.25, $28.50, or $35.75 left? **6.EE.5, MP 2**

Item	Milk	Eggs	Ham
Cost	$3.25	$2.50	$8.50

$35.75

3 Kirsten had 65 inches of wire. After she used some for a project, she had 14 inches left. The equation $x + 14 = 65$ can be used to determine the amount of wire that she used in inches. Did she use 79, 51, or 37 inches of wire? Plot the solution on the number line. **6.EE.5, MP 2**

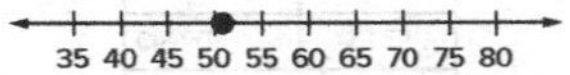

51

4 **H.O.T. Problem** George made 12 ounces of rice. He had 4 ounces left after he finished eating. Use *guess, check, and revise* strategy to solve the equation $12 - r = 4$ to find r, the number of ounces of rice that he ate. Then convert your answer to find how many cups of rice George ate. **6.EE.5, MP 6**

$r = 8$; He ate 1 cup of rice.

r	$12 - r = 4$	Are Both Sides Equal?
5	**$12 - (5) = 4$ $7 \neq 4$**	**No**
6	**$12 - (6) = 4$ $6 \neq 4$**	**No**
7	**$12 - (7) = 4$ $5 \neq 4$**	**No**
8	**$12 - (8) = 4$ $4 = 4$**	**Yes**

NAME_______________ DATE _________ PERIOD _______

Lesson 2 **Multi-Step** Problem Solving

Multi-Step Example

A bookstore has a sale on mysteries. The table shows the cost of each book format. Abigail has $70 to spend. She bought two paperbacks, one hardcover, and one audio book. Which addition equation can be used to determine how much more money Abigail still has to spend? 6.EE.7, MP 1

Book	Cost ($)
Hardcover	19
Paperback	8
E-book	10
Audio Book	25

Ⓐ $70 + 60 = x$

Ⓑ (filled) $60 + x = 70$

Ⓒ $70 + x = 60$

Ⓓ $52 + 70 = x$

Use a problem-solving model to solve this problem.

1 Understand

Read the problem. Circle the information you know.
Underline what the problem is asking you to find.

2 Plan

What will you need to do to solve the problem? Write your plan in steps.

Step 1 Determine how much Abigail spent.

Step 2 Write an equation to represent the situation.

Read to Succeed!
The total she has to spend is the sum so it will be by itself in the equation.

3 Solve

Use your plan to solve the problem. Show your steps.

Abigail spent 2($ **8**) + $ **19** + $ **25** , or $ **60** .

Let x = the amount she has left.

So, **60** $+ x =$ **70** . Choice **B** is correct. Fill in that answer choice.

4 Check

How do you know your solution is accurate?

Sample answer: The total Abigail has to spend is $70. The only equation that shows a sum of 70 is choice B.

NAME_______________ DATE _________ PERIOD _______

Lesson 2 *(continued)*

Use a problem-solving model to solve each problem.

1 Kisho has $45 to spend at a pizza shop for a pizza party. The table shows the cost of each size of pizza. He bought four small pizzas and one large pizza. Which addition equation can be used to determine how much more money he still has to spend? 6.EE.7, MP 1

Pizza Size	Cost ($)
Small	5
Medium	8
Large	10
Extra Large	15

Ⓐ $x + 45 = 30$

Ⓑ $x - 30 = 45$

Ⓒ (filled) $x + 30 = 45$

Ⓓ $x - 45 = 30$

2 Miguel has $2\frac{1}{2}$ hours to work on his homework. The table shows how much time he spent working on his English homework and his math homework. Write and solve an addition equation that can be used to find how much time, in minutes, he has left to work on his science project if he wants to take a 15-minute snack break. 6.EE.7, MP 7

Homework	Time Spent (min)
English	45
Math	28
Science project	?

$45 + 28 + 15 + s = 150$; 62 min

3 Two rectangular rugs in Winona's bedroom are shown below. The larger rug is 32 square feet bigger than the smaller rug. The equation $x + 32 = 35$ can be used to find the area of the smaller rug. What is the width of the smaller rug, in feet? 6.EE.7, MP 6

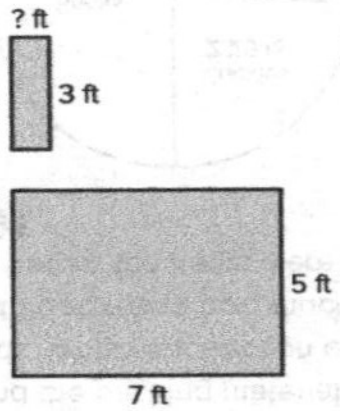

1 ft

4 **H.O.T. Problem** The perimeter of the triangle below is 58 centimeters. Write an addition equation to determine the unknown length, x. Explain how you can use the Subtraction Property of Equality to solve the equation. Write the unknown length. 6.EE.7, MP 3

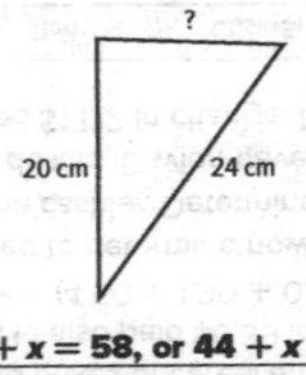

$20 + 24 + x = 58$, or $44 + x = 58$. The Subtraction Property of Equality states that if you subtract the same number from each side of an equation, the two sides remain equal. So, subtract 44 from each side. $x = 14$; 14 cm

NAME ______ DATE ______ PERIOD ______

Lesson 3 **Multi-Step** Problem Solving

Multi-Step Example

Tyson withdrew money from his savings account to go shopping. The circle graph shows how he spent the money. He has $326 left in his savings account. Which subtraction equation can be used to determine how much Tyson had in his account before he withdrew the money? **6.EE.7, MP 1**

Ⓐ $326 - 160 = m$

Ⓑ $160 - 95 = m$

Ⓒ $326 - m = 65$

Ⓓ $m - 160 = 326$

Use a problem-solving model to solve this problem.

1 Understand

Read the problem. Circle the information you know.
Underline what the problem is asking you to find.

2 Plan

What will you need to do to solve the problem? Write your plan in steps.

Step 1 Determine how much Tyson spent.

Step 2 Write an equation to represent the situation.

Read to Succeed!
The amount he has left in his account after withdrawing money to shop is the difference. So it will be by itself in the equation.

3 Solve

Use your plan to solve the problem. Show your steps.

Tyson spent $**95** + $**23** + $**42**, or $**160** in all.

Let m = the amount of money he had in his account before he withdrew the money.

So, m − **160** = **326**. Choice **D** is correct. Fill in that answer choice.

4 Check

How do you know your solution is accurate?

Sample answer: The amount Tyson has left is $326. The only equation that shows a difference of $326 is Choice D.

NAME ______ DATE ______ PERIOD ______

Lesson 3 *(continued)*

Use a problem-solving model to solve each problem.

1 Fernando weighed some items as he took them out of a carton. The circle graph shows the weight of the items in the carton. The carton and the packing material weigh 1.5 pounds. Which subtraction equation can be used to determine how much the carton weighed before the items were taken out? **6.EE.7, MP 1**

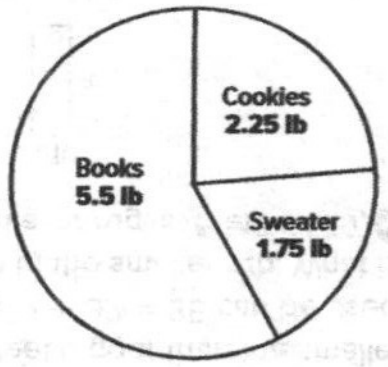

Ⓐ $9.5 - 1.5 = w$

Ⓑ $9.5 - w = 1.5$

Ⓒ $1.5 - w = 9.5$

Ⓓ $w - 9.5 = 1.5$

2 A grocery store is having a "two for the price of one" sale for several items. The cost of these items is shown in the table. Davion bought two boxes of cereal and two loaves of bread. He also paid $0.39 in tax. The equation $x - (4.50 + 1.99 + 0.39) = 13.12$ can be used to determine how much money he gave the cashier. Determine how much money, in dollars, Davion gave the cashier if he received $13.12 in change. **6.EE.7, MP 2**

Item	Cost ($)
Bread	1.99
Cereal	4.50
Orange Juice	3.00

$20

3 Nina used $12\frac{1}{3}$ yards of ribbon to make hair bows and 5 yards of ribbon to wrap gifts. She has $22\frac{2}{3}$ yards of ribbon left. Write and solve a subtraction equation that can be used to find how many yards of ribbon she had to start. **6.EE.7, MP 2**

$r - (12\frac{1}{3} + 5) = 22\frac{2}{3}$; $r = 40$

She had 40 yards of ribbon to start.

4 **H.O.T. Problem** What value of x makes the following equation true? Write a real-world problem that could be modeled with this equation. **6.EE.7, MP 7**

$$x - \frac{1}{8} = \frac{1}{12}$$

$\frac{5}{24}$; Jen had less than a pound of sunflower seeds. After she ate $\frac{1}{8}$ pound of seeds, she had $\frac{1}{12}$ pound left. How many pounds of sunflower seeds did she have to begin with?

NAME ____________ DATE ________ PERIOD ________

Lesson 4 Multi-Step Problem Solving

Chapter 7

Multi-Step Example

The table shows the number of miles Diego and some friends traveled each day and the amount of time it took. Write an equation that can be used to determine the average speed, r, at which they traveled. What property would you use to solve the equation? 6.EE.7, MP 1

Day	Distance (miles)	Time (hours)
1	110	1.9
2	90	1.5
3	105	1.8
4	120	2.1

Use a problem-solving model to solve this problem.

1 Understand

Read the problem. Circle the information you know.
Underline what the problem is asking you to find.

2 Plan

What will you need to do to solve the problem? Write your plan in steps.

Step 1 Write an equation to represent the situation.

Step 2 Determine the property used to solve the equation.

Read to Succeed!
When solving problems dealing with distance, rate, and time, the equation is $d = rt$; where d is the distance, r is the rate, and t is the time.

3 Solve

Use your plan to solve the problem. Show your steps.

The group drove a total of 425 miles in 7.3 hours.

$d = r \cdot t$ Distance, rate, time equation

$\underline{425} = r \cdot \underline{7.3}$ Substitute known values.

To solve the equation, you would use the **Division** Property of Equality.

4 Check

How do you know your solution is accurate?

Sample answer: I know the formula for distance is $d = r \cdot t$. Since the rate is not known, I used the inverse operation of multiplication, division, to solve the problem.

NAME ____________ DATE ________ PERIOD ________

Lesson 4 *(continued)*

Use a problem-solving model to solve each problem.

1 The table shows the distance Catrell biked each day and his rate. Write an equation that can be used to determine the average time, t, he spent riding his bike each day. What property would you use to solve the equation? 6.EE.7, MP 1

Day	Distance (miles)	Rate (miles/hour)
1	15	10
2	18	14
3	21	16
4	24	12

$78 = 52t$; Division Property of Equality

2 The model below shows the relationship between a gallon and pints. Use the model to write an equation to determine the number of pints given the gallons. Use the equation to convert $\frac{2}{3}$ gallon into pints. Round the answer to the nearest tenth. 6.EE.7, MP 2

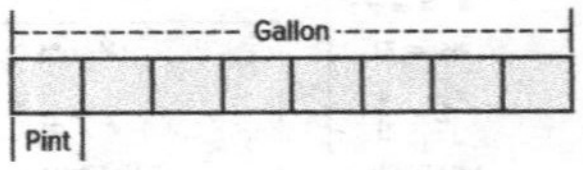

$8g = p$; 5.3

3 Linh decides to put $\frac{1}{6}$ of her paycheck into her savings account. Use the model below to write an equation that represents the amount of her paycheck in terms of the amount put into savings. Then use the equation to determine the amount of her paycheck. 6.EE.7, MP 2

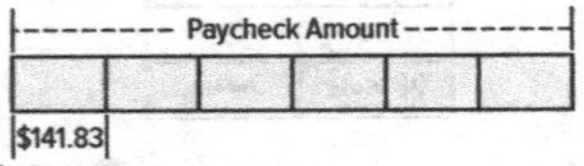

$\frac{1}{6}x = 141.83$; $850.98

4 **H.O.T. Problem** In the diagram below, $\angle ABC$ is divided into 4 angles of equal measure and $m\angle ABC = 140°$. Write and solve two equations that can be used to find the degree measure of $\angle EBF$. Identify any properties of equalities that you used to solve either equation. 6.EE.7, MP 2

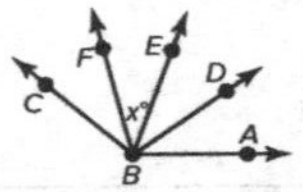

$\frac{140}{4} = x$; $4x = 140$; $x = 35$; Division Property of Equality

Answers

NAME ______ DATE ______ PERIOD ______

Lesson 5 Multi-Step Problem Solving

Multi-Step Example

Julian read a book in 4 days. The table shows the time he read each day. He read an average of 12 pages per hour. Which equation can be used to find the total number of pages p he read? What property would you use to solve the equation? 6.EE.7, MP 1

Day	Time (h)
1	1.2
2	1.7
3	2.6
4	2.5

Ⓐ $8 = p + 12$; Subtraction Property of Equality
Ⓑ $8 = p - 12$; Addition Property of Equality
Ⓒ $8 = 12p$; Division Property of Equality
Ⓓ (filled) $8 = \frac{p}{12}$; Multiplication Property of Equality

Use a problem-solving model to solve this problem.

1 Understand

Read the problem. Circle the information you know.
Underline what the problem is asking you to find.

2 Plan

What will you need to do to solve the problem? Write your plan in steps.

Step 1 Write an equation to represent the problem.

Step 2 Determine the property used to solve the equation.

Read to Succeed!
When you are writing an equation to solve a problem, it is helpful to draw a bar diagram first to model the problem.

3 Solve

Use your plan to solve the problem. Show your steps.

Julian spent a total of 8 hours reading. He read an average of 12 pages per hour.

The total time is equal to the total number of pages divided by the average number of pages.

__8__ $= \frac{p}{12}$ To solve the equation, you could use the __Multiplication__ Property of Equality and multiply each side of the equation by __12__.

Choice __D__ is the correct answer. Fill in that answer choice.

4 Check

How do you know your solution is accurate?

Sample answer: I can draw a bar diagram to check the equation. I know that the Multiplication Property of Equality is used to solve a division equation.

NAME ______ DATE ______ PERIOD ______

Lesson 5 *(continued)*

Use a problem-solving model to solve each problem.

1 It took Roberto 3 weeks to read a book for his English class. The table shows the time he spent reading the book each week. He read an average of 15 pages per hour. Which equation can be used to determine the total number of pages p he read? What property would you use to solve the equation? 6.EE.7, MP 1

Week	Time (h)
1	8.5
2	6
3	3.5

Ⓐ $18 = \frac{p}{15}$; Division Property of Equality
Ⓑ (filled) $18 = \frac{p}{15}$; Multiplication Property of Equality
Ⓒ $18 = 15p$; Division Property of Equality
Ⓓ $18 = 15p$; Multiplication Property of Equality

2 Olivia's little sister divided 40 stickers evenly among some friends. Each friend received 4 animal stickers, 2 flower stickers, and 2 plant stickers. Write and solve a multiplication equation to find how many friends received stickers. 6.EE.7, MP 2

$8f = 40$; $f = 5$

3 Latisha swam on Monday, Wednesday, and Friday for 2 weeks and on Tuesday and Thurday for 1 week. She swam an average of 25 laps each day she swam. Write and solve a division equation to find how many laps she swam in all. 6.EE.7, MP 2

$\frac{s}{8} = 25$; $s = 200$

4 **H.O.T. Problem** Timothy and Heather each solved the equation $\frac{12}{x} = 3$. Who solved the problem correctly? Explain how you know. 6.EE.7, MP 3

Timothy	Heather
$\frac{12}{x} = 3$	$\frac{12}{x} = 3$
$36 = x$	$12 = 3x$
	$4 = x$

Heather; this is a division equation, so use the Multiplication Property of Equality to solve it. Multiply both sides of the equation by x, and then divide both sides of the equation by 3.

NAME______________ DATE__________ PERIOD__________

Lesson 1 Multi-Step Problem Solving

Multi-Step Example

Sandra places wooden sculptures on top of a 3-foot-tall stand. The table can be used to compare the height of a sculpture to the height including the stand. What is the total height, in inches, of a 5-foot-tall sculpture on a stand? **6.EE.9, MP 1**

Height of Sculpture (ft) (x)	$x + 3$	Height on Stand (ft) (y)
2		
3		
5		

Ⓐ 8 in.
Ⓑ 10 in.
Ⓒ 60 in.
Ⓓ 96 in.

Use a problem-solving model to solve this problem.

1 Understand

Read the problem. Circle the information you know.
Underline what the problem is asking you to find.

2 Plan

What will you need to do to solve the problem? Write your plan in steps.

Step 1 Use the table to determine the height including the stand.

Step 2 Convert the height to inches by multiplying by 12.

3 Solve

Use your plan to solve the problem. Show your steps.

Height of Sculpture (ft) (x)	$x + 3$	Height on Stand (ft) (y)
2	**2 + 3**	**5**
3	**3 + 3**	**6**
5	**5 + 3**	**8**

Read to Succeed!
There are 12 inches in 1 foot. To convert feet to inches, multiply by 12.

So, the height on the stand is **8** feet. In inches, this is 8 × 12 or **96** inches. Choice **D** is correct. Fill in that answer choice.

4 Check

How do you know your solution is accurate?

Sample answer: Draw a bar diagram to show the height of the sculpture on the stand. Each section of the bar diagram would represent 12 inches.

NAME______________ DATE__________ PERIOD__________

Lesson 1 *(continued)*

Use a problem-solving model to solve each problem.

1 Eduardo has a $5 coupon for groceries. The amount he owes can be found by subtracting 5 from the total cost of his groceries. He buys $42 worth of groceries and pays with a $50 bill. How much change does he receive? **6.EE.9, MP 1**

Cost of Groceries (x)	$x - 5$	Amount Owed (y)
$15		
$25		
$42		

Ⓐ $8
Ⓑ $13
Ⓒ $37
Ⓓ $82

2 Sam plays a trivia game. He earns 6 points for each question that he answers correctly. He creates a table to show this relationship. What is x when $y = 30$? **6.EE.9, MP 2**

Number of Correct Answers (x)	$6(x)$	Points Earned (y)
		6
		30

5

3 Catarina invites friends to come over for breakfast. She made 50 muffins. If there are x guests, then each guest can have $50 \div x$ muffins. She creates a table to show the relationship. If $y = 2$, what is x? **6.EE.9, MP 2**

Number of Guests (x)	$50 \div (x)$	Muffins per Guest (y)
		5
		2

25

4 H.O.T. Problem If the side length of a square is x, then the perimeter of the square is $4x$. Complete the table to find the perimeters for the squares with sides lengths as shown. Describe the relationship shown in the table. **6.EE.9, MP 7**

Side Length (x)	$4(x)$	Perimeter (y)
$\frac{1}{3}$	$4\left(\frac{1}{3}\right)$	$1\frac{1}{3}$
$\frac{1}{2}$	$4\left(\frac{1}{2}\right)$	**2**
$1\frac{1}{2}$	$4\left(1\frac{1}{2}\right)$	**6**

Sample answer: As the side lengths increase, the perimeters increase. Each dependent quantity is 4 times the independent quantity.

NAME ____________ DATE ________ PERIOD ________

Lesson 2 Multi-Step Problem Solving

Chapter 8

Multi-Step Example

Autumn and Bennett painted signs for a school campaign. The table shows the total number of signs painted, based on the number of hours spent painting. How many more signs did Autumn paint than Bennett after 6 hours? **6.EE.2,** MP 1

Hours	Autumn	Bennett
1	6	4
2	9	6
3	12	8
4	15	10

Use a problem-solving model to solve this problem.

1 Understand

Read the problem. Circle the information you know.
Underline what the problem is asking you to find.

2 Plan

What will you need to do to solve the problem? Write your plan in steps.

Step 1 Determine the rule for each person.

Step 2 Use the rule to determine the number of signs made at 6 hours.

Step 3 Subtract to determine how many more signs Autumn painted.

Read to Succeed!
After determining the algebraic rule for each person, test your rule by using the independent and dependent quantities from the table.

3 Solve

Use your plan to solve the problem. Show your steps.

Autumn: 3(*x*) + 3 Bennett: 2(*x*) + 2

3(**6**) + 3 = **21** 2(**6**) + 2 = **14**

So, Autumn painted **21** − **14** or **7** more signs.

4 Check

How do you know your solution is accurate?

Sample answer: Extend the pattern in the table to determine the number of signs each person completed.

NAME ____________ DATE ________ PERIOD ________

Lesson 2 *(continued)*

Use a problem-solving model to solve each problem.

1 The table shows the number of boxes Sandra and Conisha can fill with canned food during a food drive, based on the number of hours worked. How many more boxes can Conisha fill than Sandra after 7 hours? **6.EE.2,** MP 1

Hours	Sandra	Conisha
1	4	8
2	7	16
3	10	24
4	13	32

34

2 The table shows the number of miles Lan and Bailey ran each of the last six days. How many more miles did Lan run than Bailey on the sixth day? **6.EE.2,** MP 7

Days	Lan	Bailey
1	1.10	5.0
2	2.20	4.1
3	3.30	3.2
4	4.40	2.3
5	?	?
6	?	?

6.10

3 The table shows the amount it costs to ride a go-kart, based on the number of hours. The rule to find the total cost is $a(x) + b$. What is the sum of a and b? **6.EE.2,** MP 7

Time (x)	Amount ($)
$1\frac{1}{2}$	13
2	15
3	19
5	27

11

4 H.O.T. Problem The table shows two rules. Describe each rule. What is the relationship between Rule 1 and Rule 2? **6.EE.2,** MP 7

Position	Rule 1: Value of Term	Rule 2: Value of Term
1	1	1
2	4	8
3	9	27
4	16	64

Rule 1: The value of each term is the position number times itself. Rule 2: The value of each term is the position number times itself, times itself once more. The relationship between the two rules is Position × Rule 1 = Rule 2.

NAME_____________ DATE _________ PERIOD _______

Lesson 3 Multi-Step Problem Solving

Multi-Step Example

Kya's take-home pay, y, on a given day equals her earnings, x, minus $3 for parking. The relationship can be represented by the equation $y = x - 3$. Make a table of values and plot the ordered pairs. Which ordered pair will NOT be on the graph? 6.EE.9, MP 1

Ⓐ (70, 67) Ⓒ (82, 79)
Ⓑ (51, 54) Ⓓ (93, 90)

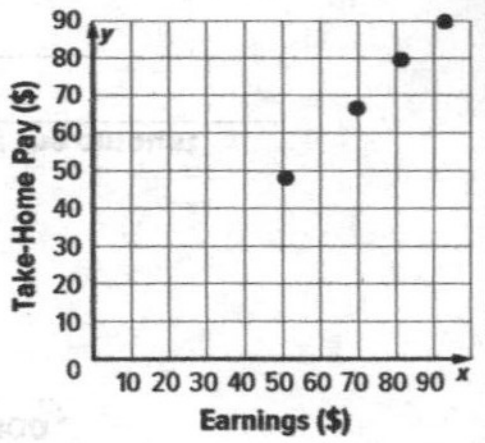

Use a problem-solving model to solve this problem.

1 Understand

Read the problem. Circle the information you know.
Underline what the problem is asking you to find.

2 Plan

What will you need to do to solve the problem? Write your plan in steps.

Step 1 Make a table to show the independent and dependent quantities.

Step 2 Write and graph the ordered pairs.

3 Solve

Use your plan to solve the problem. Show your steps.

x	$x - 3$	y
70	70 − 3	67
51	51 − 3	48
82	82 − 3	79
93	93 − 3	90

Read to Succeed!
Use the x-coordinates in the table to check the y-coordinates of the answer choices.

The ordered pair (51, 54) is not on the graph. So, choice __B__ is correct.
Fill in that answer choice.

4 Check

How do you know your solution is accurate?

Sample answer: After graphing the ordered pairs, check the x-coordinates and y-coordinates and compare to the answer choices.

NAME_____________ DATE _________ PERIOD _______

Lesson 3 *(continued)*

Use a problem-solving model to solve each problem.

1 Antonio is buying tomatoes. The equation $y = 3(x)$ represents the total number of tomatoes that he will have, y, if he buys x packs of tomatoes. Which ordered pairs will be on the graph of this equation? 6.EE.9, MP 1

Ⓐ (0, 0), (3, 1), (9, 3)
Ⓑ (0, 0), (1, 3), (3, 9)
Ⓒ (0, 3), (1, 4), (3, 10)
Ⓓ (0, 0), (2, 6), (3, 6)

2 The table shows the amount of commission, y, that Fernando earns compared to his weekly sales, x. If the equation for this situation is $y = ax$, what is the value of a? 6.EE.9, MP 7

Sales (x)	Commission (y)
$250	$75
$175	$52.50
$80	$24

0.3

3 The graph compares the number of snow cones purchased to the total cost of the snow cones. If the equation for the situation is $y = ax$, what is the value of a? As the number of snow cones increases, does the total cost increase or decrease? 6.EE.9, MP 7

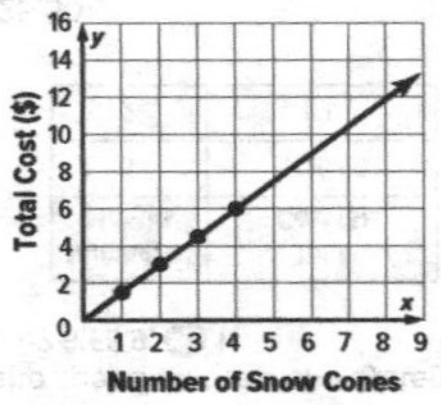

1.5; increase

4 H.O.T. Problem The perimeter of a rectangle is 30. The dimensions are unknown. Write an equation that gives the length, y, in terms of the width, x. Complete the table of values to help you. 6.EE.9, MP 7

Width (x)	Length (y)
4	11
5	10
8	7

The equation is $y = 15 - x$. Sample ordered pairs are shown in the table.

NAME ________ DATE ________ PERIOD ________

Lesson 4 Multi-Step Problem Solving

Chapter 8

Multi-Step Example

Brian's baseball team is hosting a tournament to help raise money to buy new uniforms. Each team pays $100 to compete, but some of the money will be used to pay for uniforms. Use the table to help you find the amount of money raised for uniforms if 6 teams compete. 6.EE.9, MP 1

Number of Teams Competing	Money for Uniforms ($)
1	50
2	100
3	

Ⓐ $50
Ⓑ $250
Ⓒ $300
Ⓓ $600

Use a problem-solving model to solve this problem.

1 Understand

Read the problem. Circle the information you know.
Underline what the problem is asking you to find.

2 Plan

What will you need to do to solve the problem? Write your plan in steps.

Step 1 Determine the rule that represents the situation.

Step 2 Use the rule to determine the amount of money raised if 6 teams compete.

Read to Succeed! Looking for a pattern in the table is another strategy.

3 Solve

Use your plan to solve the problem. Show your steps.

Each dependent quantity is **50** times the independent quantity.

The rule is $y =$ **50(x)**.

So, the amount raised when 6 teams compete is 50(**6**) or $**300**.

Choice **C** is correct. Fill in that answer choice.

4 Check

How do you know your solution is accurate?

Sample answer: Extend the pattern in the table to determine the amount raised when 6 teams compete.

NAME ________ DATE ________ PERIOD ________

Lesson 4 *(continued)*

Use a problem-solving model to solve each problem.

1 Felicia has an alarm system for her home. She paid $80 to have the alarm installed, and pays $45.50 each month. How much does she spend on the alarm system in the first year? 6.EE.9, MP 1

Number of Months	Total Cost ($)
1	125.50
2	171

Ⓐ $125.50
Ⓑ $216.50
Ⓒ $626.00
Ⓓ $1,005.50

2 At a craft store, Georgina buys several packs of beads. The equation $y = 1.50x$ represents her total cost y if she buys x packs of beads. If the data are graphed, what is the value of x in $(x, 16.50)$? 6.EE.9, MP 2

11

3 Montel is playing a math game. He earns points for every correct answer. The points on the graph represent the number of questions he answered correctly x, to his total score y. If the point $(7, y)$ is on the graph, what is y? 6.EE.9, MP 2

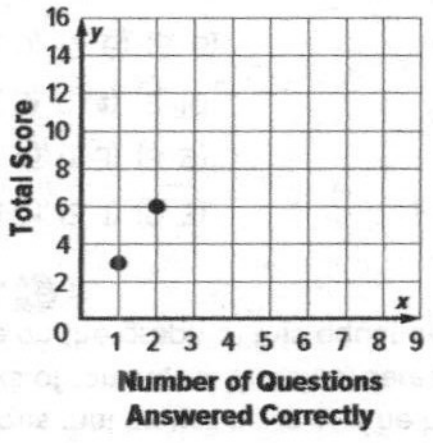

21

4 **H.O.T. Problem** Three friends are throwing a party at a park. They decide to rent a climbing wall for the party. The equation $y = 150x$ represents the total cost y for renting the wall for x hours. If they rented the wall for 5 hours and they split the cost equally, how much does each friend owe? 6.EE.9, MP 4

$250

NAME ________________ DATE ________ PERIOD ________

Lesson 5 **Multi-Step** Problem Solving

Multi-Step Example

Some friends each hope to attend a festival that costs $65. To earn money, they mowed lawns. Use the inequality $f + s \geq 65$, where f is the Friday earnings and s is the Saturday earnings, to determine who earned enough money to go to the festival. 6.EE.5, MP 1

	Friday	Saturday
Cody	$30	$10
Dominic	$60	$0
Emir	$45	$10
Fernando	$20	$55

Ⓐ Cody
Ⓑ Dominic
Ⓒ Emir
Ⓓ Fernando (answer: filled in)

Use a problem-solving model to solve this problem.

1 Understand

Read the problem. Circle the information you know.
Underline what the problem is asking you to find.

2 Plan

What will you need to do to solve the problem? Write your plan in steps.

Step 1 Substitute the values for f and s in the inequality.

Step 2 Determine whether the inequality is true.

Read to Succeed!
The inequality ≥ means greater than or equal to, so sums must be greater than or equal to 65 for the inequality to be true.

3 Solve

Use your plan to solve the problem. Show your steps.

Cody	Dominic	Emir	Fernando
$30 + 10 \geq 65$	$60 + 0 \geq 65$	$45 + 10 \geq 65$	$20 + 55 \geq 65$
40 ≥ 65	**60** ≥ 65	**55** ≥ 65	**75** ≥ 65

Since $75 \geq 65$, **Fernando** earned enough money. Choice **D** is correct. Fill in that answer choice.

4 Check

How do you know your solution is accurate?

Sample answer: Locate the values on a number line. Those values to the right of 65 on the number line are greater than 65 and make the inequality true.

NAME ________________ DATE ________ PERIOD ________

Lesson 5 *(continued)*

Use a problem-solving model to solve each problem.

1 The classmates below each have a cell phone plan that allows 200 text messages per month. Use the inequality $a + b > 200$, where a is the number of texts during Week 1 and b is the number of texts during Week 2, to determine who is already over their monthly budget of 200 texts. 6.EE.5, MP 1

	Week 1	Week 2
Olivia	25	50
Anna	150	10
Sierra	125	110
Vanesa	100	100

Ⓐ Olivia
Ⓑ Anna
Ⓒ Sierra (answer: filled in)
Ⓓ Vanesa

2 Rafael has at least 5 feet of wire. He uses 1.5 feet for a project. In the inequality $x + 1.5 \geq 5$, x represents the amount of wire that Rafael has left, in feet. What is the minimum number of inches he has left? 6.EE.5, MP 4

42 in.

3 Hannah's backpack can hold no more than 30 pounds. She has a laptop that weighs 7 pounds and books that weigh 5 pounds each. What is the maximum number of books that Hannah can carry in her backpack? 6.EE.5, MP 2

4

4 **H.O.T. Problem** If $x < 11$ and $x \geq 4$, what are the possible whole number values of x? 6.EE.5, MP 2

4, 5, 6, 7, 8, 9, 10

NAME ______________________ DATE ____________ PERIOD ________

Lesson 6 **Multi-Step** Problem Solving

Chapter 8

Multi-Step Example

The table shows the heights of 5 people and whether they were allowed to ride a certain roller coaster. Let a represent a person's height in inches. Which inequality represents the heights, in inches, of people allowed to ride this roller coaster? **6.EE.6, MP 1**

Height	Allowed
5 ft 7 in.	Yes
5 ft 3 in.	Yes
4 ft 6 in.	Yes
4 ft 5 in.	No
4 ft 3 in.	No

Ⓐ $a > 46$ Ⓒ $a \geq 54$

Ⓑ $a > 54$ Ⓓ $a < 54$

Use a problem-solving model to solve this problem.

1 Understand

Read the problem. Circle the information you know.
Underline what the problem is asking you to find.

2 Plan

What will you need to do to solve the problem? Write your plan in steps.

Step 1 Determine the minimum height allowed on the roller coaster. Convert the height to inches.

Step 2 Write an inequality that represents the situation.

Read to Succeed! When the term "minimum" is used, the inequality is greater than or equal to.

3 Solve

Use your plan to solve the problem. Show your steps.

The minimum height that was allowed on the roller coaster was **4** ft **6** in. Heights shorter than that were not allowed on the roller coaster.

4 feet **6** inches = **54** inches

So, $a \geq$ **54**. Choice **C** is correct. Fill in that answer choice.

4 Check

How do you know your solution is accurate?

Sample answer: Locate the 5 heights on a number line. The least height that was allowed on the roller coaster is the minimum.

NAME ______________________ DATE ____________ PERIOD ________

Lesson 6 *(continued)*

Use a problem-solving model to solve each problem.

1 The table shows the resting heart rate of different people and whether or not it was considered elevated. Let h represent the heart rate in beats per minute. Which inequality represents the heart rates that are considered elevated? **6.EE.6, MP 1**

Heart Rate (bpm)	Result
62	Normal
79	Normal
100	Normal
101	Elevated
105	Elevated

Ⓐ $h \leq 100$
Ⓑ $h > 101$
Ⓒ $h < 100$
Ⓓ $h \geq 101$

2 Lorenzo solved the following inequality $x + 3 < 10$ and graphed the solution on a number line. The graph contained the values −3, 2.4, 0, and 7. Which of these values is incorrect? Explain. **6.EE.6, MP 2**

7; 7 + 3 is not less than 10.

3 Marcus has $50 in his wallet. He buys 2 CDs that costs $12 each and wants to buy posters that cost $6 each. The inequality $p \leq 4$ represents the number of posters he can buy. Represent the solutions of the inequality on the number line. **6.EE.6, MP 3**

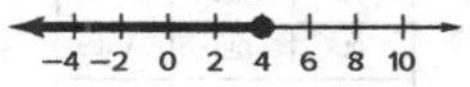

4 **H.O.T. Problem** Marianna wants to run at least 6.2 miles to train for a quarter-marathon. The inequality $m \geq 6.2$, where m is the number of miles she has run, represents the runs where she met her goal. Represent the solutions of the inequality on the number line. **6.EE.6, MP 3**

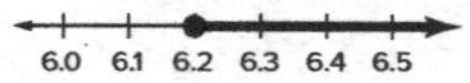

NAME__________ DATE __________ PERIOD __________

Lesson 7 **Multi-Step** Problem Solving

Chapter 8

Multi-Step Example

The table shows the costs of different size sandwiches at a sandwich shop. Ava has $30 to spend. She spends $5.25 on drinks and buys two Club sandwiches. Write and solve an inequality to find the maximum number of foot long sandwiches she can buy. **6.EE.5, MP 1**

Sandwich Size	Cost ($)
Club	2.75
6-inch	4.00
Foot long	6.25

Use a problem-solving model to solve this problem.

1 Understand

Read the problem. Circle the information you know.
Underline what the problem is asking you to find.

2 Plan

What will you need to do to solve the problem? Write your plan in steps.

Step 1 Write an inequality to represent the situation.

Step 2 Solve the inequality.

Read to Succeed!
The term "maximum" means that all values that make the inequality true will be less than the given value.

3 Solve

Use your plan to solve the problem. Show your steps.

Let x represent the number of foot long sandwiches she can buy.

$ **5.25** + 2($ **2.75**) + **6.25** x ≤ $ **30**

drinks — club sandwiches — foot long sandwiches — total she can spend

$5.25 + 2(2.75) + 6.25x \leq 30$ Write the inequality.

10.75 $+ 6.25x \leq 30$ Simplify the constants.

$6.25x \leq$ **19.25** Subtract.

$x \leq$ **3.08** Divide.

The greatest whole number that is a solution to the inequality is **3**.

4 Check

How do you know your solution is accurate?

Sample answer: Determine the amount Ava spends if she buys 3 and 4 foot long sandwiches to determine if either total is over $30.

NAME__________ DATE __________ PERIOD __________

Lesson 7 *(continued)*

Use a problem-solving model to solve each problem.

1 The table shows the costs of different bed sheet sizes at a home interior store. Phong has $100.25 to spend. He spends $45 on a blanket, and he buys one twin sheet for his brother. Write and solve an inequality to find the maximum number of full sheets he can buy. **6.EE.5, MP 1**

Sheet Size	Cost ($)
Twin	9.99
Full	10.50
Queen	20
King	28.50

$45 + $9.99 + $10.50x ≤ $100.25; 4

2 The rectangle below represents a table top that Lakita wants to cover with tiles. Each tile has an area of 7 square inches. To find the maximum number of tiles that will fit the table top, Lakita wants to use the inequality $7x \leq y$. What is the value of y? What is the maximum number of tiles she can use? **6.EE.5, MP 2**

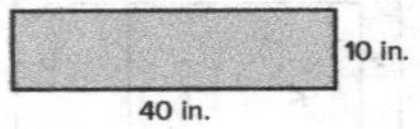

400; 57

3 Jamal wants to sell at least 50 tickets for a school raffle. He sells 10 tickets to his family. His dad is going to take at least $\frac{1}{2}$ of the remaining tickets to sell at work. Write and solve an inequality to find the minimum number of tickets his dad will take to work. Represent the solution on the number line. **6.EE.6, MP 3**

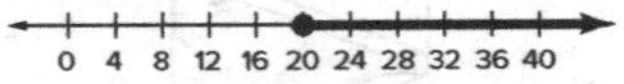

$2x \geq 40$; 20

4 **H.O.T. Problem** Write a word problem for the one-step inequality $3x \leq 15$. **6.EE.5, MP 2**

Sample answer: Heather made no more than 15 bracelets. She wants to give an equal number of bracelets to 3 friends. What is the maximum number of bracelets each friend will get? 5 bracelets

Answers

NAME ______________________ DATE ____________ PERIOD ________

Chapter 9

Lesson 1 **Multi-Step** Problem Solving

Multi-Step Example

Li is designing a flower bed for a school project. His design consists of a square inside a parallelogram. The shaded area will be planted with small shrubs. In order to get the correct number of shrubs, Li needs to determine the area of the shaded region once he has the dimensions. Which of the following formulas can be used to find the area of the shaded region? **6.G.1, MP 1**

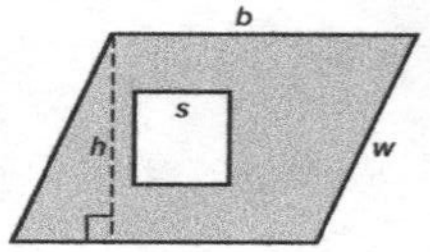

Ⓐ (filled) $A = b \cdot h - s^2$
Ⓑ $A = b \cdot w - s^2$
Ⓒ $A = b \cdot w + s^2$
Ⓓ $A = b \cdot h + s^2$

Use a problem-solving model to solve this problem.

1 Understand

Read the problem. Circle the information you know.
Underline what the problem is asking you to find.

Read to Succeed!
Be careful when finding the area of a parallelogram that the slant height is not used. The height of the parallelogram is perpendicular to the base.

2 Plan

What will you need to do to solve the problem? Write your plan in steps.

Step 1 Determine the formula for the area of the parallelogram and square.

Step 2 Subtract the area of the square from the area of the parallelogram.

3 Solve

Use your plan to solve the problem. Show your steps.

Area of Parallelogram − Area of Square = Area of Shaded Region

$b \cdot h \quad - \quad s^2 \quad = \quad b \cdot h - s^2$

The formula is $A = b \cdot h - s^2$, so choice **A** is correct. Fill in that answer choice.

4 Check

How do you know your solution is accurate?

Sample answer: Choices C and D include addition, which would not give the area of the shaded region. Choice B uses the wrong measurement of the parallelogram.

NAME ______________________ DATE ____________ PERIOD ________

Lesson 1 *(continued)*

Use a problem-solving model to solve each problem.

1 Beth is creating a flag for an art project. The flag is rectangular with two parallelograms of the same dimensions as seen in the diagram. Which formula can be used to find the area of the shaded region? **6.G.1, MP 1**

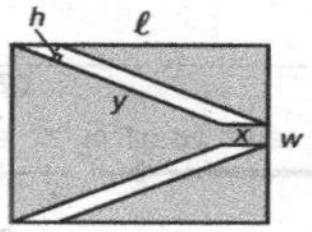

Ⓐ $A = \ell \cdot w - y \cdot h$
Ⓑ $A = \ell \cdot w - x \cdot h$
Ⓒ $A = \ell \cdot w - 2 \cdot x \cdot h$
Ⓓ (filled) $A = \ell \cdot w - 2 \cdot y \cdot h$

2 The side of an office building is made of mirrored glass panels in the shape of parallelograms. If one parallelogram-shaped piece of glass has a base of 8.5 feet and a height of 4 feet, determine how many windows there are in an 850 square foot area. **6.G.1, MP 2**

25 windows

3 An area of 861 square feet is used for 4 identical parking spaces as seen in the diagram below. Use the information in the diagram to find the height of one parallelogram-shaped parking space, *h*. **6.G.1, MP 2**

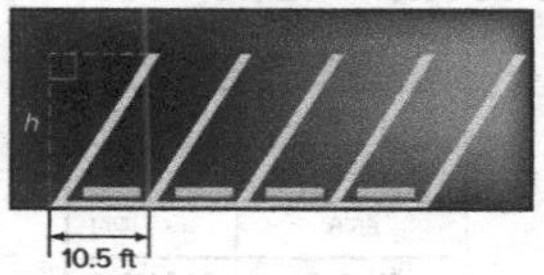

20.5 ft

4 **H.O.T. Problem** A rectangle is drawn with dimensions 6 units long and 4 units wide. Then a triangle is cut from the rectangle with dimensions shown, and placed on the left side of the rectangle to form a parallelogram. **6.G.1, MP 7**

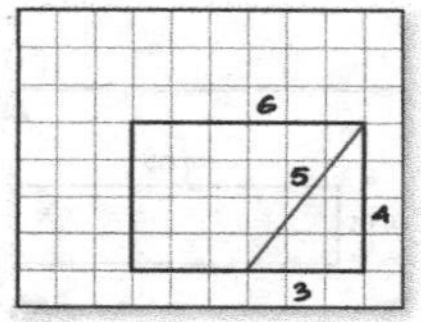

What are the area and perimeter of the original rectangle? What are the area and perimeter of the parallelogram formed by cutting and moving the triangle?

rectangle: $A = 24$ sq units, $P = 20$ units,

parallelogram: $A = 24$ sq units,

$P = 22$ units

NAME ________________ DATE ________ PERIOD ________

Lesson 2 Multi-Step Problem Solving

Chapter 9

Multi-Step Example

The triangle on the grid represents a triangular-shaped pillow. What is the area of the triangle? 6.G.1, MP 1

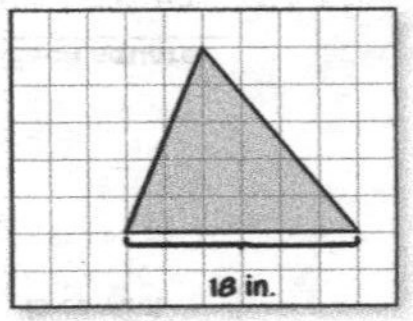

Ⓐ 270 in^2
Ⓑ 135 in^2 (filled)
Ⓒ 45 in^2
Ⓓ 15 in^2

Use a problem-solving model to solve this problem.

1 Understand

Read the problem. Circle the information you know.
Underline what the problem is asking you to find.

2 Plan

What will you need to do to solve the problem? Write your plan in steps.

Step 1 Determine the height of the triangle.

Step 2 Use the formula $A = \frac{1}{2}bh$ to find the area of the triangle.

3 Solve

Use your plan to solve the problem. Show your steps.

Each square on the grid represents 18 ÷ [6] or [3] inches.

The height of the triangle is [5] × [3] or [15] inches.

The area of the triangle is $\frac{1}{2}bh$ or $\frac{1}{2}$ · [18] · [15].

The area is [135] square inches.

Choice [B] is correct. Fill in that answer choice.

Read to Succeed! Remember to determine the scale of the grid.

4 Check

How do you know your solution is accurate?

Sample answer: Each square on the grid represents 9 square inches. The triangle covers about 15 squares. Since 15 × 9 is 135, the answer is accurate.

NAME ________________ DATE ________ PERIOD ________

Lesson 2 *(continued)*

Use a problem-solving model to solve each problem.

1 The figure below represents a swimming pool. A rope is attached from point *A* to *C*, and triangle *ABC* will be a designated adult swimming area. What is the area of the adult swimming region? **6.G.1, MP 1**

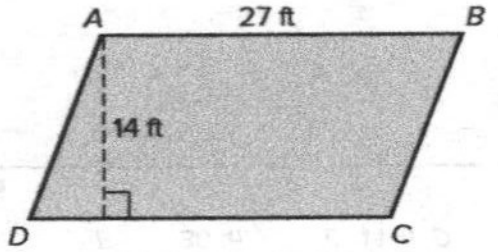

Ⓐ 41 ft^2
Ⓑ 82 ft^2
Ⓒ 189 ft^2 (filled)
Ⓓ 378 ft^2

2 The triangle on the grid outlines the border of a town. Each square on the grid represents a side length of 1.5 miles. What is the area of the town in square miles? **6.G.1, MP 4**

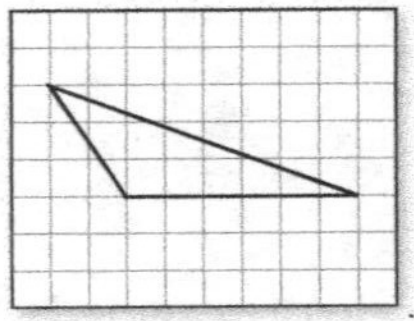

20.25 mi^2

3 The table shows the dimensions of three triangles. How much greater is the area of triangle C than triangle A, in square centimeters? **6.G.1, MP 7**

Triangle	Base	Height
A	8.5 cm	6 cm
B	7 cm	7 cm
C	9 cm	6.5 cm

3.75 cm^2

4 H.O.T. Problem Marco is going to paint 30% of the triangle. How many square inches will he paint? **6.G.1, MP 2**

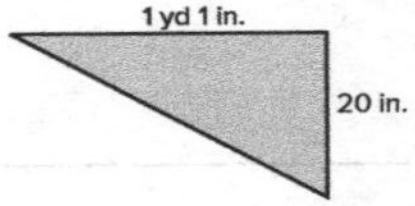

111 in^2

Answers

NAME ______________________ DATE ____________ PERIOD ____________

Lesson 3 Multi-Step Problem Solving

Multi-Step Example

The figure on the grid represents a parking lot. Asphalt for the parking lot costs $8.95 per square foot. How much will it cost to asphalt the parking lot, to the nearest cent? 6.G.1, MP 1

Ⓐ $290.88
Ⓑ $805.50
Ⓒ $18,616.00 (filled)
Ⓓ $37,232.00

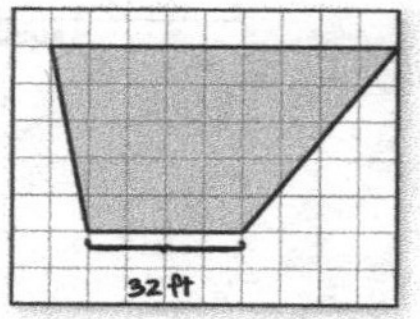

Use a problem-solving model to solve this problem.

1 Understand

Read the problem. Circle the information you know.
Underline what the problem is asking you to find.

2 Plan

What will you need to do to solve the problem? Write your plan in steps.

Step 1 Determine the area of the trapezoid.

Step 2 Multiply to determine the cost of the asphalt.

Read to Succeed!
The height of a trapezoid is perpendicular to the two bases.

3 Solve

Use your plan to solve the problem. Show your steps.

Each square on the grid represents a length of 32 ÷ 4 or 8 feet.
The two bases are 32 feet and 72 feet. The height is 40 feet.

$A = \frac{1}{2}h(b_1 + b_2)$ Area of a trapezoid

$A = \frac{1}{2}(40)(72 + 32)$ Replace h with 40, b_1 with 72, and b_2 with 32.

$A = 2{,}080$ Multiply.

So, the cost of the asphalt is $8.95 × 2,080 or $18,616. Choice __C__ is correct.
Fill in that answer choice.

4 Check

How do you know your solution is reasonable?

Sample answer: The area of the trapezoid is about 30 squares. Each square is 64 square feet. So, the area is about 1,920 sq ft. The cost of the asphalt is about $10. So, the cost for the parking lot is about 1,920 × $10 or $19,200.

The answer is reasonable.

NAME ______________________ DATE ____________ PERIOD ____________

Lesson 3 *(continued)*

Use a problem-solving model to solve each problem.

1 The figure on the grid represents the floor of an office. Each square on the grid represents 2 units. Which expression represents an area of a floor that is twice this size? 6.G.1, MP 1

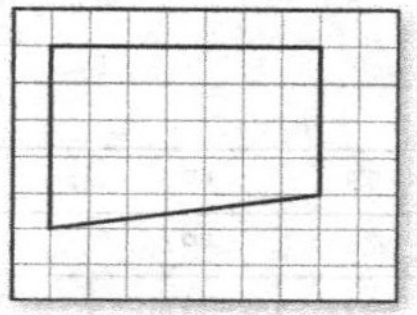

Ⓐ $\frac{1}{2}(7)(4 + 5)$
Ⓑ $7(4 + 5)$
Ⓒ $\frac{1}{2}(14)(8 + 10)$
Ⓓ $(14)(8 + 10)$ (filled)

2 A farmer spread fertilizer onto the plot of land shown below. He uses 4 scoops of fertilizer per square meter. If he used 312 scoops, what is the height of the trapezoid in meters? 6.G.1, MP 2

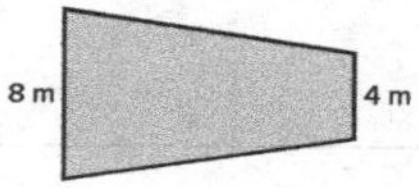

13 m

3 In the figure below, the area inside *ABEF* will be colored red. What will be the red area, in square feet? 6.G.1, MP 2

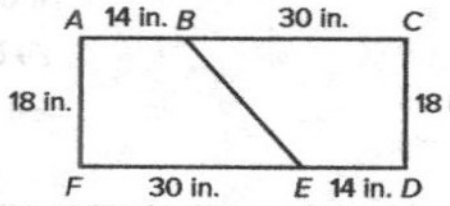

2.75

4 H.O.T. Problem Suppose the bases and height of a trapezoid are all multiplied by 3. How will the area change? 6.G.1, MP 7

The area will be multiplied by 9.

NAME ______________ DATE ________ PERIOD ________

Lesson 4 **Multi-Step** Problem Solving

Chapter 9

Multi-Step Example

The side lengths of the smaller triangle are multiplied by the same number to create a larger triangle with a base of 1 foot and height of $\frac{1}{2}$ foot. How many times greater is the area of the larger triangle? **6.G.1, MP 1**

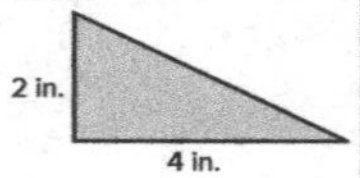

Ⓐ 3 Ⓒ 9 (filled)
Ⓑ 4 Ⓓ 36

Use a problem-solving model to solve this problem.

1 Understand

Read the problem. Circle the information you know.
Underline what the problem is asking you to find.

2 Plan

What will you need to do to solve the problem? Write your plan in steps.

Step 1 Determine the **area** of the smaller triangle.

Step 2 Convert the dimensions of the larger triangle to **inches**, determine the area, then compare the **areas**.

3 Solve

Use your plan to solve the problem. Show your steps.

Determine the area of the smaller triangle.

$\frac{1}{2} \times 2 \times 4 =$ **4 in²**

Determine the area of the larger triangle in inches.

1 ft = **12 in.** $\frac{1}{2}$ ft = **6 in.** $\frac{1}{2} \times 12 \times 6 =$ **36 in²**

36 ÷ **4** = **9**

The area of the larger triangle is **9** times larger.

So, the correct answer is **C**. Fill in that answer choice.

Read to Succeed!
Remember to convert the units of the larger triangle to inches before comparing the area of each triangle.

4 Check

How do you know your solution is accurate?

Sample answer: You can use the area of the smaller triangle, 4 square inches, to determine the area of the larger triangle by multiplying by 3² or 9. The area of the larger triangle is 36 square inches, which is 9 times greater.

NAME ______________ DATE ________ PERIOD ________

Lesson 4 *(continued)*

Use a problem-solving model to solve each problem.

1 The side lengths of the rectangle below are multiplied by $2\frac{3}{4}$. What effect would this have on the perimeter? **6.G.1, MP 1**

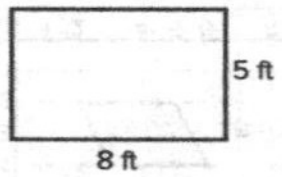

Ⓐ The perimeter will be $1\frac{3}{4}$ times greater.
Ⓑ The perimeter will be $2\frac{3}{4}$ times greater. (filled)
Ⓒ The perimeter will be $2\frac{5}{8}$ times greater.
Ⓓ The perimeter will be 3 times greater.

2 The dimensions of the rectangles listed in the table will all be multiplied by 3. What is the combined area of the enlarged rectangles in square meters? **6.G.1, MP 8**

Rectangle	Length (cm)	Width (cm)
Rectangle A	4	5
Rectangle B	6	6
Rectangle C	7	9

0.1071 m²

3 The Pentagon in Washington, D.C. is a regular pentagon. Juan made two scale models of the Pentagon. The perimeter of the larger model is how many times greater than the perimeter of the smaller model? (*Hint:* 1 inch ≈ 2.54 cm) **6.G.1, MP 8**

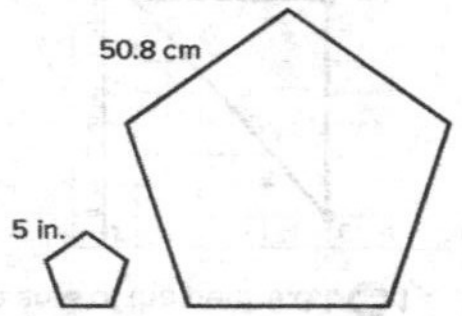

4 times

4 **H.O.T. Problem** The area of Rectangle B is 5 times greater than the area of Rectangle A. Give possible dimensions for each rectangle. Justify your answer. **6.G.1, MP 3**

Sample answer: Rectangle A is 2 by 5, and Rectangle B is 10 by 5. When both sides of a rectangle are multiplied by the same number, the area is multiplied by that number squared. But if only one side of a rectangle is multiplied by a number, the area is multiplied by that number.

Answers

NAME ______________________ DATE ____________ PERIOD ________

Lesson 5 Multi-Step Problem Solving

Multi-Step Example

Madeline drew this diagram of her rectangular vegetable garden. What are the coordinates of the vertices of the rectangle she drew? If the length of each grid square on the diagram is 2 yards, what is the area of Madeline's garden? 6.G.1, MP 1

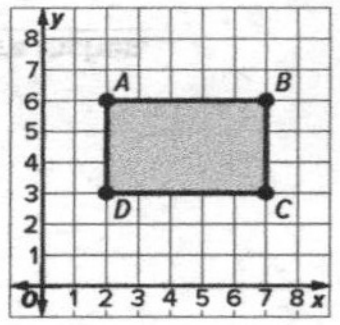

Use a problem-solving model to solve this problem.

Read to Succeed!
Remember, when x-coordinates are the same, you can subtract y-coordinates to find distance. And, when y-coordinates are the same, you can subtract x-coordinates to find distance.

1 Understand

Read the problem. Circle the information you know.
Underline what the problem is asking you to find.

2 Plan

What will you need to do to solve the problem? Write your plan in steps.

Step 1 Write the coordinates of the vertices of the rectangle and then determine the length of each side.

Step 2 Multiply the length and the width of the rectangle by 2 yards to find the length and the width of the garden.

Step 3 Multiply the length by the width to find the area of the garden.

3 Solve

Use your plan to solve the problem. Show your steps.

The coordinates of the vertices are A(**2** , **6**), B(**7** , **6**), C(**7** , **3**), D(**2** , **3**).

Length: Subtract x-coordinates → $\overline{AB}$: 7 − 2 = **5** and $\overline{CD}$: **7 − 2 = 5**

Width: Subtract y-coordinates → $\overline{AD}$: 6 − 3 = **3** and $\overline{BC}$: **6 − 3 = 3**

Multiply the length and the width by 2 yards: 5 × 2 = **10** . 3 × 2 = **6**

Find the area: **10 × 6 = 60** . The area of the garden is **60** square yards.

4 Check

How do you know your solution is accurate?

Sample answer: I know that the area of each grid square is 4 square yards. There are 15 squares in the rectangle. 4 × 15 = 60 square yards.

NAME ______________________ DATE ____________ PERIOD ________

Lesson 5 *(continued)*

Use a problem-solving model to solve each problem.

1 Hudson drew this diagram of the triangular pen he wants to build for his pet rabbits. What are the coordinates of the vertices of the triangle he drew? If the length of each grid square on the diagram is 1.5 feet, what is the area of the pen? 6.G.1, MP 1

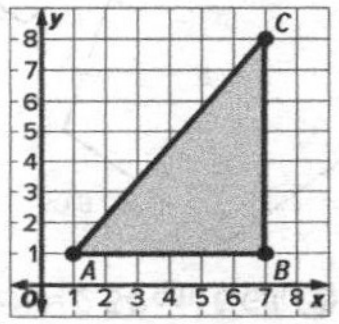

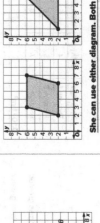

A(1, 1), B(7, 1), C(7, 8); 47.25 ft²

2 Montrel drew a diagram of his rectangular pool. What are the coordinates of the vertices of the pool he drew? The length of each grid square on the diagram is 1 foot. Montrel wants to put a deck around the outside of his pool that is 5 feet wide on all sides. What will be the perimeter of the outside of the deck? 6.G.3, MP 2

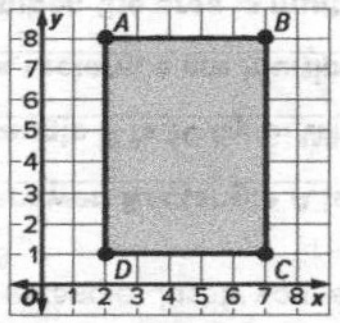

A(2, 8), B(7, 8), C(7, 1), D(2, 1)

64 ft

3 The diagram shows the shape and size of two different tiles that Charl wants to use to cover a wall. He wants to cover 72 square feet with each kind of tile. If the length of each grid square on the diagram is 1 foot, how many of each tile does Charl need? 6.G.1, MP 2

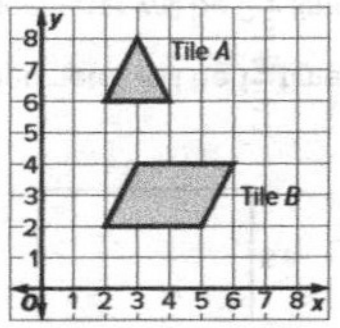

Tile A: 36; Tile B: 12

4 **H.O.T. Problem** Vivian drew two diagrams for flower gardens. The length of each grid square on the diagrams is 2 feet. Vivian wants a flower garden that is 64 square feet. Which diagram should she use for her garden? Explain. 6.G.1, MP 3

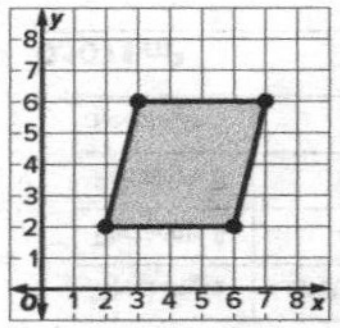

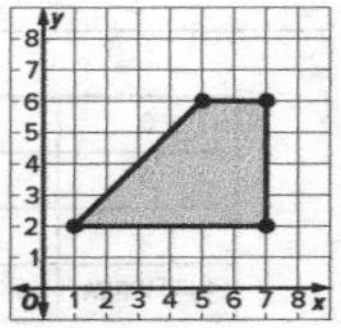

She can use either diagram. Both diagrams show gardens that are 64 ft².

NAME ______________________ DATE ____________ PERIOD ________

Lesson 6 Multi-Step Problem Solving

Multi-Step Example

The diagram shows a wall in Javier's living room that he wants to paint. Find the total area to be painted. 6.G.1, MP 1

Ⓐ 165 square feet
Ⓑ 189 square feet
Ⓒ 207 square feet
Ⓓ 231 square feet

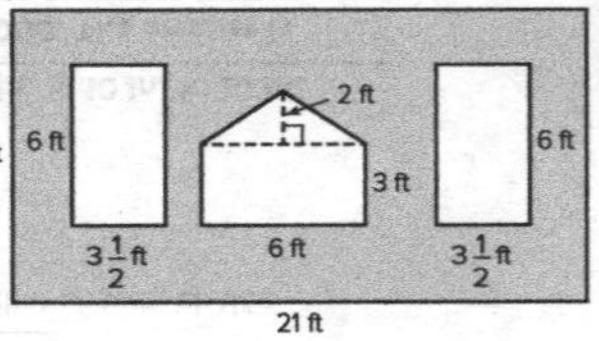

Use a problem-solving model to solve this problem.

Read to Succeed! The window in the center of the wall is a composite figure. You can find the area of a composite figure by separating it into figures for which you know how to find the area.

1 Understand

Read the problem. Circle the information you know.
Underline what the problem is asking you to find.

2 Plan

What will you need to do to solve the problem? Write your plan in steps.

Step 1 Find the area of the wall including the windows.

Step 2 Find the areas of the three windows.

Step 3 Subtract the areas of the windows from the total area of the wall.

3 Solve

Use your plan to solve the problem. Show your steps.

$21 \times 11 =$ **231** The area of the wall including the windows is **231** square feet.

Window on the left	Middle window	Window on the right
$6 \times 3\frac{1}{2} =$ **21**	$3 \times 6 = 18$; $\frac{1}{2}(6)(2) = 6$ $18 + 6 =$ **24**	$6 \times 3\frac{1}{2} =$ **21**

$21 + 24 + 21 =$ **66** The total area of the windows is **66** square feet.

$231 - 66 =$ **165** The total area to be painted is **165** square feet.

So, **A** is the correct answer. Fill in that answer choice.

4 Check

How do you know your solution is accurate?

Sample answer: I can work backward and use addition to check:

$165 + 21 + 24 + 21 = 231$.

NAME ______________________ DATE ____________ PERIOD ________

Lesson 6 *(continued)*

Use a problem-solving model to solve each problem.

1 The diagram shows where Jada plans to plant flowers and vegetables and place sod in her rectangular backyard. How many square meters of sod does she need? 6.G.1, MP 1

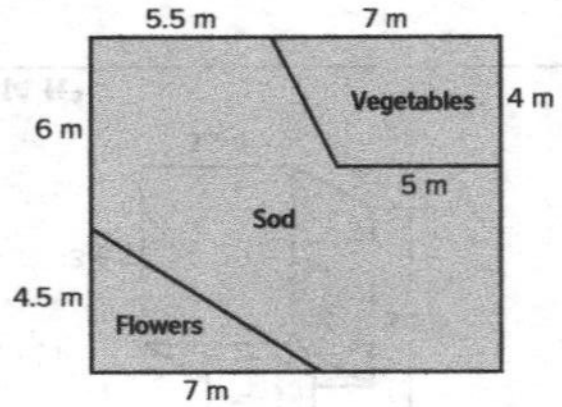

Ⓐ 131.25 square meters
Ⓑ 107.25 square meters
Ⓒ 91.5 square meters
Ⓓ 75.5 square meters

2 There are doors into the closet from both Ella's bedroom and Bethany's bedroom. How many square yards of carpet will it take to cover the floors in the two bedrooms and the closet? 6.G.1, MP 2

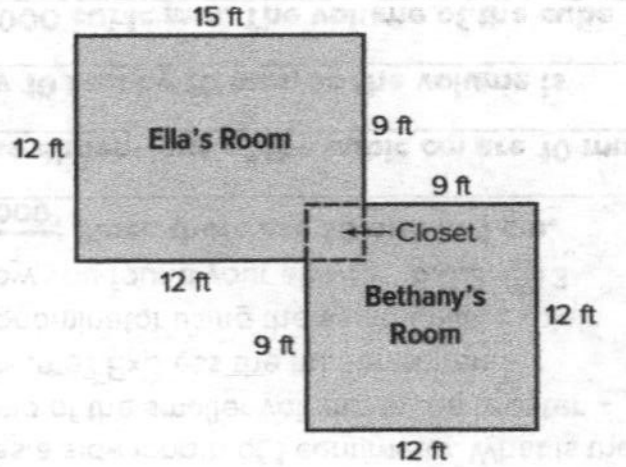

35 yd²

3 Ellery drew a diagram of a pen she fenced in for her rabbits. What is the area of the pen? 6.G.1, MP 2

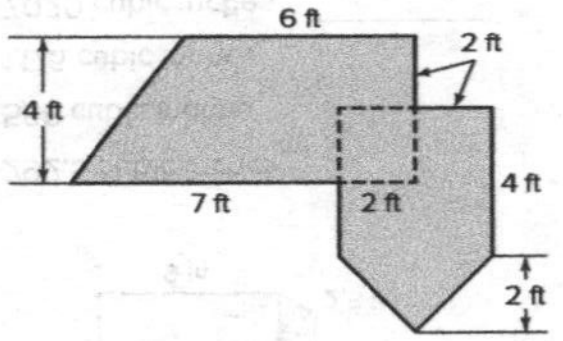

46 ft²

4 H.O.T. Problem To find the area of the shaded composite figure below, you can find the sum of the areas of the two shaded triangles and the shaded square. Describe another way you can find the area of the shaded composite figure. Then use one of the ways to find the area. 6.G.1, MP 3

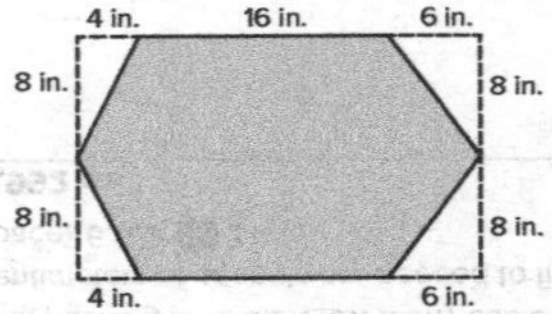

Sample answer: Find the area of the rectangle that encloses the shaded figure and then subtract the sum of the areas of the four small white triangles. $26 \times 16 = 416$; $416 - (16 + 16 + 24 + 24) = 336$. So the area of the shaded figure is 336 square inches.

NAME ____________ DATE ________ PERIOD ________

Lesson 1 **Multi-Step** Problem Solving

Multi-Step Example

If the fish tank shown is 80% filled with water, how much water is in the tank? **6.G.2,** MP 1

Ⓐ 5,772 cubic inches
Ⓑ 4,617.6 cubic inches (filled)
Ⓒ 1,154.4 cubic inches
Ⓓ 384.8 cubic inches

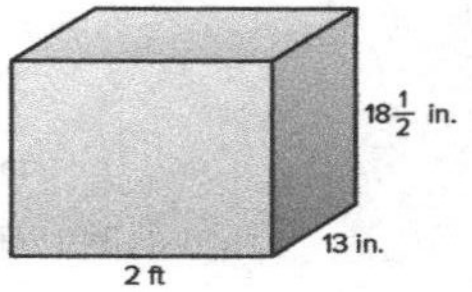

Use a problem-solving model to solve this problem.

1 Understand

Read the problem. Circle the information you know.
Underline what the problem is asking you to find.

2 Plan

What will you need to do to solve the problem? Write your plan in steps.

Step 1 Determine the volume of the tank.

Step 2 Multiply to determine the volume of water in the tank.

Read to Succeed!
The dimensions of the fish tank are given in feet and inches. Convert 2 feet to inches before finding the volume.

3 Solve

Use your plan to solve the problem. Show your steps.

$V = \ell \cdot w \cdot h$ Volume of a rectangular prism

$V = 24 \cdot 13 \cdot 18.5$ $\ell = 24$ in., $w = 13$ in., $h = 18.5$ in.

$V =$ **5,772** Multiply.

To find 80% of the volume, multiply by 0.80. The volume of water is **5,772** × 0.80 or **4,617.6** cubic inches. Choice **B** is correct. Fill in that answer choice.

4 Check

How do you know your solution is reasonable?

Sample answer: The volume of the tank is about 25 in. × 10 in. × 20 in., or 5,000 cubic inches. 80% is $\frac{4}{5}$ and $\frac{4}{5}$ of 5,000 is 4,000. The answer is reasonable.

NAME ____________ DATE ________ PERIOD ________

Lesson 1 *(continued)*

Use a problem-solving model to solve each problem.

1 The figure is a box full of cereal. If a case of 24 boxes are filled, how much cereal is there in all? **6.G.2,** MP 1

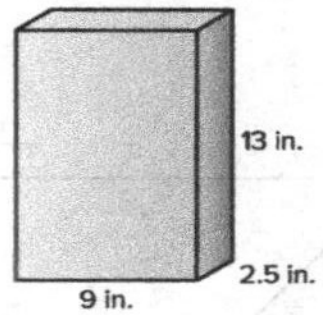

Ⓐ 292.5 cubic inches
Ⓑ 588 cubic inches
Ⓒ 1,176 cubic inches
Ⓓ 7,020 cubic inches (filled)

2 A storage cube that has an edge length of 16 centimeters is being packed in a cardboard box with a length of 28 centimeters, a width of 18 centimeters, and a height of 22 centimeters. The extra space is being filled with packing peanuts. How many cubic centimeters of peanuts are needed to fill the space? **6.G.2,** MP 2

6,992 cm^3

3 What is the volume of the statue in cubic feet? Round to the nearest hundredth. **6.G.2,** MP 7

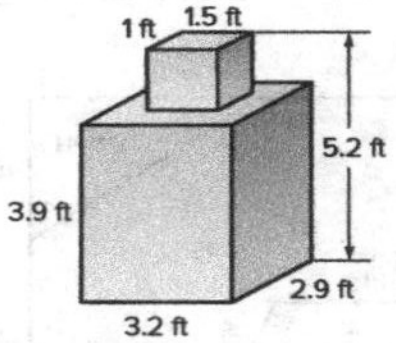

38.14 ft^3

4 **H.O.T. Problem** One cube has a side length of 1 millimeter, and another cube has a side length of 1 centimeter. What is the ratio of the smaller volume to the greater volume? Express the numerator and denominator using the same units. Explain how you found your answer. **6.G.2,** MP 3

$\frac{1}{1,000}$; Since there are 10 mm in 1 cm, the dimensions of the cubic cm are 10 mm by 10 mm by 10 mm, so the volume is 1,000 cubic mm. The volume of the cube with side length 1 mm is 1 cubic mm.

NAME________________ DATE __________ PERIOD __________

Lesson 2 Multi-Step Problem Solving

Multi-Step Example

Angela has a candle in the shape of a triangular prism with the dimensions shown in the drawing. If she burns the candle and reduces the volume by 25%, what will be the volume of the candle that is left?
Extension of 6.G.2, MP 1

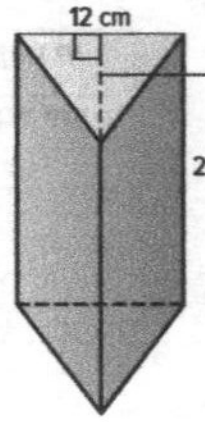

Use a problem-solving model to solve this problem.

Read to Succeed!
Remember, the formula for the volume of a triangular prism is V = Bh, where B is the area of the base and h is the height.

1 Understand

Read the problem. Circle the information you know.
Underline what the problem is asking you to find.

2 Plan

What will you need to do to solve the problem? Write your plan in steps.

Step 1 Determine the volume of the candle.

Step 2 Subtract the percent that has burned from 100% to find the percent that will be left.

Step 3 Multiply to determine the volume of the candle that will be left.

3 Solve

Use your plan to solve the problem. Show your steps.

$V = Bh$
$V = \left(\frac{1}{2} \cdot 12 \cdot 8\right)20$
$V =$ __960__ cm^3

$100 - 25 =$ __75__
__75__ % of the volume will be left.

To find 75% of the volume, multiply by 0.75.
$960 \cdot 0.75 = 720\ cm^3$

So, __720__ cubic centimeters will be left.

4 Check

How do you know your solution is reasonable?
Sample answer: The volume of the candle is $\left(\frac{1}{2} \cdot 10 \cdot 10\right)20$, or 1,000 cubic centimeters. 75% is $\frac{3}{4}$, and $\frac{3}{4}$ of 1,000 is 750. The answer is reasonable.

NAME________________ DATE __________ PERIOD __________

Lesson 2 *(continued)*

Use a problem-solving model to solve each problem.

1 Jamaal has a carton in the shape of a triangular prism with the dimensions shown in the diagram. He packs a gift in the carton that takes up $\frac{2}{3}$ of the volume of the carton. What is the volume of the space that is left in the carton after the gift is packed inside?
Extension of 6.G.2, MP 1

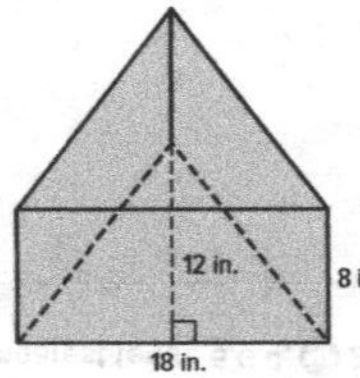

Ⓐ 216 cubic inches
Ⓑ 288 cubic inches (circled)
Ⓒ 648 cubic inches
Ⓓ 864 cubic inches

2 The diagram shows the dimensions of a fish pond that is in the shape of a triangular prism. Tom wants to build a fish pond with a depth that is $1\frac{1}{2}$ times greater than the depth of the fish pond shown in the diagram. How many cubic meters of water will be needed to fill Tom's pond to the top?
Extension of 6.G.2, MP 2

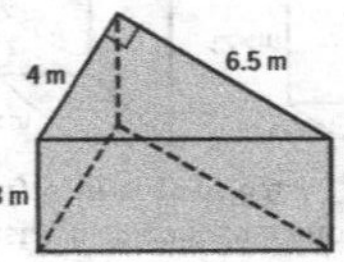

58.5 m³

3 The diagram shows the dimensions of two vases. Which vase holds the greater volume of water? How much greater?
Extension of 6.G.2, MP 2

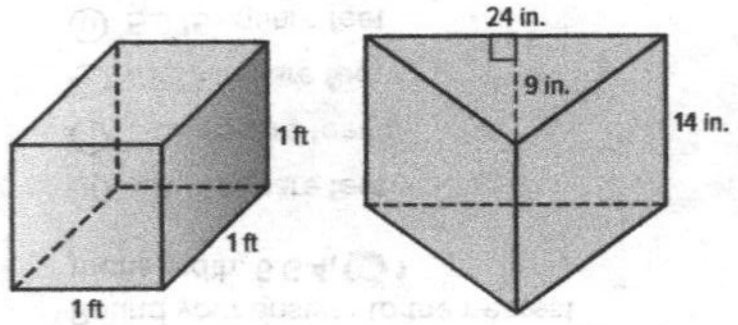

Vase A; 216 in³ or $\frac{1}{8}$ ft³

4 **H.O.T. Problem** The diagrams show the dimensions of two sheds. If both sheds have the same volume, what is the missing dimension on the shed on the right? Explain.
Extension of 6.G.2, MP 3

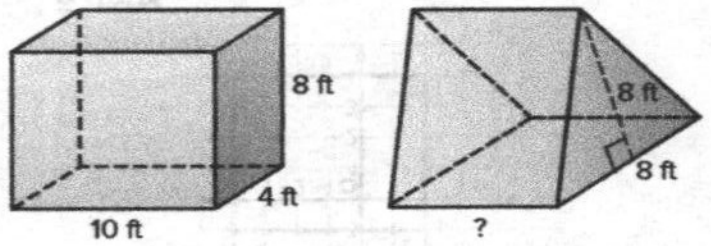

10 ft; The volume of the shed that is a rectangular prism is 10 × 4 × 8, or 320 cubic feet. Divide the volume of the shed on the left by the area of the base of the triangular prism to find the height of the triangular prism. 320 ÷ 32 = 10.

NAME_______________ DATE _________ PERIOD _________

Lesson 3 **Multi-Step** Problem Solving

Chapter 10

Multi-Step Example

Determine the surface area for each package. How much greater is the surface area of package B? **6.G.4, MP 1**

Ⓐ 10 square inches
Ⓑ 12 square inches (filled)
Ⓒ 20 square inches
Ⓓ 24 square inches

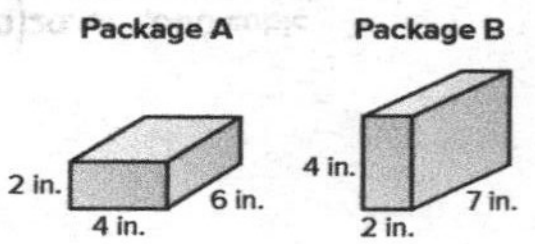

Use a problem-solving model to solve this problem.

1 Understand

Read the problem. Circle the information you know.
Underline what the problem is asking you to find.

2 Plan

What will you need to do to solve the problem? Write your plan in steps.

Step 1 Determine the __surface area__ for each prism.

Step 2 __Subtract__ the surface areas.

Read to Succeed! You may need to draw a net of each prism to help you visualize the area of each face.

3 Solve

Use your plan to solve the problem. Show your steps.

Package A surface area:
front and back: 2(2 • 4) = __16__
top and bottom: 2(6 • 4) = __48__
sides: 2(2 • 6) = __24__
__16__ + __48__ + __24__ = __88__ Add.

Package B surface area:
front and back: 2(2 • 4) = __16__
top and bottom: 2(2 • 7) = __28__
sides: 2(7 • 4) = __56__
__16__ + __28__ + __56__ = __100__ Add.

__100__ – __88__ = __12__ square inches greater Subtract.

So, the correct answer is __B__. Fill in that answer choice.

4 Check

How do you know your solution is accurate?

Sample answer: Draw a net of each rectangular prism to determine the area of each face. The surface area for package A is 88 square inches, which is 12 square inches less than package B, or 100 square inches.

NAME_______________ DATE _________ PERIOD _________

Lesson 3 *(continued)*

Use a problem-solving model to solve each problem.

1 Taro wants to build a storage box that will exactly fit his 6 reference books that are each 8 inches wide and 11 inches long. If half of his books are 1 inch thick, and half are 2 inches thick, how much material, in square feet, will he need to make the storage box? Round your answer to the nearest thousandth. **6.G.4, MP 1**

Ⓐ 3.597 square feet (filled)
Ⓑ 3.65 square feet
Ⓒ 4.735 square feet
Ⓓ 5.375 square feet

2 The coordinate grid shows the base of a rectangular prism. If the prism has a surface area of 170 units, what is its height, in units? **6.G.4, MP 7**

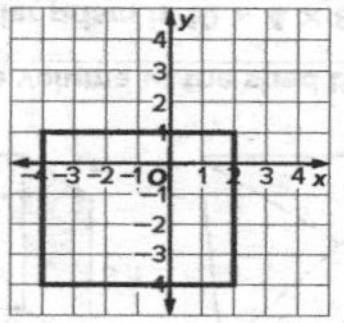

5 units

3 Each side length of a unit cube measures 2 units and increases by 50% every minute. What is the ratio of the surface area after 3 minutes to the original surface area? Write your answer as a decimal rounded to the nearest tenth. **6.G.4, MP 2**

11.4

4 H.O.T. Problem A chemical company wants to reduce the cost of their shipping containers. The measurements of the containers are shown. They pay for the containers by the amount of material required to make them. If they want to ship the greatest volume of chemicals at the lowest cost, which container should they use? Justify your answer. **6.G.4, MP 3**

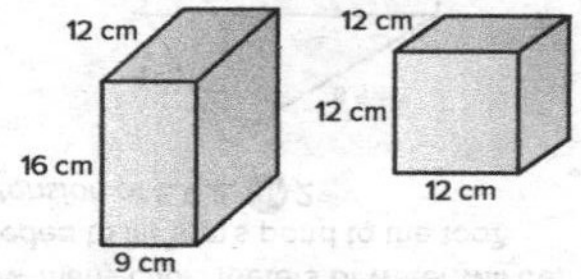

Sample answer: The volumes of both containers are 1,728 cm³. The surface area of the left container is 888 cm² and the surface area of the right container is 864 cm². Since 864 < 888, they should use the right container.

NAME ______________ DATE __________ PERIOD __________

Lesson 4 **Multi-Step** Problem Solving

Multi-Step Example

Two play houses at a children's gym are shown at the right. How much greater, in square feet, is the surface area of the larger house than the smaller house? **6.G.4, MP 1**

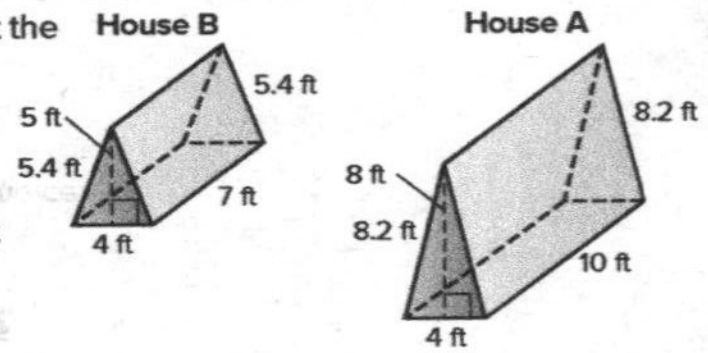

Ⓐ 112.4 square feet Ⓑ 123.6 square feet
Ⓒ 224 square feet Ⓓ 236 square feet

Use a problem-solving model to solve this problem.

1 Understand

Read the problem. Circle the information you know.
Underline what the problem is asking you to find.

2 Plan

What will you need to do to solve the problem? Write your plan in steps.

Step 1 Determine the **surface area** for each triangular prism.

Step 2 **Subtract** the surface areas.

Read to Succeed!
You may need to draw a net of each prism to help you visualize the area of each face.

3 Solve

Use your plan to solve the problem. Show your steps.

House A surface area:
triangular bases: $\frac{1}{2}(4 \cdot 8)(2) = \underline{32}$
faces: $2(8.2 \cdot 10) = \underline{164}$
$4 \cdot 10 = \underline{40}$

House B surface area:
triangular bases: $\frac{1}{2}(4 \cdot 5)(2) = \underline{20}$
faces: $2(5.4 \cdot 7) = \underline{75.6}$
$4 \cdot 7 = \underline{28}$

$\underline{32} + \underline{164} + \underline{40} = \underline{236}$ Add. $\underline{20} + \underline{75.6} + \underline{28} = \underline{123.6}$ Add.

$\underline{236} - \underline{123.6} = \underline{112.4}$ square feet Subtract.

So, the correct answer is **A**. Fill in that answer choice.

4 Check

Sample answer: Draw nets to determine the area of each face. The surface area for tunnel A is 236 ft², which is 112.4 ft² more than tunnel B.

NAME ______________ DATE __________ PERIOD __________

Lesson 4 *(continued)*

Use a problem-solving model to solve each problem.

1 In science class, Marco compares the two light prisms shown below. How much larger, in square inches, is the surface area of the larger light prism than the smaller light prism? **6.G.4, MP 1**

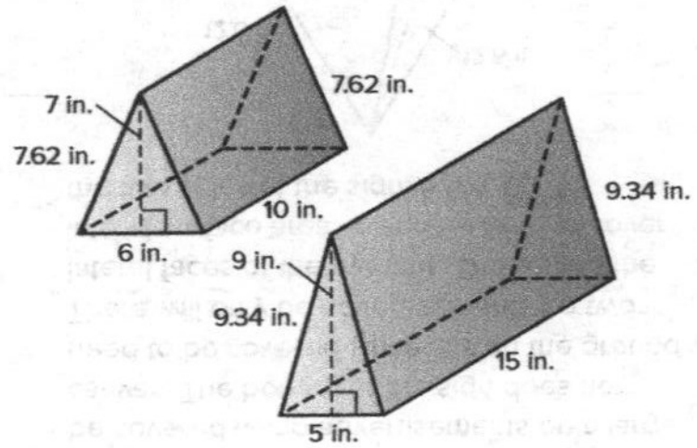

Ⓐ 142.8 square feet
Ⓑ 145.8 square feet
Ⓒ 212.4 square feet
Ⓓ 567.6 square feet

2 The net below represents a portion of a mural on a park sidewalk. If the dimensions are doubled, how many times greater is the surface area of the similar net, in square yards? **6.G.4, MP 7**

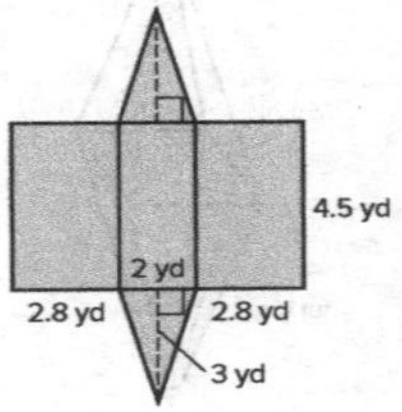

4 times

3 Gloria purchased a wedge pillow as shown below. She wants to make a pillow case for it. She has 500 square inches of fabric. How many more square inches of fabric does she need for the pillow case? **6.G.4, MP 7**

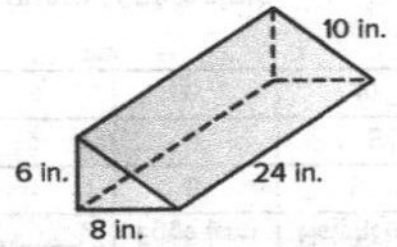

124 square inches

4 **H.O.T. Problem** The rectangular prism shown is cut in half diagonally to create the triangular prism. Is the surface area of the right triangular prism equal to one-half the surface area of the rectangular prism? Explain. **6.G.4, MP 3**

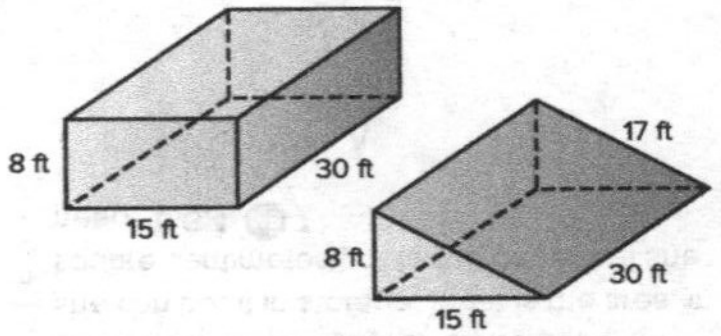

No; Sample answer: The surface area of the triangular prism is 1,320 square feet, and the surface area of the rectangular prism is 1,620 square feet.

NAME________________ DATE ________ PERIOD ________

Lesson 5 **Multi-Step** Problem Solving

Chapter 10

Multi-Step Example

The table shows the dimensions of three different square pyramids. What is difference between the greatest and least surface area, in square inches? **6.G.4, MP 1**

Pyramid	(Base Edge (in.)	Slant Height (in.)
A	2	5
B	5	12
C	3.5	9

Ⓐ 51.25 square inches
Ⓑ 69.75 square inches
Ⓒ 96 square inches
Ⓓ 121 square inches

Use a problem-solving model to solve this problem.

1 Understand

Read the problem. Circle the information you know.
Underline what the problem is asking you to find.

2 Plan

What will you need to do to solve the problem? Write your plan in steps.

Step 1 Determine the **surface area** for each square pyramid.

Step 2 **Subtract** the least from the greatest surface areas.

3 Solve

Use your plan to solve the problem. Show your steps.

Determine the area of each base and lateral face:

A: $(2)(2) = \mathbf{4}$ B: $(5)(5) = \mathbf{25}$ C: $(3.5)(3.5) = \mathbf{12.25}$

A: $\frac{1}{2}(2)(5) = \mathbf{5}$ B: $\frac{1}{2}(5)(12) = \mathbf{30}$ C: $\frac{1}{2}(3.5)(9) = \mathbf{15.75}$

A: **4** + **5** + **5** + **5** + **5** = **24** Add.

B: **25** + **30** + **30** + **30** + **30** = **145** Add.

C: **12.25** + **15.75** + **15.75** + **15.75** = **75.25** Add.

145 − **24** = **121** square inches Subtract.

So, the correct answer is **D**. Fill in that answer choice.

Read to Succeed!
A square pyramid has four triangular sides. Determine the area of one side, then add the area four times to determine the lateral surface area.

4 Check

How do you know your solution is accurate?

Sample answer: Draw a net of each square pyramid to determine the surface area. The surface area of the greatest pyramid is 145 square inches, which is 121 square inches greater than the least surface area, which is 24 square inches.

NAME________________ DATE ________ PERIOD ________

Lesson 5 *(continued)*

Use a problem-solving model to solve each problem.

1 The table shows the dimensions of three different square pyramids. What is difference between the greatest and least surface area, in square centimeters? **6.G.4, MP 1**

Pyramid	Base Edge (cm)	Slant Height (cm)
1	5	8.5
2	8	5
3	6	10

Ⓐ 12 square centimeters
Ⓑ 34 square centimeters
Ⓒ 46 square centimeters
Ⓓ 55 square centimeters

2 The net of Alana's crystal square pyramid is shown below. She wants to wrap the pyramid in three layers of tissue paper so she can put it in storage. What is the area, in square centimeters, of tissue paper will she need? **6.G.4, MP 7**

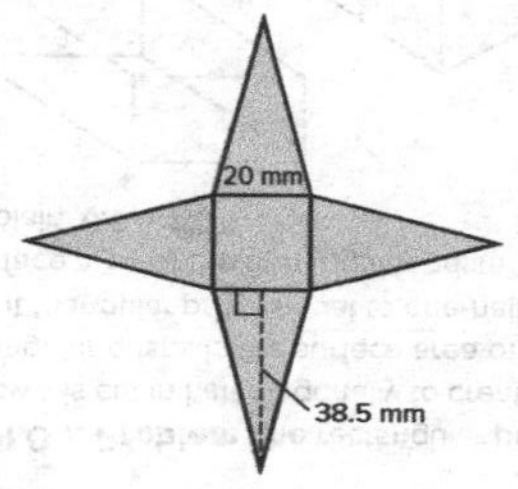

58.2 cm²

3 The pyramid below represents a sign at the entryway to a state park. The sign is going to be covered using advertisements on a large canvas. The bottom of the sign does not need to be covered since it is on the ground. There will only be advertisements on two lateral faces of the pyramid. Determine the lateral surface area, in square feet, to cover the two sides of the sign. **6.G.4, MP 7**

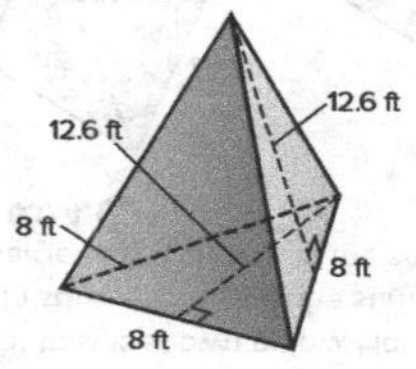

100.8 ft²

4 H.O.T. Problem The square pyramids below are congruent. What is the surface area of the composite figure? Explain. **6.G.4, MP 3**

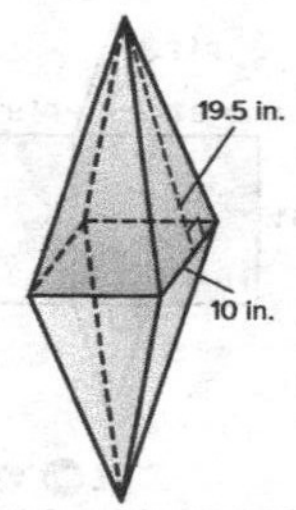

780 in²; Sample answer: The two pyramids share a base, which is not on the surface and is not needed. $\frac{1}{2}(10)(19.5) = 97.5$ and $97.5(8) = 780$ in²

NAME ______________ DATE ________ PERIOD ________

Lesson 1 **Multi-Step** Problem Solving

Chapter 11

Multi-Step Example

Edward's family owns a tree farm, which is open every day of the week except Monday. Edward kept track of how many trees were sold each day for two weeks. How much greater was the mean number of trees for Week 2 than for Week 1? **6.SP.3, MP 1**

Week 1	
Day	Trees
Tuesday	7
Wednesday	12
Thursday	6
Friday	14
Saturday	22
Sunday	17

Week 2	
Day	Trees
Tuesday	10
Wednesday	8
Thursday	12
Friday	17
Saturday	31
Sunday	18

Use a problem-solving model to solve this problem.

1 Understand

Read the problem. Circle the information you know.
Underline what the problem is asking you to find.

2 Plan

What will you need to do to solve the problem? Write your plan in steps.

Step 1 Determine the mean for each week.

Step 2 Subtract to find how much greater the mean for Week 2 is than Week 1.

Read to Succeed! To determine the mean, add each value in the set and divide by the number of values in the set.

3 Solve

Use your plan to solve the problem. Show your steps.

Week 1: $\frac{7 + 12 + 6 + 14 + 22 + 17}{6} = \underline{13}$

Week 2: $\frac{10 + 8 + 12 + 17 + 31 + 18}{6} = \underline{16}$

So, Week 2's mean is 16 − 13 or __3__ trees greater.

4 Check

How do you know your solution is reasonable?

Sample answer: The values for Week 2 are generally greater than the values for Week 1. The answer is reasonable.

NAME ______________ DATE ________ PERIOD ________

Lesson 1 *(continued)*

Use a problem-solving model to solve each problem.

1 The dot plot shows how many minutes Mr. Elliot's piano students said they practiced on the day before their lessons. Imala practiced 31 minutes but forgot to tell Mr. Elliot. If Imala's time were included, by how much time (in minutes) would the mean increase? **6.SP.3, MP 1**

Minutes of Practice

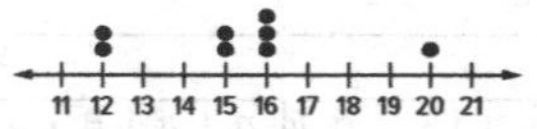

1.75 min

2 The graph shows how many rides a group of friends went on at the fair. Each ride costs $2.75. What was the mean amount of money, in dollars, spent per person to go on the rides? **6.SP.3, MP 2**

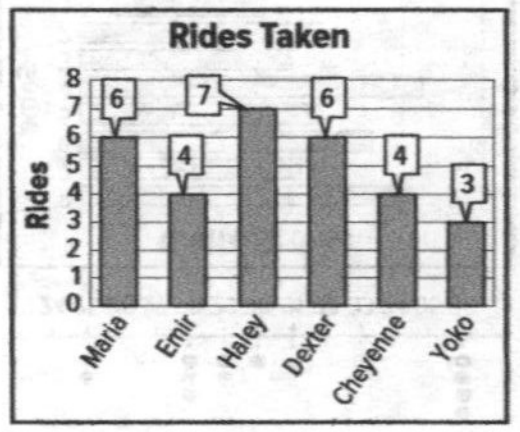

$13.75

3 For Quon's first six quizzes, he had a mean score of 33 points. After the seventh quiz, his mean score was 32 points. After the eighth quiz, the mean was 34. What was the difference in scores between his seventh and eighth quizzes? **6.SP.3, MP 8**

22 points

4 **H.O.T. Problem** Create a list of 8 values with a mean of 26. Justify your response. **6.SP.3, MP 3**

Sample answer: 14, 20, 22, 24, 30, 31, 33, 34; 14 + 20 + 22 + 24 + 30 + 31 + 33 + 34 = 208; 208 ÷ 8 = 26

Answers

NAME________ DATE ________ PERIOD ________

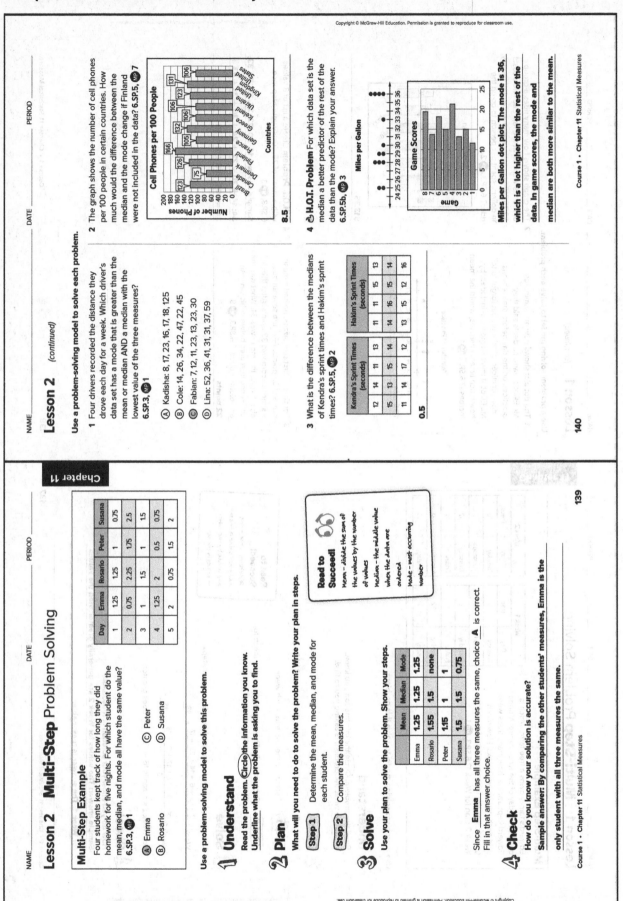

Lesson 2 **Multi-Step** Problem Solving

Chapter 11

Multi-Step Example

Four students kept track of how long they did homework for five nights. For which student do the mean, median, and mode all have the same value? 6.SP.3, MP 1

Day	Emma	Rosario	Peter	Susana
1	1.25	1.25	1	0.75
2	0.75	2.25	1.75	2.5
3	1	1.5	1	1.5
4	1.25	2	0.5	0.75
5	2	0.75	1.5	2

Ⓐ Emma (filled in)
Ⓑ Rosario
Ⓒ Peter
Ⓓ Susana

Use a problem-solving model to solve this problem.

1 Understand

Read the problem. Circle the information you know. Underline what the problem is asking you to find.

2 Plan

What will you need to do to solve the problem? Write your plan in steps.

Step 1 Determine the mean, median, and mode for each student.

Step 2 Compare the measures.

Read to Succeed!
Mean – divide the sum of the values by the number of values
Median – the middle value when the data are ordered
Mode – most occurring number

3 Solve

Use your plan to solve the problem. Show your steps.

	Mean	Median	Mode
Emma	**1.25**	**1.25**	**1.25**
Rosario	**1.55**	**1.5**	**none**
Peter	**1.15**	**1**	**1**
Susana	**1.5**	**1.5**	**0.75**

Since **Emma** has all three measures the same, choice **A** is correct. Fill in that answer choice.

4 Check

How do you know your solution is accurate?

Sample answer: By comparing the other students' measures, Emma is the only student with all three measures the same.

NAME________ DATE ________ PERIOD ________

Lesson 2 *(continued)*

Use a problem-solving model to solve each problem.

1 Four drivers recorded the distance they drove each day for a week. Which driver's data set has a mode that is greater than the mean or median AND a median with the lowest value of the three measures? **6.SP.3,** MP 1

Ⓐ Kadisha: 8, 17, 23, 16, 17, 18, 125
Ⓑ Cole: 14, 26, 34, 22, 47, 22, 45
Ⓒ Fabian: 7, 12, 11, 23, 13, 23, 30 (filled in)
Ⓓ Lina: 52, 36, 41, 31, 31, 37, 59

2 The graph shows the number of cell phones per 100 people in certain countries. How much would the difference between the median and the mode change if Finland were not included in the data? **6.SP.5,** MP 7

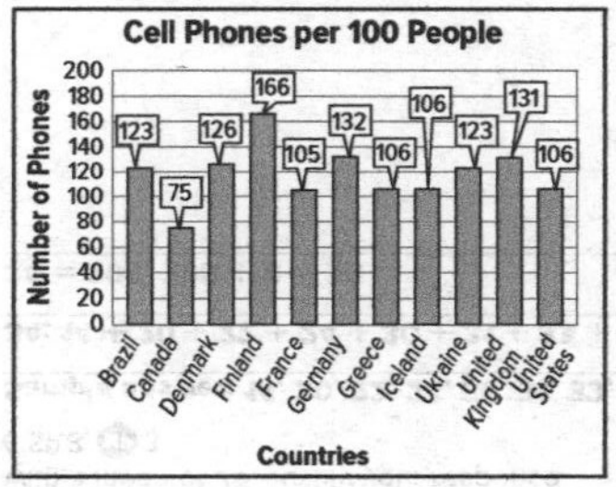

8.5

3 What is the difference between the medians of Kendra's sprint times and Hakim's sprint times? **6.SP.5,** MP 2

Kendra's Sprint Times (seconds)			
12	14	11	13
15	13	15	14
11	14	17	12

Hakim's Sprint Times (seconds)			
11	11	15	13
14	16	15	14
13	15	12	16

0.5

4 H.O.T. Problem For which data set is the median a better predictor of the rest of the data than the mode? Explain your answer. **6.SP.5b,** MP 3

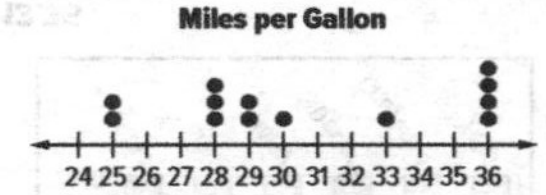

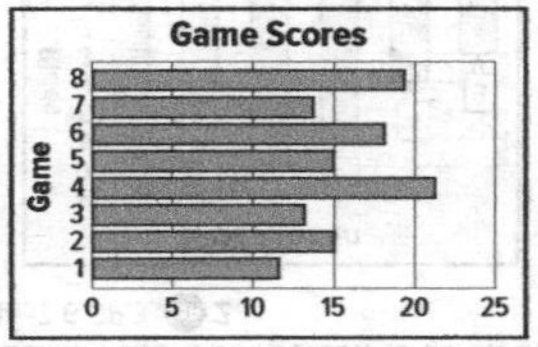

Miles per Gallon dot plot; The mode is 36, which is a lot higher than the rest of the data. In game scores, the mode and median are both more similar to the mean.

NAME_______________ DATE _________ PERIOD _______

Lesson 3 **Multi-Step** Problem Solving

Multi-Step Example

Carmen and Noah are running for president of the middle school student government. Votes are counted by classroom. What is the difference between the interquartile ranges for the two candidates? **6.SP.3, MP 1**

Voting Results					
Room Number	Number of Votes		Room Number	Number of Votes	
	Carmen	Noah		Carmen	Noah
1	12	9	5	8	17
2	6	18	6	2	13
3	14	8	7	18	7
4	20	6	8	12	12

Use a problem-solving model to solve this problem.

1 Understand

Read the problem. Circle the information you know.
Underline what the problem is asking you to find.

2 Plan

What will you need to do to solve the problem? Write your plan in steps.

Step 1 Order the values for each person from least to greatest.

Step 2 Determine the IQR for each data set.

Step 3 Subtract to find the difference.

Read to Succeed!
Interquartile range is the difference between the third quartile and the first quartile.

3 Solve

Use your plan to solve the problem. Show your steps.

Carmen: 2, 6, 8, 12, 12, 14, 18, 20 IQR: **16 − 7 or 9**

Noah: 6, 7, 8, 9, 12, 13, 17, 18 IQR: **15 − 7.5 or 7.5**

The difference between the interquartile ranges is **9 − 7.5 or 1.5**.

4 Check

How do you know your solution is reasonable?

Sample answer: Represent each data set on a dot plot to check the median, first, and third quartiles.

NAME_______________ DATE _________ PERIOD _______

Lesson 3 *(continued)*

Use a problem-solving model to solve each problem.

1 Melissa is keeping track of the temperature in her town at noon each day. She has recorded the temperature for six days so far. How much greater will the interquartile range be if Saturday's temperature is 70°F than if it is 58°F? **6.SP.3, MP 1**

Temperature at Noon	
Day	Temperature (°F)
Sunday	64
Monday	72
Tuesday	58
Wednesday	54
Thursday	60
Friday	62
Saturday	?

6

2 Jamal cut out these shapes from construction paper. What is the interquartile range for the areas of the shapes? **6.SP.3, MP 7**

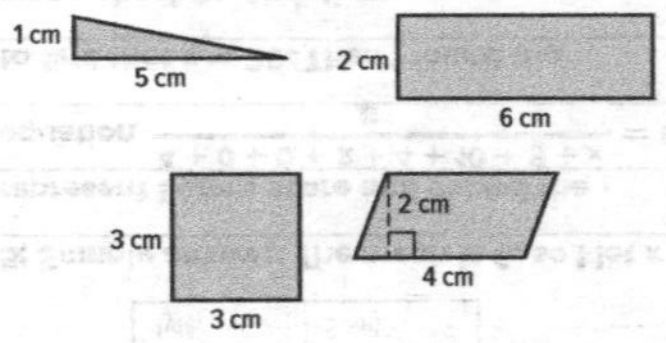

5.25

3 Tamiko is training for a bicycle race. She made a graph of the number of miles she rode each week for 6 weeks. If the median at Week 7 increases by 1.5, how many miles did she ride in Week 7? **6.SP.5, MP 7**

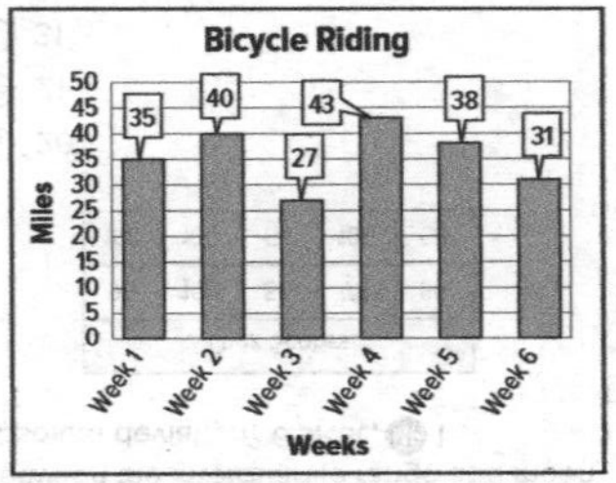

38 (Accept any number 38 or greater)

4 **H.O.T. Problem** Describe the measures of spread for the data set. Change the data set by adding an outlier. Describe the new measures of spread. **6.SP.5c, MP 3**

1,124	465
650	976
840	711
712	925

range: 659; median: 776; first quartile: 680.5; third quartile: 950.5; IQR: 270. Answers will vary for new data set; the new data value should be less than 275.5 or greater than 1,355.5. The range will increase, the median will be 712 or 840, the first quartile will be 557.5 or 680.5, the third quartile will be 950.5 or 1,050, and the IQR will be 393 or 369.5.

NAME_______________ DATE _______ PERIOD _______

Lesson 4 Multi-Step Problem Solving

Multi-Step Example

The table shows the number of hours various students worked on a school project. The students want to compare measures of spread. What is the difference between the interquartile range and mean absolute deviation? **6.SP.5c, MP 1**

Hours Worked			
Luke	0.5	Aria	3.5
Erin	2.5	Donte	1.0
Kim	2.5	Chante	3.5
Sierra	2.0	Kya	0.5

Ⓐ 0.25
Ⓑ 1.25 (filled)
Ⓒ 2.00
Ⓓ 5.75

Use a problem-solving model to solve this problem.

1 Understand

Read the problem. Circle the information you know.
Underline what the problem is asking you to find.

2 Plan

What will you need to do to solve the problem? Write your plan in steps.

Step 1 Determine the **mean absolute deviation** and **interquartile** range.

Step 2 Determine the **difference** between the two values.

3 Solve

Use your plan to solve the problem. Show your steps.

The mean absolute deviation is **1**. The interquartile range is **2.25**. The difference between the interquartile range and the mean absolute deviation is **2.25** − **1**, or **1.25**.

So, the correct answer is **B**. Fill in that answer choice.

Read to Succeed!
Remember to take the absolute value of the differences between each value and the mean when finding the mean absolute deviation.

4 Check

How do you know your solution is accurate?

I used technology tools to check my calculations. I used the graphing calculator stat menu to check the interquartile range and an online calculator to check the mean absolute deviation.

NAME_______________ DATE _______ PERIOD _______

Lesson 4 *(continued)*

Use a problem-solving model to solve each problem.

1 The table shows the quiz scores of various students. The students want to compare the measures of spread. What is the difference between the interquartile range and mean absolute deviation? **6.SP.5c, MP 1**

Quiz Scores				
95	100	50	75	60
100	100	60	100	60

Ⓐ 20
Ⓑ 21 (filled)
Ⓒ 31
Ⓓ 40

2 Eight students were asked how many persons live in their home. The results of the survey are shown in the line plot. What is the mean absolute deviation of the data? **6.SP.5c, MP 2**

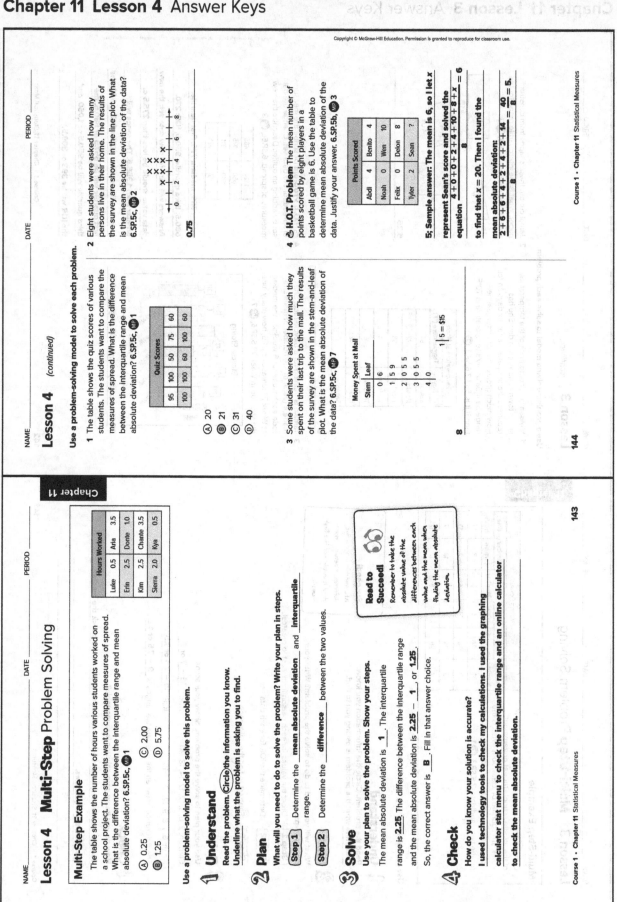

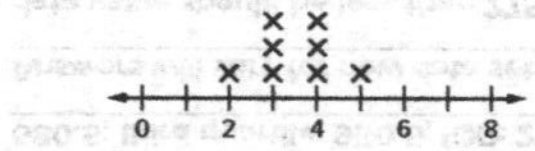

0.75

3 Some students were asked how much they spent on their last trip to the mall. The results of the survey are shown in the stem-and-leaf plot. What is the mean absolute deviation of the data? **6.SP.5c, MP 7**

Money Spent at Mall

Stem	Leaf
0	6
1	5 9
2	0 5 5
3	0 5 5
4	0

1 | 5 = $15

8

4 **H.O.T. Problem** The mean number of points scored by eight players in a basketball game is 6. Use the table to determine mean absolute deviation of the data. Justify your answer. **6.SP.5b, MP 3**

Points Scored			
Abdi	4	Benito	4
Noah	0	Wen	10
Felix	0	Delon	8
Tyler	2	Sean	?

5; Sample answer: The mean is 6, so I let x represent Sean's score and solved the equation $\frac{4+0+0+2+4+10+8+x}{8} = 6$ to find that $x = 20$. Then I found the mean absolute deviation: $\frac{2+6+6+4+2+4+2+14}{8} = \frac{40}{8} = 5$.

NAME______________________ DATE ____________ PERIOD ________

Lesson 5 **Multi-Step** Problem Solving

Chapter 11

Multi-Step Example

The table shows the ages of the people at a family dinner. Identify the outlier in the data set. Then determine how the outlier affects the mean of the data. **6.SP.5c, MP 1**

Age of Family Members (years)						
39	48	47	41	38	84	39

Ⓐ outlier: 48; mean age with the outlier decreased by 48

Ⓑ outlier: 84; mean age with the outlier decreased by 6

Ⓒ outlier: 84; mean age with the outlier increased by 6

Ⓓ outlier: 48; mean age with the outlier increased by 48

Use a problem-solving model to solve this problem.

1 Understand

Read the problem. Circle the information you know.
Underline what the problem is asking you to find.

2 Plan

What will you need to do to solve the problem? Write your plan in steps.

Step 1 Determine the outlier, or deviation from the majority of the data set.

Step 2 Determine the mean for the data set both with and without the outlier.

Step 3 Subtract to compare the mean age with and without the outlier.

Read to Succeed!
Measures of center are used to summarize a data set. Outliers often make one measure more appropriate to use than others.

3 Solve

Use your plan to solve the problem. Show your steps.

Compared to the other ages, 84 is very old. So, **84** is an outlier.

Mean with the outlier: 39 + 47 + 38 + 39 + 48 + 41 + 84 ÷ 7 = **48**

Mean without the outliner: 39 + 47 + 38 + 39 + 48 + 41 ÷ 6 = **42**

Compare: 48 − 42 = 6. So, the correct answer is **C**. Fill in that answer choice.

4 Check

How do you know your solution is reasonable?

Sample answer: Because the outlier is a greater number than the other numbers in the data set, the mean with the outlier is going to increase. The answer is reasonable.

NAME______________________ DATE ____________ PERIOD ________

Lesson 5 *(continued)*

Use a problem-solving model to solve each problem.

1 The table shows the weekly deposits Malcolm made in his savings account. Identify the outlier in the data set. Then determine how the outlier affects the mean of the data. **6.SP.5c, MP 1**

Deposits in Savings Account ($)								
41	41	28	38	26	26	5	36	32

Ⓐ outlier: 41; mean with the outlier increased by about 3.2

Ⓑ outlier: 5; mean with the outlier increased by about 3.2

Ⓒ outlier: 41; mean with the outlier decreased by about 3.2

Ⓓ outlier: 5; mean with the outlier decreased by about 3.2

2 The scores Miriam received on the science tests are 95, 80, 95, 85, 45, 95, 75, 85, and 90. Identify the outlier in the data set. Determine the mean, median, and mode without the outlier. Then tell which measure of center best describes the data without the outlier. **6.SP.5d, MP 7**

outlier: 45; mean: 87.5; median: 87.5; mode: 95; since the mean and the median are both 87.5, either of these measures of center best describes the data.

3 List eight data values for which the median is the best measure of center for the data set. Explain. **6.SP.5d, MP 3**

Sample answer: 1, 3, 4, 5, 6, 7, 22, 25; This data set has two extreme values, which will make the mean, 9.125, too great, and there is no mode. So, the median, 5.5, is the best measure of center.

4 H.O.T. Problem The table shows the lengths of some rivers in the United States. Identify the outlier. Find the measures of center with and without the outlier. Tell which measure of center best describes the data with and without the outlier. **6.SP.5c, MP 3**

River	Length (mi)
Columbia	1,243
Mississippi	2,340
Ohio-Allegheny	1,250
Peace	1,210
Red	1,290

outlier: 2,340; With the outlier, the mean is 1,466.6, the median is 1,250, and there is no mode. Without the outlier, the mean is 1,248.25, the median is 1,246.5, and there is no mode. With the outlier the best measure is the median; without the outlier the best measure is either the mean or the median.

Answers

NAME ______ DATE ______ PERIOD ______

Lesson 1 **Multi-Step** Problem Solving

Chapter 12

Multi-Step Example

Which description matches the data in the line plot of U.S. presidents' years in office? (Round decimals to the nearest whole number.) **6.SP.4, MP 1**

Presidents' Years in Office 1901–2009

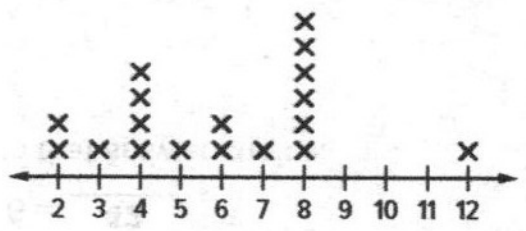

Ⓐ mean: 5, mode: 8, median: 7, range: 10, interquartile range: 4, outlier: 12

Ⓑ mean: 6, mode: 8, median: 6, range: 8, interquartile range: 6, outlier: 12

Ⓒ mean: 6, mode: 8, median: 6, range: 10, interquartile range: 4, no outlier

Ⓓ mean: 7, mode: 8, median: 5, range: 12, interquartile range: 4, no outlier

Use a problem-solving model to solve this problem.

1 Understand

Read the problem. Circle the information you know.
Underline what the problem is asking you to find.

2 Plan

What will you need to do to solve the problem? Write your plan in steps.

Step 1 Determine the measures of center.

Step 2 Determine the measures of spread.

Read to Succeed!
Determine the measures of center and spread and compare your answers to the answer choices, making sure to account for each measure.

3 Solve

Use your plan to solve the problem. Show your steps.

mean: about **6** range: 12 – 2 or **10** mode: **8**

IQR: 8 – 4 or **4** median: **6** outlier: **none**

Choice **C** lists the correct measures. Fill in that answer choice.

4 Check

How do you know your solution is reasonable?

Sample answer: After finding mean, only choices B and C will work. After finding the range, choice C is correct.

NAME ______ DATE ______ PERIOD ______

Lesson 1 *(continued)*

Use a problem-solving model to solve each problem.

1 Which line plot matches Noshi's description of his quiz scores? **6.SP.4, MP 1**

The mean and range are both 30.
The mode is 35. The interquartile range is 15.

Ⓐ
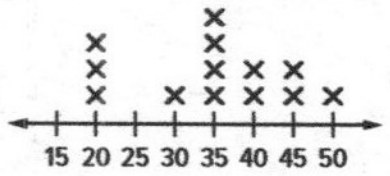

Ⓑ
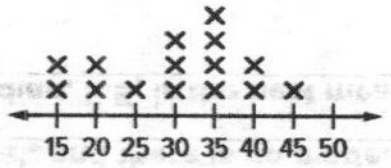

Ⓒ
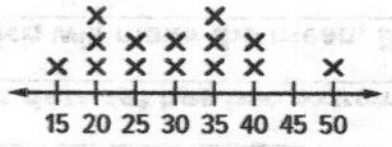

Ⓓ
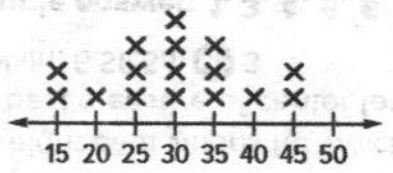

2 Liliana is shopping for pencils that are sold by the dozen (12). She finds out how much one pencil costs at each price and makes a line plot. Show the line plot. What price should have three marks above it? **6.SP.5a, MP 2**

Pencil Prices (per dozen)				
4 for $4.80	4 for $9.60	1 for $2.40	4 for $14.40	2 for $6.00
2 for $3.60	12 for $14.40	4 for $7.20	1 for $3.60	6 for $14.40

Price for One Pencil ($)

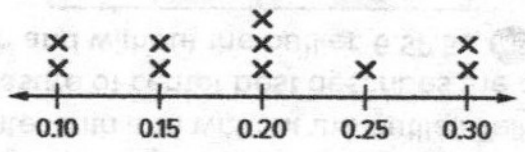

0.20

3 Joseph asked his friends how many pages of their history book they had read. He made a line plot to show his data and he said that the mean was 6. Then he noticed that he'd forgotten to include one number on the line plot. What number did he forget? **6.SP.5c, MP 4**

Number of Pages Read

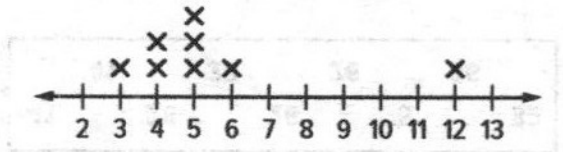

10

4 **H.O.T. Problem** The Math Club has been selling cookies during lunch. Make a line plot to show the number of cookies sold each day. Describe the data's measures of spread and center. **6.SP.4, MP 3**

Number of Cookies Sold
12, 16, 10, 12, 15, 13, 14, 20, 15, 15, 13, 11, 13, 14, 15, 14, 11, 12, 13, 15

Number of Cookies Sold

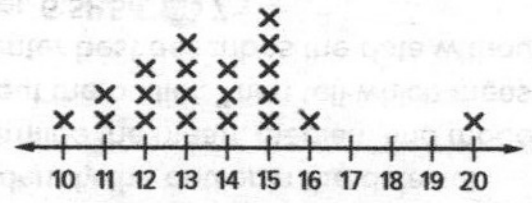

mean: 13.65; median: 13.5; mode: 15; first quartile: 12; third quartile: 15; IQR: 3; range: 10; outlier: 20

NAME_________________ DATE _________ PERIOD _______

Lesson 2 **Multi-Step** Problem Solving

Chapter 12

Multi-Step Example

The histogram shows the distances a volleyball team travels to their games. What percent of the games did they travel more than 24 miles? Round to the nearest tenth. 6.SP.5a, MP 1

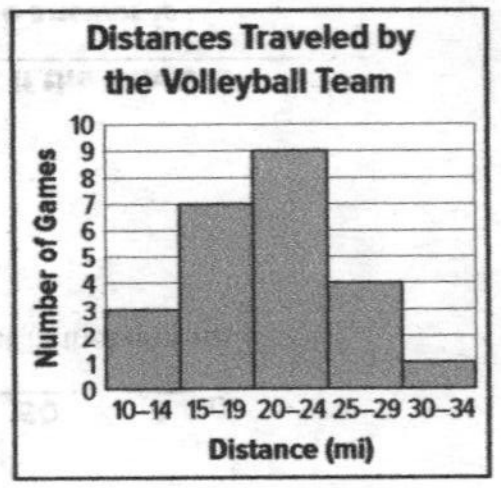

Use a problem-solving model to solve this problem.

1 Understand

Read the problem. Circle the information you know.
Underline what the problem is asking you to find.

2 Plan

What will you need to do to solve the problem?
Write your plan in steps.

Step 1 Determine how many games were more than 24 miles away. Determine the total number of games.

Step 2 Express as a percent.

Read to Succeed!
The percentage is the number of games greater than 24 miles divided by the total number of games. The decimal is then expressed as a percent.

3 Solve

Use your plan to solve the problem. Show your steps.

Number of games greater than 24 miles away: **5**

Total number of games: **24**

So, **5** out of **24** games were played greater than 24 miles away.

This is **5** ÷ **24** or **0.208** or **20.8** % of the games.

4 Check

How do you know your solution is reasonable?

Sample answer: Draw a bar diagram to represent the total number of games. Label a section as those games traveled greater than 24 miles. Determine the percent.

NAME_________________ DATE _________ PERIOD _______

Lesson 2 *(continued)*

Use a problem-solving model to solve each problem.

1 The histogram shows the average monthly temperature for cities in the United States for the month of August. What percent of cities have a monthly temperature of less than 80°F? Round to the nearest tenth. 6.SP.5a, MP 1

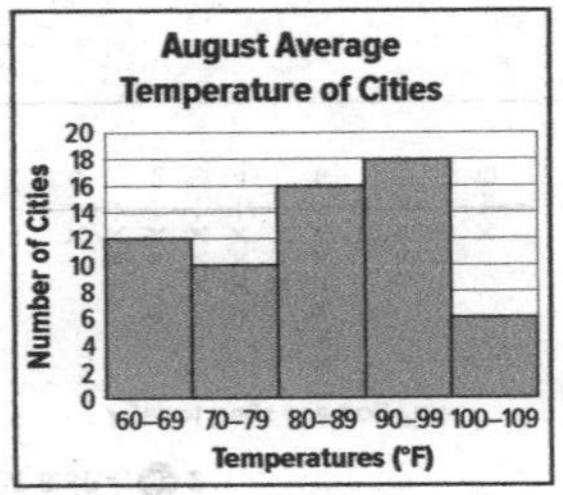

35.5%

2 The students in Mrs. Sanchez's class recorded their heights. The histogram shows the heights of the students. What fraction of the students are taller than 55 inches? Simplify your answer. 6.SP.5a, MP 2

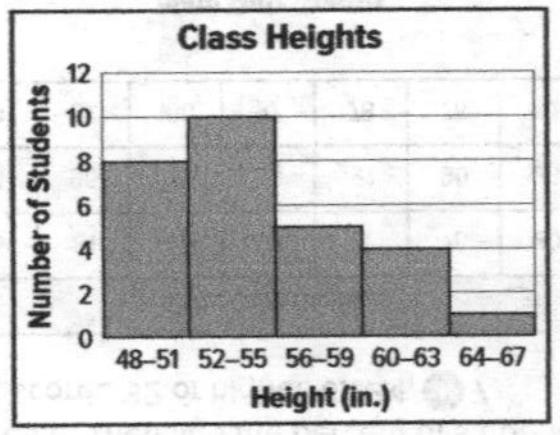

$\mathbf{\frac{5}{14}}$

3 A government program plants small trees in parks. The histogram shows the number of trees planted in 48 different parks. What is the difference in the number of parks that had the least trees planted and the most trees planted? 6.SP.5a, MP 2

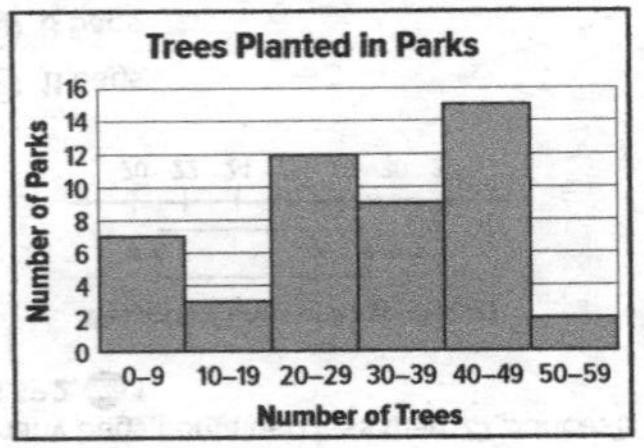

5 parks

4 **H.O.T. Problem** Valley View Middle School is holding a fundraiser for a local charity. The table shows the number of classes that raised money. Create a histogram for the data. Then find the percent of classes that raised $75 or more. 6.SP.4, MP 4

Amount ($)	Number of Classes
25–49	|||
50–74	卌 ||||
75–99	卌 卌
100–124	卌 |||

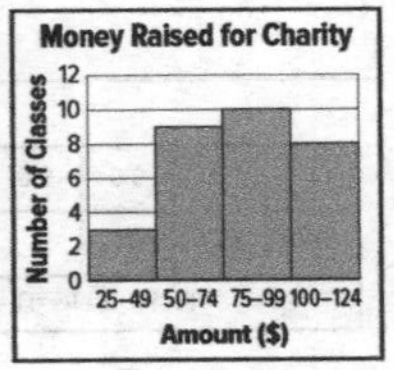

60%

NAME ____________ DATE ____________ PERIOD ____________

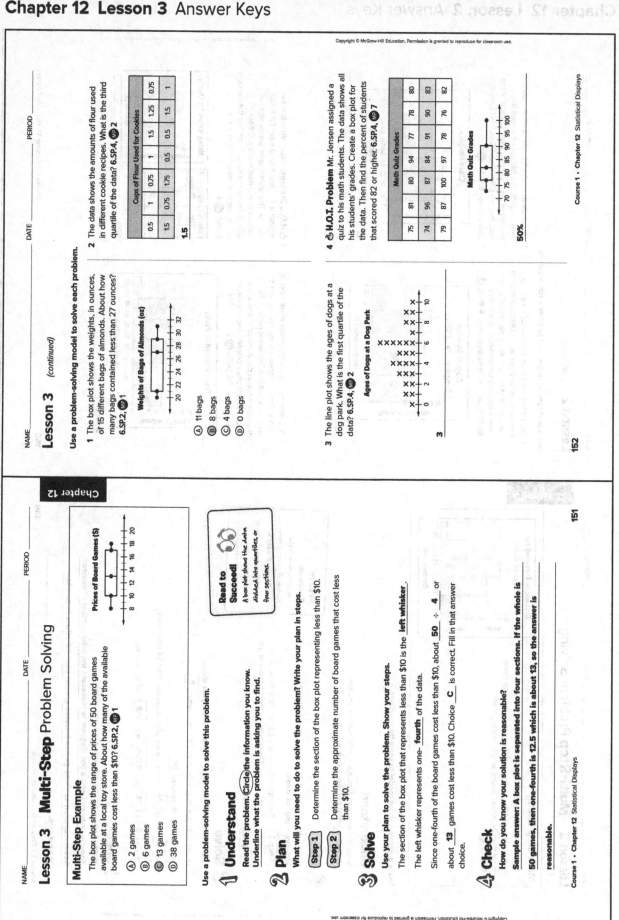

Lesson 3 **Multi-Step** Problem Solving

Chapter 12

Multi-Step Example

The box plot shows the range of prices of 50 board games available at a local toy store. About how many of the available board games cost less than $10? **6.SP.2, MP 1**

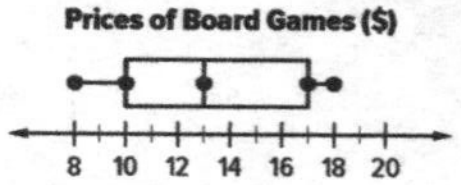

Ⓐ 2 games
Ⓑ 6 games
Ⓒ 13 games (filled in)
Ⓓ 38 games

Use a problem-solving model to solve this problem.

1 Understand

Read the problem. Circle the information you know.
Underline what the problem is asking you to find.

Read to Succeed!
A box plot shows the data divided into quartiles, or four sections.

2 Plan

What will you need to do to solve the problem? Write your plan in steps.

Step 1 Determine the section of the box plot representing less than $10.

Step 2 Determine the approximate number of board games that cost less than $10.

3 Solve

Use your plan to solve the problem. Show your steps.

The section of the box plot that represents less than $10 is the **left whisker**.

The left whisker represents one- **fourth** of the data.

Since one-fourth of the board games cost less than $10, about **50** ÷ **4** or about **13** games cost less than $10. Choice **C** is correct. Fill in that answer choice.

4 Check

How do you know your solution is reasonable?

Sample answer: A box plot is separated into four sections. If the whole is 50 games, then one-fourth is 12.5 which is about 13, so the answer is reasonable.

NAME ____________ DATE ____________ PERIOD ____________

Lesson 3 *(continued)*

Use a problem-solving model to solve each problem.

1 The box plot shows the weights, in ounces, of 15 different bags of almonds. About how many bags contained less than 27 ounces? **6.SP.2, MP 1**

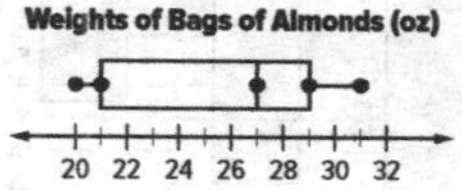

Ⓐ 11 bags
Ⓑ 8 bags (filled in)
Ⓒ 4 bags
Ⓓ 0 bags

2 The data shows the amounts of flour used in different cookie recipes. What is the third quartile of the data? **6.SP.4, MP 2**

Cups of Flour Used for Cookies						
0.5	1	0.75	1	1.5	1.25	0.75
1.5	0.75	1.75	0.5	0.5	1.5	1

1.5

3 The line plot shows the ages of dogs at a dog park. What is the first quartile of the data? **6.SP.4, MP 2**

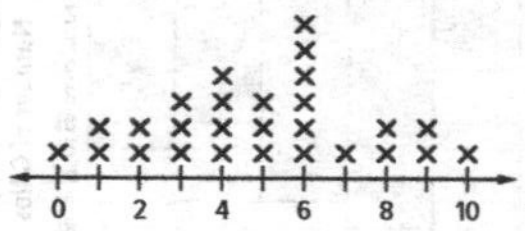

3

4 **H.O.T. Problem** Mr. Jensen assigned a quiz to his math students. The data shows all his students' grades. Create a box plot for the data. Then find the percent of students that scored 82 or higher. **6.SP.4, MP 7**

Math Quiz Grades						
75	81	80	94	77	78	80
74	96	87	84	91	90	83
79	87	100	97	78	76	82

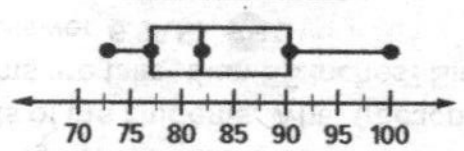

50%

NAME______________ DATE ________ PERIOD ________

Lesson 4 Multi-Step Problem Solving

Multi-Step Example

The line plot shows the ages of the students in Session 1 of Mr. Garcia's martial arts class. For Session 2, the students were the same except that a 12-year old dropped out and a 9-year old enrolled. Which shows the best measures of center and spread for the **Session 2** data distribution? **6.SP.5d, MP 1**

Ages in Martial Arts Class Session 1

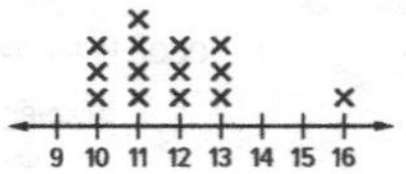

Ⓐ mean ≈ 11.6, range = 7

Ⓑ mean ≈ 11.8, range = 6

Ⓒ median = 11, interquartile range = 3

Ⓓ median = 11.5, interquartile range = 2

Use a problem-solving model to solve this problem.

1 Understand

Read the problem. Circle the information you know.
Underline what the problem is asking you to find.

2 Plan

What will you need to do to solve the problem?
Write your plan in steps.

Step 1 Determine the data set with the values adjusted.

Step 2 Determine the mean, median, range, and interquartile range of the new data set.

Read to Succeed!
List the values in order after replacing the value of 12 with the value of 9.

3 Solve

Use your plan to solve the problem. Show your steps.

The new data set is 9, 10, 10, 10, 11, 11, 11, 11, 12, 12, 13, 13, 13, 16.

The mean is **11.6**, range is **7**, median is **11**, and interquartile range is **3**. Choice **C** is correct. Fill in that answer choice.

4 Check

How do you know your solution is reasonable?

Sample answer: Check each part of the choices. Each measure must be correct for the entire answer choice to be correct.

NAME______________ DATE ________ PERIOD ________

Lesson 4 *(continued)*

Use a problem-solving model to solve each problem.

1 The line plot shows the amount of time students recorded engaging in physical activity for Week 1. For Week 2, $\frac{2}{3}$ as many students recorded 5 hours, $\frac{2}{3}$ as many students recorded 6 hours, and twice as many students recorded 10 hours.

Hours of Physical Activity Week 1

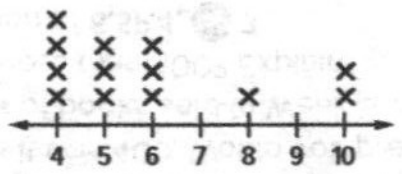

Which option shows the best measures of center and spread for the Week 2 data distribution? **6.SP.5d, MP 1**

Ⓐ mean ≈ 5.9, range = 6

Ⓑ mean ≈ 6.6, range = 6

Ⓒ median = 5, interquartile range = 3

Ⓓ median = 6, interquartile range = 6

2 The stem-and-leaf plot shows students' quiz scores in Ms. Warren's math class.

Quiz Scores

Stem	Leaf
7	5 5 5
8	0 0 0 5 5 5 5
9	0 0 0 0 5 5 5
10	0 0 0

7 | 5 = 75%

Lola found the range and the best measure of center for the data distribution. What two measures did Lola find? What is the sum of the two numbers Lola found? **6.SP.5d, MP 7**

Sample answer: range and median; 112.5

3 Donte recorded the number of repairs his bicycle shop completed each day for a month. The box plot shows a summary of the data.

Describe the shape of the distribution **6.SP.2, MP 4**

Sample answer: The shape of the distribution is symmetric. The left side of the data looks like the right side. There are no outliers.

4 H.O.T. Problem The histogram summarizes the players' heights on Rachel's volleyball team.

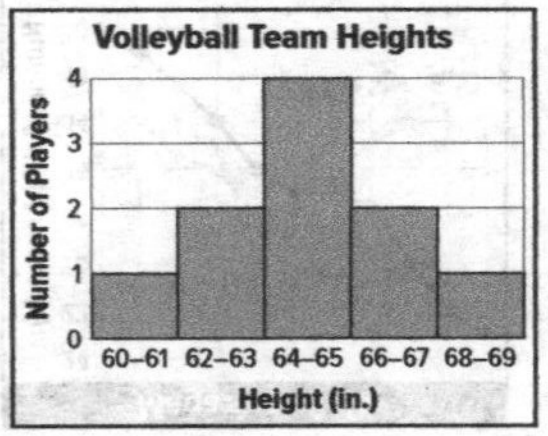

a. What are the range and best measure of center if all the heights are even integers?

b. What are the range and best measure of center if all the heights are odd integers? **6.SP.2, MP 2**

a. range = 8, mean = median = 64

b. range = 8, mean = median = 65

NAME________________ DATE ________ PERIOD ________

Lesson 5 **Multi-Step** Problem Solving

Multi-Step Example

The line graph shows the enrollments at Midway Middle School and Oakhill Middle School between the years 1980 and 2010. Based on the graph, predict which school will have a greater enrollment in 2020? Explain. *Extension of* **6.SP.4, MP 1**

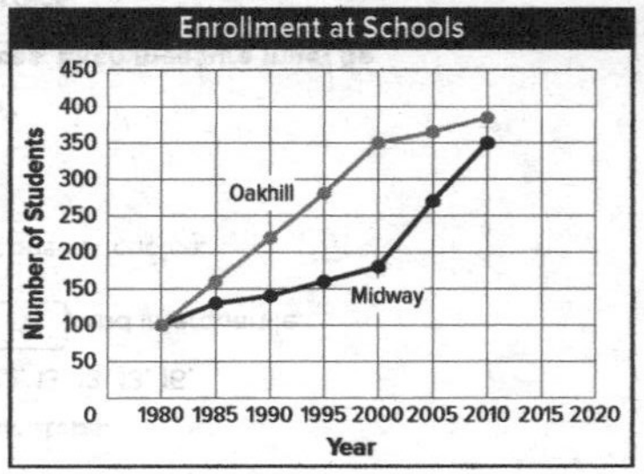

Use a problem-solving model to solve this problem.

Read to Succeed!
Line graphs show change over time. You can look at trends shown in a line graph to predict future data.

1 Understand

Read the problem. Circle the information you know.
Underline what the problem is asking you to find.

2 Plan

What will you need to do to solve the problem? Write your plan in steps.

Step 1 Observe the slant and the steepness of each line.

Step 2 Look for trends and make predictions based on the trends.

3 Solve

Use your plan to solve the problem. Show your steps.

Midway's enrollment increased slowly between 1980 and 2000, and more quickly between **2000** and **2010**. Oakhill's enrollment increased quickly between 1980 and 2000, and more slowly between **2000** and **2010**.

Midway's enrollment in recent years has increased more quickly than Oakhill's enrollment. If the trend continues, the enrollment at **Midway** will be greater in 2020.

4 Check

How do you know your solution is reasonable?

Sample answer: Extend each line using the same slant as from 2000 to 2010.

Observe which line reaches the horizontal grid line for 450 first.

NAME________________ DATE ________ PERIOD ________

Lesson 5 *(continued)*

Use a problem-solving model to solve each problem.

1 The line graph shows the number of minutes Tyson and Kindra exercised each week. Based on the graph, predict who will spend more time exercising in Week 7? Explain. *Extension of* **6.SP.4, MP 1**

Tyson; Sample answer: Tyson has continued to exercise more each week than the week before. Kindra exercised more for two weeks but since then has exercised less. So, if the trends continue, Tyson will spend more time exercising in Week 7 than Kindra.

2 The line graph shows the circulation of two magazines for ten years. For what percent of these ten years did *The Voice* have a greater circulation than *The Star*? *Extension of* **6.SP.4, MP 2**

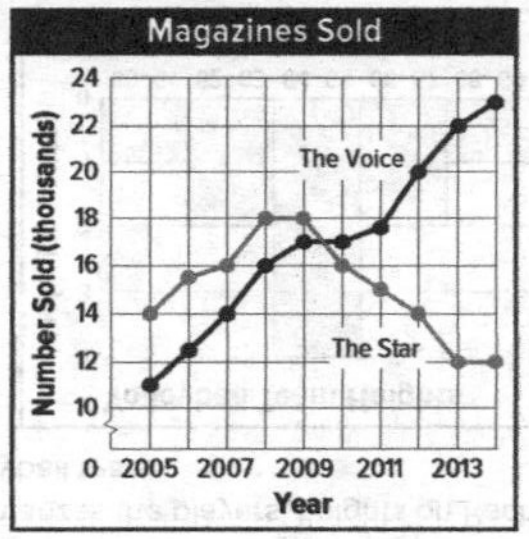

50%

3 **H.O.T. Problem** Make a line graph of the data in the table. Suppose the number of books sold in Week 6 was 350, in Week 7 it was 375, and in Week 8 it was 400. Would you predict that the number of books sold in Week 9 will be greater than or less than 400? Explain. *Extension of* **6.SP.4, MP 3**

Books Sales at Bob's Book Barn

Week	1	2	3	4	5
Number Sold	100	125	175	250	350

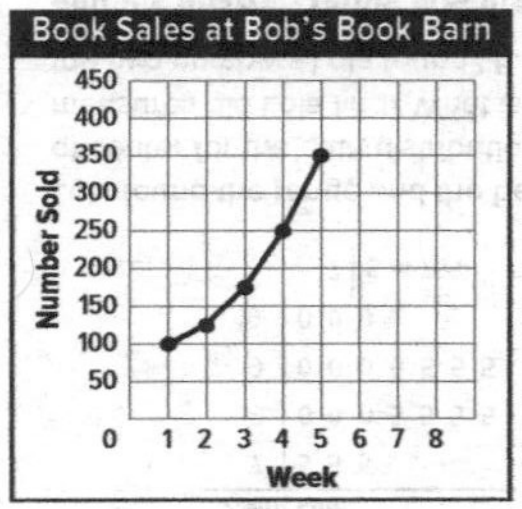

greater than 400; Sample answer: Sales have continued to increase each week except for one week when they remained the same as for the previous week.

NAME________________ DATE ________ PERIOD ________

Lesson 6 Multi-Step Problem Solving

Chapter 12

Multi-Step Example

The table shows the distance 12 students on the track team ran one week. The next week, each student ran exactly twice as far as they did Week 1. Which statement is true about finding the mode of the data for Week 2? *Extension of* **6.SP.4, MP 1**

Number of Miles Run Week 1			
5	7	4	9
10	4	7	4
8	4	6	4

Ⓐ A dot plot will best show that the mode is 8. (filled in)
Ⓑ A box plot will best show that the mode is 8.
Ⓒ A dot plot will best show that the mode is 11.
Ⓓ A box plot will best show that the mode is 11.

Use a problem-solving model to solve this problem.

1 Understand

Read the problem. Circle the information you know.
Underline what the problem is asking you to find.

2 Plan

What will you need to do to solve the problem? Write your plan in steps.

Step 1 Determine the values for Week 2.

Step 2 Determine the mode of the Week 2.

3 Solve

Use your plan to solve the problem. Show your steps.

The values for Week 2 are 10, 20, 16, 14, 8, 8, 8, 14, 12, 18, 8, 8.

The mode is **8**.

Since a box plot does not show the mode, a dot plot is the best representation to show the mode. Choice **A** is correct. Fill in that answer choice.

Read to Succeed!
Remember to read each statement before deciding on a choice.

4 Check

How do you know your solution is accurate?

Sample answer: Since the new mode is 8, choices C and D are incorrect.

A box plot does not show mode, so choice B is incorrect.

NAME________________ DATE ________ PERIOD ________

Lesson 6 *(continued)*

Use a problem-solving model to solve each problem.

1 The table shows the number of books each student in an afterschool club read in January. In February, each student met the goal of reading one more book than they did in January. Which statement is true about the mode of the data for February? *Extension of* **6.SP.4, MP 1**

Number of Books Read in January			
7	1	4	4
5	6	8	9
4	5	7	3

Ⓐ A box plot will best show that the number of books read most is 6 books.
Ⓑ A box plot will best show that the number of books read most is between 5 and 8.
Ⓒ A histogram will best show that the number of books read most is between 4 and 7.
Ⓓ A dot plot will best show that the number of books read most is 5. (filled in)

2 Dana created a histogram and a box plot using the data showing the number of minutes she exercised each day for 10 days.

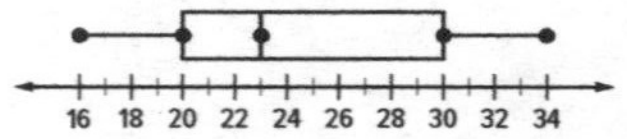

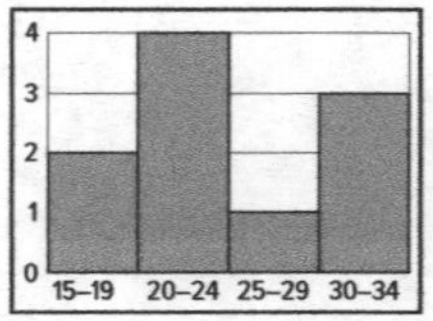

Which representation can she use to determine how many days she exercised 25 minutes or more? How many days did she exercise for 25 minutes or more? *Extension of* **6.SP.4, MP 2**

histogram; 4 days

3 **H.O.T. Problem** A real estate agent wants to create a display to show the trend in the median sales of houses over the past 6 months. He has 50 data entries for each month for the past 6 months. Explain what two types of displays he can use, one to show the median and one to show the trend in the median over the past 6 months. *Extension of* **6.SP.4, MP 3**

Sample answer: The median for a large number of data entry items can be easily seen in a box plot. A line graph shows change over time and can be used to show the trend of the median over six months.

Answers